Europeans Globalizing

Making Europe: Technology and Transformations, 1850–2000

Series editors: Johan Schot (SPRU – Science Policy Research Unit, University of Sussex, UK) and Phil Scranton (Rutgers University, USA)

Book series overview:
Consumers, Tinkerers, Rebels: The People Who Shaped Europe
by Ruth Oldenziel (Eindhoven University of Technology, the Netherlands) and Mikael Hård (Darmstadt University of Technology, Germany)

Building Europe on Expertise: Innovators, Organizers, Networkers
by Martin Kohlrausch (KU Leuven, Belgium) and Helmuth Trischler (Deutsches Museum, Germany)

Europe's Infrastructure Transition: Economy, War, Nature
by Per Högselius (Royal Institute of Technology (KTH), Sweden), Arne Kaijser (Royal Institute of Technology (KTH), Sweden) and Erik van der Vleuten (Eindhoven University of Technology, the Netherlands)

Writing the Rules for Europe: Experts, Cartels, and International Organizations
by Wolfram Kaiser (University of Portsmouth, UK) and Johan Schot (SPRU – Science Policy Research Unit, University of Sussex, UK)

Communicating Europe: Technologies, Information, Events
by Andreas Fickers (University of Luxembourg, Luxembourg) and Pascal Griset (Paris-Sorbonne University, France)

Europeans Globalizing: Mapping, Exploiting, Exchanging
by Maria Paula Diogo (NOVA University of Lisbon, Portugal) and Dirk van Laak (University of Giessen, Germany)

Initiator: Foundation for the History of Technology (Eindhoven University of Technology, the Netherlands)

The Foundation for the History of Technology (SHT) seeks to develop and communicate knowledge that increases our understanding of the critical role that technology plays in the history of the modern world. Established in 1988 in the Netherlands, SHT initiates and supports scholarly research in the history of technology. This includes large-scale national and international research programs, as well as numerous individual projects, many of which are in collaboration with Eindhoven University of Technology. SHT also coordinates Tensions of Europe (TOE), an international research network of more than 250 scholars from across Europe and beyond who are studying the role of technology as an agent of change in European history. For more information visit www.histech.nl.

Europeans Globalizing

Mapping, Exploiting, Exchanging

Maria Paula Diogo

and

Dirk van Laak

First published 2016 by
PALGRAVE MACMILLAN

Palgrave Macmillan in the UK is an imprint of Macmillan Publishers Limited, registered in England, company number 785998, of 4 Crinan Street, London N1 9XW.

Palgrave Macmillan in the US is a division of St Martin's Press LLC, 175 Fifth Avenue, New York, NY 10010.

Palgrave Macmillan is the global academic imprint of the above companies and has companies and representatives throughout the world.

Palgrave® and Macmillan® are registered trademarks in the United States, the United Kingdom, Europe and other countries.

ISBN: 978–0–230–27963–6

This book is printed on paper suitable for recycling and made from fully managed and sustained forest sources. Logging, pulping and manufacturing processes are expected to conform to the environmental regulations of the country of origin.

A catalogue record for this book is available from the British Library.

A catalog record for this book is available from the Library of Congress.

Printed and bound by CPI Group (UK) Ltd, Croydon, CR0 4YY

Contents

Making Europe:
An Introduction to the Series

In a typical conversation about twentieth-century European history, the subject of war will almost certainly arise—whether it is the Great War, the Second World War, or the Cold War. Similarly, historians who write about contemporary European history often view war as the twentieth century's iconic event. In fact, many scholars rely on Europe's political history, rife as it is with military conflict, to set the timeframe for their work. The influential historian Eric Hobsbawm, for example, defined the twentieth century as beginning with the First World War and ending with the collapse of the Soviet Union; Hobsbawm named this period—1914 to 1991—The Short Twentieth Century. Indeed, the topic of war and rupture has dominated the discourse on Europe in the twentieth century—and understandably so.

We, the editors and authors of the *Making Europe* series, however, have taken an alternative approach to our subject. We offer a European history viewed through the lens of technology rather than war. We believe that a European history with technology at its core can help to understand the continuities that have endured despite the rupture of wars. *Making Europe* places continuities—from the rise of institutions like CERN to the evolution of hacker networks—in a longer-term perspective. The *Making Europe* narrative suggests that recent European history is as much about building connections across national borders as it is about playing out conflicts between nation-states. This view of technology from a transnational perspective has proven to be felicitous. As a phenomenon, technology has always been particularly mobile; this mobility has allowed new technologies to help shape international relations between countries, companies, organizations, and people.

To understand the role of technology in this history, we required ourselves to rethink the very meaning of technology: referencing far more than machines alone, technology also embraces people and values; ideas, skills, and knowledge. Technological change, in our view, is a deeply human process. Technology was—and still is—central to the creation of Europe. And given its centrality, technology has been hotly contested—politically, economically, and culturally—in the making of Europe.

Technology's role in shaping Europe coalesced around 1850, when a new era began, an era from 1850 to 2000 that we refer to as The Long Twentieth Century. It was during the mid-nineteenth century that a newly globalizing world began to emerge. This was a world in which the many new transportation and communication technologies played a decisive part. At this time, technology became a reference point for European superiority—both within and beyond Europe. Cross-border connections and institutions thrived; the knowledge-sharing practices that fostered these connections were widely circulated and adopted. This circulation of knowledge led to a worldwide imagining, negotiating, and experiencing of Europe that still exists today. This was also the foundation for the formal process of European integration that gained traction in the 1950s. Our perspective simultaneously decenters the European Union and its direct predecessors—which, after all, comprised only one force of Europeanization among many—and places the process of European integration in long-term historical context. Acknowledging that this dynamic of integration continues today, *Making Europe* presents and interprets a history that is still in the making.

That said, it is clear to us as historians that the decade 1990–2000 marks another watershed: it was in this period that the digital revolution gained new momentum, as did shifting power relationships at the global level. This spurred the European Union to become a hegemonic force of Europeanization, and it helped globalization to enter a new phase. Simultaneously, however, the processes of integration and globalization in this apparent new phase have proven to be fragile: in light of the global economic crisis, Europe's future, called into doubt, has become a pressing issue, and one with a sharp political edge. Accordingly, Europe's past has also come under fresh scrutiny. We contend that technology will continue to play a central role in defining Europe; that the politics of Europe is the politics of technology as much as anything else; and that now is the opportune time to explore technology's historical role in the creation of Europe.

Making Europe provides a perspective on European history that transcends borders. The volumes in the series examine the linking—and, in some cases, the disruption—of infrastructures and knowledge networks that operate beyond nations and states. Also mapped here is the transnational circulation—and appropriation—of people, products, and ideas. The people and organizations featured in this series employed particular notions of Europe in building their cross-border connections. Indeed, they imagined and invented new Europes, often making clear distinctions between which people and places belonged and which were alien to the concept and the reality of Europe. *Making Europe* asks: Who projected their ideas of Europe? When did these projections take place—how, and why? The series looks at the people and the organizations that perceived themselves as central—and peripheral—to Europe, its colonies, and the transatlantic crossings that were part of the European imagination. Examined here are migrants and experts, foods and inventions, markets and regulations—virtually everything that was identified, experienced, and communicated as "European." This Europeanization, we find, had significant—and sometimes unintended—consequences: some connections between people and institutions were lasting, others broken, these continuities and ruptures shaping Europe as both an imagined place and a living community. *Making Europe* explores the stability and fragility of these European connections, communities, and institutions.

The majority of existing studies of Europe have been based on one of two approaches. First is the, often massive, single-author narrative. Second is the essay collection, which presents many voices, in some cases edited to align the authors' themes. In the field of European history, single-author volumes have tended to be broad-ranging and to address different timeframes and regions. Often, single-author volumes are a compilation of national stories; at their best, compilations transcend their individual stories to posit a complete European picture. Essay collections, for their part, have generally assumed a sharper focus—on particular communities, ethnicities, and empires, for example. These usual approaches point to a distinctive feature of *Making Europe*: in this series, five of the six volumes have two authors; one book has three writers. These voices, thirteen in all, create multiple narratives. The six sets of *Making Europe*'s co-authors have worked as a team to draft a series of volumes with coordinated yet individual themes (see www. makingeurope.eu). These six volumes contain six distinct points of view; as editors, we have imposed neither uniformity nor the pressure to harmonize narratives. In our opinion, the most informative new contributions to European history embrace diverse actors and diverse meanings, a range of purposes and understandings. *Making Europe* captures this diversity, reflecting a dynamic European history that continues to unfold.

All of the authors in the series have drawn on the European Science Foundation's "Inventing Europe" collaborative research initiatives as well as the Foundation for the History of Technology's "Tensions of Europe" project, begun in 1998 (see www.tensionsofeurope.eu). They have profited from an intensive period of discussion and joint research and writing at the Netherlands Institute for Advanced Study in the Wassenaar dunes in 2010–11. The fruits of these initiatives include the *Making Europe* book series as well as a web-based exhibit "Inventing Europe, European Digital Museum for Science and Technology" that encompasses a dozen of Europe's technology and science museums (see www.inventingeurope.eu) and scores of scholarly publications. All aim to promote creativity in fostering a more inclusive understanding of technology's role in refashioning Europe—an ongoing process that is as fascinating as it is contentious. The authors of *Making Europe* have asked themselves what shape an open-ended European history of technology would take. They provide their answers in the form of this book series.

The first volume in the *Making Europe* series, entitled *Consumers, Tinkerers, Rebels: The People who Shaped Europe,* is written by Ruth

Oldenziel and Mikael Hård. This volume spotlights the people who "made" Europe by appropriating and consuming a wide range of technologies—from the sewing machine to the bicycle, the Barbie doll to the personal computer. What emerges is a fascinating portrait of how Europeans lived during The Long Twentieth Century. Explored here are the questions of who, exactly, decided how Europeans dressed and dwelled? Traveled and dined? Worked and played? Who, in fact, can be credited with shaping the daily lives of Europeans? The authors argue that, while inventors, engineers, and politicians played their parts, it was consumers, tinkerers, and rebels who have been the unrecognized force in the making of Europe.

The second volume in the series, entitled *Building Europe on Expertise: Innovators, Organizers, Networkers*, is written by Martin Kohlrausch and Helmuth Trischler. Here the focus shifts from consumers of technology to a new breed of professionals: the technical and scientific experts whose influence soared from around 1850 onward. The authors show how these experts created, organized, and spread knowledge—enabling them to shape societies, create cross-border connections, and set political agendas. During Europe's Long Twentieth Century, technoscientific experts became a strategic resource for serving national, international, and transnational interests, the authors argue. They revisit experts' visions of Europe, showing how these visions manifested in the dictatorships of Nazi Germany and Stalinist Russia—as well as helping to build Europe's vast research networks during the Cold War. *Building Europe on Expertise* ends with today's efforts to reinvent the European Union—as a knowledge-based society defined by experts.

The third volume in the series, *Europe's Infrastructure Transition: Economy, War, Nature*, is written by Per Högselius, Arne Kaijser, and Erik van der Vleuten. This book elaborates on the first two volumes by introducing a new cast of historical actors: system-builders. These individuals and organizations helped to transform Europe by envisioning, constructing, and manipulating large-scale transport, communications, and energy systems. Their efforts reshaped Europe as a geographical entity by forming massive new material interconnections—and divisions—between places. This had far-reaching implications for European integration; for peaceful economic exchange; for military planning and logistics. System-builders challenged Europe's natural barriers, from the Alps to northern Europe's forests and the vast marshlands to the east. But Europe's water, air, and land were not only connected, they were transformed radically, sometimes destroyed. In response, system-builders eventually turned much of Europe's environment itself into infrastructure, interlinking isolated ecosystems via human-made corridors and networks.

The fourth volume, *Making the Rules for Europe: Experts, Cartels, International Organizations* is written by Wolfram Kaiser and Johan Schot. Here, the focus becomes the norms and standards of technological innovation—discussed in depth for transport and heavy industry. Featured are the people and organizations that debated, negotiated, and regulated the cross-border issues raised by innovation. Presented here are individuals with special—and often interdisciplinary—expertise in technology, business, and law. Often, these experts sought to de-politicize issues by deeming them technical; this yielded workable solutions to shared problems. It also paved experts' way in rule-making for multiple, distinct yet overlapping, and frequently competing "Europes." In the pursuit of finding technological solutions, many institutions' transnational practices survived ruptures, including the two World Wars. After the Second World War, the European Union was obliged to accommodate—and to compete with—other institutions' established practices in order for the EU to gain greater influence in shaping Europe.

The fifth volume, *Communicating Europe: Technologies, Information, Events,* analyzes Europe's information and communication systems from roughly 1850 onward. Authors Andreas Fickers and Pascal Griset place these technologies at the very heart of European society. Presented here is a global vision of media, telecommunications, and computers that reveals the tensions inherent in designing and appropriating electrical and electronic devices. The authors argue that the control in the material realm by research and entrepreneurship and the emergence of new forms of creativity and new ways of life are two sides of the same coin, mostly driven by political and cultural forces. Examined in this volume are the political, economic, and cultural realities and meanings of information and communication technologies on a European level. This perspective, which extends over the long term, provides the tools for a new critical understanding of the digital revolution.

How did today's globalized, thoroughly mapped-out world emerge? What part did technology play in Europe's international encounters, colonial and otherwise? *Europeans Globalizing*, written by Maria Paula Diogo and Dirk van Laak, concludes the *Making Europe* series with a study of how Europe interacted with the rest of the world from 1850 until the close of the twentieth century. The volume details how technologies were applied and creatively adopted–from India to Argentina, South Africa to the Arctic. From the turn of the twentieth century onwards, we witness assumptions about Europe's technologically-based superiority being continuously challenged. And we discover that globalized Europe in its present form looks quite different from what Europeans once imagined.

Consumers and tinkerers; engineers and scientists; system-builders and inventors. Experts in technology, law, and business; communicators and entrepreneurs; politicians and ambassadors. This is a cross-section of the actors represented on *Making Europe*'s pages. These actors, through the institutions and organizations they cultivated, the connections they created, the rules and practices they fostered, co-created Europe. Narrated from contrasting as well as complementary viewpoints, the six volumes in the series create a collage of co-existent portraits that depict Europe's Long Twentieth Century; its technologies; and its meanings. Together, these histories form the view of modern Europe that we and the authors wish to contribute to the historical record at this time.

Johan Schot & Philip Scranton
Making Europe Series Editors
Amsterdam, the Netherlands & Camden, New Jersey, USA
July 2013

Acknowledgements

This book is the result of a difficult, sometimes turbulent, yet extremely enjoyable journey of collaborative writing between two authors who did not know each other before embarking in this adventure. During the first weeks at the Netherlands Institute for Advanced Study in Wassenaar, we used our time to weave common ground, to understand and bridge our academic backgrounds and to build a solid friendship that would allow us to enjoy a pizza together while discussing the table of contents and, later, to write the chapters while hundreds of kilometers apart.

Many people have helped us to bring this book to fruition. Our friends and colleagues, who authored the other five books of the *Making Europe* series, were our most enthusiastic and critical audience during the long period of gestation of the manuscript. Andreas Fickers, Pascal Griset, Mikael Hård, Per Högselius, Arne Kaijser, Wolfram Kaiser, Martin Kohlrausch, Ruth Oldenziel, Johan Schot, Helmut Trischler, and Erik van der Vleuten discussed with us a significant number of hypotheses both concerning the organization and the content of the book and helped us to focus our lens of analysis and to choose our case studies. During the first stage of our book, these long and fruitful discussions also included Matthias Middell, whose insights were especially helpful and inspiring, and he contributed much to the book's conceptual design.

We would like to thank the anonymous reviewers for their comments and suggestions, which we tried to accommodate as much as possible in terms of our own view of what the book should be. We also thank the suggestions that arose from numerous debates on the various topics included in this book with our PhD students and post-doctoral researchers, and the valuable assistance of João Miguel Machado in preparing the final version of the manuscript according to the editors' stylesheet. Our thanks also go to Freya Buechter-Greiner for translating major parts of the text from German into English and to Ana Carneiro for revising the first translation from Portuguese into English.

A special word of gratitude to Katherine Kay-Mouat and to Jan Korsten, who guided us through the world of picture editing, helping us to find the images we had in mind and deal with the practicalities of having them in our book, and to Jenny McCall, Jade Moulds and Holly Tyler at Palgrave.

This book would have not been possible without the strong support and superb advice of Phil Scranton and Johan Schot, the series editors. Even in the darkest hours, they were there for us and always found the right words of encouragement and the necessary solutions. We are also in deep debt to Nil Disco, who took the responsibility for editing the manuscript in terms of both its language and organization. His *plan de campagne* (Nil's words) proved to be both efficient and diplomatic, and Nil always made sure that there was a consensus among all three of us, whenever a more substantial change was needed.

We also wish to thank the organizations that contributed financially to making this book possible: the Foundation for the History of Technology and the Netherlands Institute for Advanced Study. We also extend thanks to: the Portuguese Fundação para a Ciência e a Tecnologia for funding the research project PTDC/ HIS-HCT/118359/2010 that allowed for part of the original research included in this book; the Centre for the History of Science and Technology-CIUHCT for financial support for some of the traveling expenses; and the Faculdade de Ciências e Tecnologia/ Universidade NOVA de Lisboa (Portugal) and the Justus-Liebig-Universität for granting sabbatical leave for the authors' research.

Maria Paula Diogo and Dirk van Laak
Lisbon and Giessen, April 2016

Introduction

For many the "modern world" first became a palpable reality with the Great Exhibition of the Works of Industry of All Nations held in London in 1851. The multitudes that flocked to the Crystal Palace were shown a new world ruled by steam, machines, speed, and railroads. Contemporaries were impressed with the idea that this modern civilization would soon turn global and envelop the rest of the world. Satirical commentators pictured a civilizing steam engine that within minutes converted "savages" into educated Europeans. Though the satire harbored a kernel of truth, the actual story was of course much more complex—and certainly far less mechanical. On the following pages we will explore some of this complexity.

We propose to do this by re-examining Europe and Europe's interaction with the rest of the world over the past 150 years. Our images of this historical landscape are focused through what we call the "lens of technology," a way of seeing that brings technology to the forefront of the narrative. The book therefore combines colonial, post-colonial, global, and European history with a fresh perspective on the key roles of technology and knowledge systems.

Civilisations-Dampf-Maschine auf der Londoner Industrie-Ausstellung.

AW.

Die wilden Völkerstämme werden oben in die Maschine geworfen und erscheinen nach zwei Minuten vollständig civilisirt als gebildete Europäer! —

Fig. 0.1 Civilizing Steam Engine: *During the first World's Fair, taking place at the Crystal Palace in London from May 1 to October 11, 1851, German magazine* Kladderadatsch *ironically conceived a civilizing steam engine. The caption explained: "The wild peoples are thrown into the machine from above, and within two minutes they re-emerge as completely civilized and educated Europeans!" However satirical, the drawing revealingly captured the central position of technology for European self-confidence.*

Meeting & Dominating the "Other": From Christianity to Technology

In *Prisoners of the Sun*, Tintin, the young hero of a comic strip probably familiar to most of us, applies European science in a ruse to escape from immanent death at the hands of Inca emperor. Knowing from his almanac that an eclipse was about to occur, Tintin invokes Pachacamac (the Sun) to display his awesome power by blackening his own face, a dodge which frightens the Incas out of their wits and allows him and his friends, Captain Haddock and Professor Calculus, to escape the impeding human sacrifice commanded by the emperor.

This fictional episode, which nicely demonstrates the ideology of European superiority based on scientific and technical knowledge, is based on real events that took place in the course of European contacts with non-European cultures in the eighteenth and nineteenth centuries. John Lander, assistant to the Scottish traveler and

Fig. 0.2 *Deus ex Machina:* *The idea that European techno-scientific rationality is the ultimate worldview became a contemporary* doxa, *insofar it was accepted as self-evident. In* Prisoners of the Sun *Tintin used his knowledge that an eclipse was about to occur to show the Inca emperor that he had the power to command Pachacamac (the Sun). Hergé was inspired by accounts claiming that Columbus subdued a 1503 revolt of indigenous tribes in Jamaica using knowledge of a solar eclipse that had been predicted by Giovanni Muller's 1474 calendar. The "Oh Wow!" effect, that is, astounding non-Europeans by using European science and technology, was largely used during early encounters between Europeans and the "others."*

explorer Hugh Clapperton during the latter's expedition to western Africa in 1830–31, described a very similar incident:

> So peculiarly unearthly, wild, and horrifying was the appearance of the dancing group and the clamour which they made. It was perhaps fortunate for us that we had an almanac with us, which foretold the eclipse: for although we neglected to inform the king of this circumstance, we were yet enabled to tell him and his people the exact time of its disappearance. This succeeded in some measure in suppressing their fears, for they would believe anything we might tell them; and perhaps, also, it has procured for us a lasting reputation and a name.[1]

That said, science and technology had not always been the matrix for European self-identity nor had they always been the royal road to domination. Contacts between Europeans and the "other" (of course entailing the ongoing construction of the very definition of Europe in a geographical and cultural sense) were for many centuries mediated by more or less militant professions of Christianity, rather than by scientific knowledge or technical prowess. During the Middle Ages, Europe turned in on itself as a way of countering the Muslim threat. The Crusades of the twelfth and thirteenth centuries were mounted in order to restore the holy city of Jerusalem to the Christian (that is, European) world. From the fifteenth century onwards, the expansion of various European countries, first Portugal and Spain, and later France, Britain, and the Netherlands, incorporated the mission of bringing the world's "heathens" into the Christian faith.

The practical Christianization of the "other" worked out rather differently according to the degree of formal structuring of local religions. It was most effective, although often hybridized with native magical practices (*cadomblé*, voodoo), in populations whose beliefs were more rarefied, like the Indians of Brazil and the black African tribes. Christianization fared rather less well in regions with more formal and authoritative religious structures such as Islam in North Africa, Hinduism in India, or Buddhism in China and Japan.

Regardless of the success of such proselytization, up until the eighteenth century the contours of European identity had for the most part been religiously inspired. During the Enlightenment, however, a new element was added to the European genome: science and technology. This new cultural matrix first flexed its muscles in a confrontation between the "enlightened space" of the European center and the peripheral regions of the European South and East, vast Roman and Orthodox Catholic lands whose inhabitants seemed impervious to reason, science, and technology. The new line of division rested on the Enlightenment "religion" of science and technology, replacing the older one based on a unitary Christian faith. In this context, British, German, and French intellectuals traveling to Spain, Portugal, or Hungary saw themselves as emissaries of the "civilized world" to exotic lands still in thrall to religious obscurantism. Voltaire, for example, leaves us in no doubt about his deep aversion as he approaches Spain. In his *Précis du siècle de Louis XV* (1768), he roundly condemns this country, dominated by the Inquisition and where he could still sense the barbaric smell of the burning fires of the *autos-da-fé*.

Beyond Europe, the encounter with the "other," the "savage," the "barbarian" was thenceforth governed by Enlightenment logic. The measure of civilization was transferred from the sphere of the sacred to the profane, to the material world of science and technology. The missionary was replaced first by the explorer and later, in the nineteenth century, by the engineer. Although these characters and their rhetoric are not hermetic categories, and were in fact often conflated in the same individual, their relative contributions to the "civilization" of non-European spaces undergoes a clear transformation in the eighteenth century. Despite digressions on the wondrous fauna, flora, and human societies encountered on exotic coasts and in impenetrable hinterlands, the reports of European explorers are above all minute accounts of the economic potentials of the unexplored regions. In fact, the romanticized idea of the explorer as an intrepid adventurer belongs more to popular imagery than to actual history. Although certainly courageous, the eighteenth-century explorer was still in essence a reporter, a note-taker in the service of his patrons. His aim was to observe, gather, and record information in a professional way. Usually, these missions were funded by the state or by private patrons and had a precise program and specific tasks to be carried out in order to justify the initial investment. By the end of the eighteenth century, the Enlightenment was taking a decidedly techno-economic turn in the context of a nascent Industrial Revolution. New values had been assimilated to the original rationalist core, values built on industrial capitalism and on the Saint-Simonian concept of progress, based on the trinity of science, technology, and industry.

In the colonial territories, the relation with the "other" as well as the imperialist exploitation of resources was guided by techno-scientific logics. The axis of the *mission civilisatrice* was the integration of the colonized peoples, the "other," into the matrix of European civilization, organized around scientific and technical knowledge and the concepts of productivity and labor. It was a powerful mechanism of domination and control which entailed, either by imposition, diffusion, or appropriation, assimilation of "natives" into an alien culture that considered itself superior and was engaged in civilizing the native by turning him into a de facto European citizen:[2]

> A few years ago men knowledgeable of the African wilderness became aware of the fundamental help provided by railways in transforming and civilizing this immense continent, for the most part still in a barbaric stage, and almost exclusively populated by a primitive and savage race devoted to hunting and war and living in complete darkness.

Inspired by new approaches to the Earth and Mankind by European engineers, scientists, historians, and philosophers, resistance to European expansion came to be regarded as a violation of the laws of history and progress. The often extreme imbalance of power vis à vis native populations and the inclination of the Europeans to measure almost everything in terms of historical "progress" led to the colonialists' conviction that they were the ordained tools of history, "burdened" with spreading civilization to all corners of the globe. Globalization, an inevitable attribute of European industrial economy, depended in a deep sense on technological mastery. Hence the imperialism that was the core of globalization was fundamentally a "technopolitical" and, we add, a "technoeconomic" phenomenon, insofar as technology was used to achieve political and economic goals.[3]

Technology allowed European nations to first define and then take control of the globe by imposing their worldviews and their economic institutions on non-Europeans. Territorial domination to secure resources and markets and the control of multidirectional flows of goods, people, and information were critical to building an efficient global market. Technology took Europeans almost everywhere in the world, by ship, steam power or aircraft. Technology in a wider sense allowed them to survive in "uncivilized" settings, e.g., by using medicines to combat tropical diseases or equipment for opening up land and for constructing settlements. It was technology that enabled relatively small numbers of Europeans to conquer and control huge territories, by means of rifles, machine guns, or railroads. Technology was applied to exploit resources and to develop the foundations of industrial production in the colonies. And after Europeans had been forced out and power shifted towards "indigenous" peoples, it was European technology that the ex-colonized adopted: for example, telephone lines, traffic facilities, broadcasting and television infrastructures.[4] Weapons, steamships and harbors, railroads and roads, new agricultural techniques—all were tools to achieve domination and ultimately efficient exploitation. Though this domination was often challenged, European nations managed to impose their rules. The "Scramble for Africa" and the "Open Door Policy" in China and Japan were paradigmatic for Europe's indisputable hegemony on the eve of the twentieth century.

What is Europe?

All this still begs the question of what precisely we understand by "Europe." The terms "Europe" and "European" have been topics of chronic debate. In this book, we are more interested in Europe and Europeans as a performative category, that is, as a global actor with specific proclivities rather than in any of the static historical and geographical definitions that have been current in different times and spaces. The Greek historian Herodotus coined the term "Europe" in ancient times along geographical lines, designating it as one of the three parts of the world (the others were Asia and Libya/Africa). It resurfaced again in the Middle Ages to designate the lands of Christendom, partly defined in opposition to the Byzantine and the Islamic Empires, which threatened Europe at its southeastern borders. This concept of Europe, mostly cultural and religious, was at the heart of the Carolingian Renaissance and remained unchallenged up through the Late Middle Ages.

The fifteenth and sixteenth centuries, the headiest phase of the Renaissance, remade the concept of Europe: new nation states consolidated their identities and their frontiers, and they began to project themselves into the world with tentative overseas expansion. The Europe which emerged in the Renaissance was a diversified and asymmetrically-built space, encompassing different political and cultural realities: from a more Western Europe, built around crystallizing nation states—Britain, France, Portugal, and Spain—to a more (South) Eastern Europe, composed of atomized structures—Germany, Italy, or the Balkans. But in either case it is clear that there was a pan-European drift away from the tradition of the unified and centralized empire that had been embodied earlier in the churchly system of caesaropapism and that lived forth in the emasculated guise of the Holy Roman Empire.

The Renaissance's unfolding and mathematization of uniform maritime space—epitomized in chartmaking and navigational technologies—provided Europeans with access to the world's coasts and hinterlands and radically transformed the place of Europe in the world and its self-perception. It placed Europe at the apex of a new civilizational framework, with the rest of the world relegated to second-class status. With long-distance trade, colonialism, imperialism, and the beginnings of globalization, a new concept of Europe emerged, based on a widely-shared and internally-competitive European ambition to extend itself outside its traditional geographical space. Thenceforth, the concept of Europe was neither historical, nor cultural, nor geographical, but *civilizational*, premised on a radically new way of perceiving nature and its relationship with Humanity. As the Renaissance gave way to the Enlightenment, science and technology became the core of this new concept of Europe and its self-imposed global "mission" to disseminate science and technology as practices to observe, measure, explain, and dominate nature and peoples. The world was no longer conceived as ruled by the discretionary will of God, but rather seen as a deeply-organized, hierarchical, and mathematized machine that could be ruled by those who purported to understand how the machine worked.

Europeans had found a new vocation and Europe became the dominant actor in a world-transforming network. Science and technology, having become pillars of European civilization, were now called upon to play their part in constructing a new global society. They therefore deserve a central position in any account of globalization and that is precisely what we aim to do in this book. In so doing, however, we reject both technological determinism as well as a regression to the Marxist concept of first instance. Both of these approaches simplify and make a caricature of the shaping force of technics in society. Nonetheless, it remains clear that economic, political, religious, and cultural factors have too long taken pride of place in the literature on European relationships with the rest of the world, and that it will not do to relegate science and technology to a place in the wings. Our narrative aims to bring these formative forces to center stage, emphasizing their role as structuring elements in the making of Europe and a globalized world.

Throughout the 150 years that will occupy us in this volume, European globalization exhibited distinct geometries, from an initial phase of bold exploration to systematic domination over the rest of the world and finally to the definitive loss of its global leadership after the Second World War. The legacy at the threshold of the twenty-first century was a mixed one: while politicians still waxed optimistic, self-doubt lurked in the shadows, a gnawing fear that Europe's glorious and prosperous past was now definitively over, leaving only an insecure and fearful future nearly devoid of hope. To analyze these transformations, we shall focus on the use and circulation of agents and technical objects, always bearing in mind that technology and science are inevitably intertwined in complex ways with economic, political, social, and cultural processes. In doing so, we hope to assess not only how Europe shaped the rest of the world, but also to illuminate the extent to which Europe itself was reshaped by these contacts.

The Give & Take of Technological Imperialism

In tracing the paths of European science and technology through the world we emphasize circulation and interaction rather than diffusion because it is evident that European science and technology, in transforming the globe, also transformed Europe. More to the point, the objects and human victims of European technologies and scientific judgements were not simply passive recipients but rather scientific and technological agents in their own right. More often than not their resistance to and interventions in—and their adoption, adaptation, and piracy of—European knowledge and technologies transformed not only the substance of European science and technology, but even the way Europeans thought about their own techno-scientific culture and about themselves.

   This puts us at odds with approaches that portray the construction of modern knowledge and particularly the spread of global science and technology as a process of diffusion from an imperial center to colonial peripheries.[5] This also applies to classical authors in this field, who build their narratives mostly from the European perspective, thus approaching imperial relationships as one-directional.[6] More recently, the focus has changed to analyzing the diversity of intercultural encounters, moving from the concept of imposition, in which one of the elements is passive, to the concept of negotiation and appropriation, which implies an interaction between two or more active elements. The older, limited concept of science and technology transfer, in which "non-scientific cultures" were simply subjected to the modern European science-driven world view, has shifted to a new framework that highlights indigenous forms of knowledge, as well as connections between indigenous and European knowledge systems. This perspective, which came to fruition within the fields of colonial and imperial studies, today extends to spatial analysis within Europe, where clear asymmetries between centers and peripheries are evident, both positions generating specific identities.[7] To Voltaire, traveling in Spain or Portugal embodied a contact with "the other," the "barbarian," as exotic as Mungo Park facing the Niger tribes. To West Europeans, traveling eastwards from Budapest was as radical an adventure as penetrating the African hinterland.[8]

Scholars in the fields of Subaltern Studies and Postcolonial Studies, focusing on a bottom-up approach that on the one hand amplified the voices of the masses and on the other deployed a narrative of how colonized elites produced their own non-Westernized selves, have been emphasizing the formation of colonial and postcolonial identities.[9] The New Imperial History has stressed the indivisibility of imperial and domestic colonial histories to examine the links between imperialism and globalization, where both metropolis and colonies are critical to understanding the nature, forms, and legacies of imperialism.[10] Very challenging concepts have emerged from these recent historiographical trends, namely the "contact zone," "cultural brokering," "go-betweens," "portals of globalization," and "international junctions and sites," all highlighting the multifarious forms used to build up and circulate knowledge among different groups and geographies.[11]

The key point is that the colonized were not passive receptors of colonization. But neither was there any kind of uniform response to the imposition of European technological and scientific regimes—nor for that matter to the seductiveness of European technologies for native elites. Responses to European technological imperialism varied widely, depending on the situation, on the technology, and on who was purveying the goods. Not only do we find immediate responses, as for example armed resistance, but also more subtle forms of neutralizing European technological dominance by preserving local indigenous technologies or creating hybridized European–native technologies, or "creole technologies."[12] Ancestral forms of irrigation were successfully used in India alongside the pro-industrial Nehru government's huge program to build irrigation canals and dams. Both the Dutch East Indies railroad tracks on Java and the German Berlin–Baghdad railroad tracks were used not only by locomotives, but also by water buffalos and mules pulling traditional wagons.

Hybrid solutions mixing local and Western technologies are probably the most common, but more extreme cases including total rejection or complete adoption of Western technology are also part of the story. Although Europeans were confident that "techno-dazzling" the natives would lead them to embrace Western technology, many tribes in Sub-Saharan Africa just ignored European technology and continued to rely exclusively on their traditional techniques. Often arrested and dragged to unpaid work in non-specialized tasks (for example, earth-moving and stone-crushing for building railroad lines), they escaped and disappeared into the jungle whenever they had the chance.

In Japan, at the other extreme, the appropriation of Western technology was a conscious state-driven option. The Japanese government chose deliberately to emulate Western technology as the main instrument to achieve modernity and political and economic power. Based on the principles *of Datsua-nyuuou* (Out of Asia, into the West) and the *Wakon Yousai* (Japanese spirit and Western technology), Meiji leaders (1868–1912) were able to bridge the tensions between traditional Japanese values and those embodied by Western technology. There was no significant resistance to the Meiji modernization agenda, because it was presented as a way of strengthening Japan and not as a threat to national and cultural identities.

In India, despite the integration of large communities of Indian technicians in the process of building up infrastructures,[13] the enforcement of European techno-scientific modernity led to harsh opposition, well illustrated by the Bhadralok Debate (1890–1915).[14] This debate, which took place on the pages of the newspaper *Dawn*, opposed two extreme standpoints: the revivalists, who rejected Western knowledge, and the westernizers, who defended the full adoption of Western science and technology. In between were the moderates, known as the revitalists, who tried to reconcile Indian traditional knowledge and Western knowledge.

In Northern Africa, various local powers vehemently opposed the European presence by refusing to recognize its alleged superiority and inflicting military defeats, if only occasionally and without much consequence. The conquest of Khartoum by Muhammad Ahmad, who called himself the Mahdi (the Expected One or the Redeemer) in 1885, subduing by force both the Egyptian Khedive and the British representative, General Charles Gordon, exemplifies of such military resistance.

Regarding Black Africa, as already noted, the situation was totally distinct. The multiplicity of groups and cultures, on the one hand, and its low technical index from the European perspective, on the other, led to a climate of deregulated imposition of European culture, an unfettered display of technical superiority. To "bring civilisation to the natives"[15] was an ideological imperative and the best way to enhance their "desire for progress" was to astound them with technological devices:[16]

> In the century of steam engines and electricity Europe does not have to use old methods to civilise Africa....A gun, a sophisticated fire engine, a steam machine, a large road, a railway, the whistle and movement of an engine, etc., produce in the inhabitants of Africa a deeper stimulus to their intellectual development than masses and sermons preached by the most eloquent missionary.

While Japanese and also Indian governments used European technology as an explicit modernization strategy, the European Enlightenment also provided a more generic road map for organizing new nation states with clear borders, governing and administrating institutions, and the rule of law—at least in theory. For many, European political culture and material technology became a Holy Grail, initially for indigenous elites, then gradually for non-European populations in general. The call of European technologies was exceedingly seductive, holding out the promise of "getting connected" and "being integrated" into the new world order. In this respect, Europe was as much an "irresistible empire" as the United States later proved to be with its more ostentatious consumer culture. Technology and active technology transfer acted as pr-eminent agents of change and eventually led to a "shared history" between Europe and a globalizing world.[17]

If a "religion of technology" indeed existed, colonial techniques were certainly a prominent mode of worship.[18] The "religious mission" was materialized in colonial infrastructures: harbors, roads, railroads, telegraphs, telephones, electricity lines, irrigation systems, and airports—accompanied by the knowledge and expertise that enabled their use, that is, writing and education, scientific knowledge, organizational and management skills. But if technologies had thus become "tools of empire,"[19] the empires also struck back. The imperialized peoples of Africa and Asia inevitably found their way to Europe, sometimes in great numbers, with often unsettling effects on traditional societal patterns. So, European societies were inevitably connected to their non-European territories, albeit rarely on an equal footing.

This double process of "Europeanizing" the world and "globalizing" Europe exhibited multiple tensions.[20] Distance and the difficulties of transferring and/or adapting European technologies in tropical latitudes often set the stage for dramatic conflicts and conquests, often accompanied by grandiloquent colonial rhetoric. The metamorphoses of Charles Marlow as he goes upriver searching for Mr. Kurtz in Joseph Conrad's *Heart of Darkness* evoke the detachment often felt towards Europe by those who experienced the darker sides of colonial life. The 1899 novel reflected on the complex ambivalence inherent in the "civilizing mission" and its latent inhuman brutality. Taking the Congo as an example, Conrad illuminated the sudden transition of benevolent colonialism into an attitude that had the protagonist exclaiming: "Exterminate all the brutes!" This novel experienced a renaissance of sorts when postcolonial authors began to reflect on the long tradition of European colonial brutality that began with the first conquistadors and that was still business as usual for the French army during the Algerian war.

The other side of the coin was the use of colonial territories by metropolitan engineers to experiment with novel technologies and technological solutions and the use of colonies by European scientists as field laboratories. Investments in colonial technology and science not only provided novel sorts of expertise, it also augmented the professional and political status of its practitioners. Moreover, colonial territories offered a dynamic job market that flourished thanks to the circulation of technologies, experts, and expertise both between European nations and their colonies and among colonial powers themselves. A number of scientific and technological institutions were created in support of this new global science and technology, the paradigm being Imperial College in London. State administrations themselves were often overhauled to accommodate the existence of a colonial corps of engineers and physicians. The career trajectories of many of these engineers reveal how their colonial experience reshaped their practice in European settings, making it hard for them to distinguish between colonizing Africa and Asia and colonizing their own countries.

Nationalism, Imperialism, & the European Technological Identity

Despite the persistence of the rhetoric of the "civilizing mission" throughout the nineteenth and into the twentieth century and the apparently-cohesive ideology of the "white man's burden," Europe remained an entity divided amongst itself. In fact, tensions among European powers were a major driving force in the effective occupation of the colonies and the opening up of new territories. Colonial history reveals an intricate network of competing interests, which involved the most powerful countries no less than the peripheral ones. Great Britain and France vested their global leadership by becoming imperial powers. Germany revamped its national agenda by demanding to be part of the "division of the world." Spain envisaged its territories in Northern Africa as a pillar of its national prestige. Smaller countries such as Portugal, Belgium, or the Netherlands claimed a superior position in the European arena partly on the basis of their imperial status. The history of European imperialism and colonialism is indeed a history of nation states, fiercely defending their own interests vis-à-vis the non-European world, actively building their own spheres of influence and always struggling to maintain a place in the European arena.

Standardization of time and space, for example, was as much a tool of imperial control as of inter-imperial strife. We can appreciate why European empires and their global expansion would profit from the overarching coordination provided by such institutions as the International Telegraph Union or the International Meteorological Organization. But we would be missing the obvious not to perceive, in the laying down of submarine cables, the material expression of *national* imperial policies. It would be obtuse not to recognize the conversion of the Eiffel Tower into a monumental antenna broadcasting time signals and aligning clock hands across Europe as also a tactical move in the intense rivalry between the British and French empires. The encouragement of wireless technology by Henri Poincaré as head of the French Bureau of Longitude (who in that capacity was also the moving force behind transforming the Eiffel Tower into the greatest time synchronizer in the world) was an understandable technological strategy given the monopoly of English companies over the global

Fig. 0.3 Taming Time:
The French created the
Bureau des Longitudes
*to respond to and defy
British maritime control.
On May 1, 1910, the
first time signal was
broadcast from the
antenna placed at the top
of the Eiffel Tower, which
anchored France's new
radio service, considered
essential to the country's
defense. The frequency
of the original broadcast
was near 150 kHz and
the radiated power was
about 40 kW, spanning
the Atlantic Ocean and
reaching North America.
French leadership in
standardizing time was
secured by the* Ponts
et Chaussées *engineer*
Charles Lallemande,
*who led the adoption
of Greenwich Time
in France and pushed
forward a plan for
creating an international
time bureau.*

LES VINGT-CINQ ANS DE LA TOUR EIFFEL
Son utilisation actuelle pour la télégraphie sans fil

telegraph network—a monopoly which enabled them at will to cut all communications between Paris and its colonies. By 1907 France's army was able to celebrate its success in using radio when its forces fighting Moroccan rebels proved able to communicate

with their commanders in France. This is a significant example of
how a technology developed for colonial rule enabled one empire,
the French, to escape from the technological clutches of another
European empire, the British.[21]

But all these national rivalries notwithstanding, colonizers still
felt a common "Europeanness" in sharing a comparable set of
scientifically- and technologically-based knowledge and skills, in
contrast with the others, the "uncivilized" or colonized peoples. It
was the Europeans who appeared to be the rational, the "technical
race."[22] Technology therefore is at the very heart of what, outside
Europe, was meant by "Europeanization."

The colonial world structured itself in accordance with logics
of domination and integration, that is, by imposing a "European"
way of seeing and understanding.[23] Its techno-scientific rationality
operated on space and economies, aiming to incorporate local
production into a world economy dominated by Europe, and it
operated on people through the "domestication" of the natives, by
reformatting their ways of living and thinking. As we shall see in
pursuing the story of Europeanization, these vectors of domination
exhibited different outcomes and varied levels of success.

From Guns to Railroads

Of all the European technologies mustered to control and exploit
foreign populations, none was so ubiquitous and consequential
as the initial monopoly on and lasting superiority of European
firepower: from blunderbusses, to rifles, gunboats, gatling guns,
automatic pistols, tanks, bombs, military aircraft, and rockets.
Certainly the initial stages of colonization in the sixteenth and
seventeenth centuries were only possible thanks to the liberal
use—or the threat of use—of firearms and musket balls. Yet by
the early nineteenth century, during the expeditions that prepared
infrastructural interventions, firearms had lost none of their former
prominence. François Le Vaillant, a French bird fancier educated
in the Dutch East Indies, in his dealings with the famous and
feared Kaffirs describes the awe caused by his "double-barrelled
fuzees and pistols." With them he hit "two swallows cleaving the
air before me...and they fell at the distance of a few paces from

us."[24] According to Heinrich Barth, in his monumental *Travels and Discoveries in North and Central Africa*, Mungo Park, a Scottish explorer, was known among the Niger tribes for his ferocity and indiscriminate use of firearms, as *táwakast*, that is, wild beast, a name which the natives ultimately extended to all Europeans.[25] The massive use of firearms by Henry Morton Stanley against the natives, throughout his multiple campaigns in Africa was infamous, as was the way in which he described his military power as a "valuable service in helping civilization to overcome barbarism."[26]

European colonial "gunslingers" of this ilk were also popular fictional characters, as witness the astounding success of H. Rider Haggard's series of adventure books featuring the redoubtable big-game hunter, adventurer, and "go-between" Alan Quatermain.[27] Quatermain, like Le Vaillant and Park, based his fighting reputation on frequent displays of gunslinging prowess. In a comic book version of *The Place of Skulls*, although he shows himself a redoubtable fighter with African weaponry, he leaves little doubt about how the white man will fight and what the inevitable outcome will be. This was, after all, the nineteenth century.

Defeating the Mahdi's warriors and the solid Chinese empire was just as assuredly the result of superior European weaponry. But whereas firearms were invaluable for short-term control, the long-term domination and occupation—and certainly the systematic exploitation—of non-Western territories were possible only thanks to newer technologies that controlled space and time: railroads, telegraphs, steamboats, harbors, and roads. They became the keys to the effective exploitation of the colonies and the growing integration of new territories in a Western-driven world economy. Despite the differences between the Raj and Angola, China and the Congo, or Indonesia and Vietnam, railroads, roads, telegraphs, and seaports everywhere proved critical to the European powers in domesticating and exploiting their imperial territories. The concept of modernity was itself driven by these infrastructures, which therefore became the paradigm of success and the main tool of economic power.

Railroads were especially effective for opening up "the interior" once seaports had been established as colonial beachheads. "Railroad imperialism" mirrored the development of Europe itself: railroads were pivotal for integrating European national states, for modernizing peripheral countries, and for the expansion of

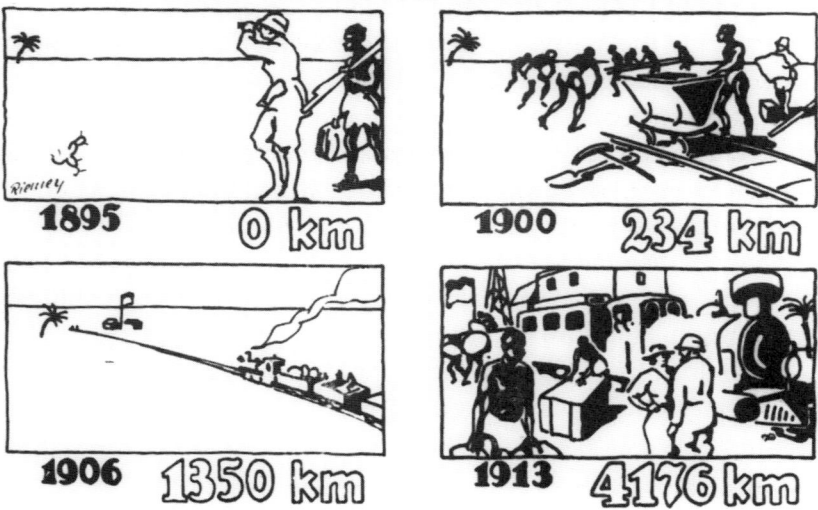

Die Entwicklung des Verkehrs
in unseren afrikanischen Kolonien

1895   0 km

1900   234 km

1906   1350 km

1913   4176 km

Fig. 0.4 Colonial Development: A series of pictures, used in German revisionist propaganda after the First World War, illustrated "the development of traffic in our African colonies" from exploration, the building of railroads, the networking of destinations, up to the ensuing establishment of trade and industry. This was how the effects of "imperial infrastructures" usually were assessed.

Europe's eastern flank, the Russian Empire, even further East. Imperialists of all nations were quite obsessed with what railroads could achieve in integrating even the last unknown territories. On this understanding, the French planned a Trans-Saharan railroad; the British conqueror Cecil Rhodes pursued the building of a railroad line from Cape Town to Cairo; Great Britain's colonial agenda for India centered on a railroad network; the Portuguese struggled to build a coast-to-coast railroad line linking Angola to Mozambique; the Dutch wanted to build the first railroad line in Asia across the island of Java. Summarizing this "railroad fever," British explorer Henry Morton Stanley stated that the African continent would not have any value unless railroads opened it up like a nutshell.

By the end of the nineteenth century, the hierarchic reorganization of the spaces was completed, technical superiority becoming central and equivalent to superior civilization. The word "savage," recurrent in narratives by Europeans, encompassed all those who were alien to the European worldview, that is, civilization *tout-court*; it was semantically equivalent to the concept of barbarian in the Roman Empire. This conviction that technology provided the scale for evaluating "the other" soon led to the belief that techno-logical superiority was synonymous with cultural and even racial superiority. All these narratives relate in one way or another to the legacy of positioning Europe in world history as the superior power. But what really drives our analysis is the investigation of how the very interaction of Europeans with developments in other parts of the world actually produced the notion of *European* structures and patterns—in the sense that actors became ready and willing to perceive European techniques and practices as distinct and different from anything else on the planet.

Organization of the Book

Faced with the challenge of telling the story of how a number of dissimilar European nations spun a great variety of technological webs across the globe over a period of 150 years, it was obvious to us that this volume would either have to be extremely thick or that we would have to be very selective and at times only suggestive. We have opted for the latter, abandoning all pretense at writing anything like a comprehensive history and choosing instead to make our points about the technological underpinnings of Europe's global world on the basis of a limited set of case studies presented as vignettes. Although this is a book on European history written through the lens of European technology, it is not a Eurocentric narrative inasmuch as it focuses on how "Europe" was built as a consequence of its contacts and interactions with other parts of the global world. In the same vein, we do not see this volume as any kind of worldwide capstone to the *Making Europe* series. Like its predecessors, it is a stand-alone volume that explores a specific aspect—in this case globalization—of Europe's centuries-long enchantment with science and technology.[28]

Our guiding research questions are these: first, how and to what extent has technology been pivotal for European–non-European relations since the nineteenth century? How did local and temporal environments, military or civilian purposes, or the developmental momentum of specific technologies influence this interaction? Did technologies inform the images that Europe drew of itself and that set it apart from "the other"? Are there specific patterns of imposing technological regimes outside of Europe? And how did non-Europeans react to these challenges, how was European knowledge and technology absorbed, creatively adapted, or declined?

We did feel it was incumbent upon us to impose some order on this *mer à boire*: at the very least some sense of chronology. It does, for example, make a real difference to the actual performance of (neo)imperialism if we are writing about an era in which communication and transport between the imperial metropolis and the peripheries are a matter of months or a matter of hours or milliseconds, even though the political and economic dynamics might seem quite similar. And of course religious, cultural, political, and economic landscapes have been radically transformed through time as well, producing historically-specific contexts for the enactment of Europe's technologically-driven conquest of the globe.

By training and inclination, historians are not prone to generalizing or abstracting—with the significant exception of imposing a more or less explicit periodization on their chronologies. This narrative ploy suggests a fundamental shift in the order of things, some new historical configuration that presents new types of challenges and opportunities for actors and in which outcomes are distinctly different from the time before and after. We see three such periods in the history of Europeans globalizing.

The first comprises what is commonly called the Age of Discovery conventionally lasting up to the mid-eighteenth century. But it is clear that "discovering" the globe (and "discovering" Europe in the bargain) went on long after that. So we would prefer to speak of "ages" of discovery in the plural. These start somewhere around 1400 but go on well into the twentieth century. Our focus is of course on the last 150 years, but much of the globalizing that went on in that period—and for that matter the globalization that we ourselves are witness to—cannot be adequately understood without some appreciation of what happened in previous centuries. This was the *Gründerzeit* of European globalization, a period in which some basic rules of engagement were worked out, trade and supply routes established, and spheres of influence set out.

The Ages of Discovery are closely related to the notion of *mapping*, as the title of chapter 1 suggests. The core pursuit here is the observation, collation, and schematization of near and distant places, objects, living beings, and human communities for the purpose of orientation and the guidance of practices. It is about

Fig. 0.5 An Upside Down Europe:
Sebastian Münster was one of the most influential geographers of the sixteenth century. The Moderna Europae Descriptio, *included in Münster's work* Geographia (1540) is *one of the first printed maps of the continent of Europe and the first to be included in a set of maps of the four continents. These wood engraving maps are famous for their decorative elements. The map of Europe, which shows a sailing ship off the Coast of Spain, is unique and unusual by modern standards, as it is oriented with South at the top and the Iberian Peninsula on the right of the map.*

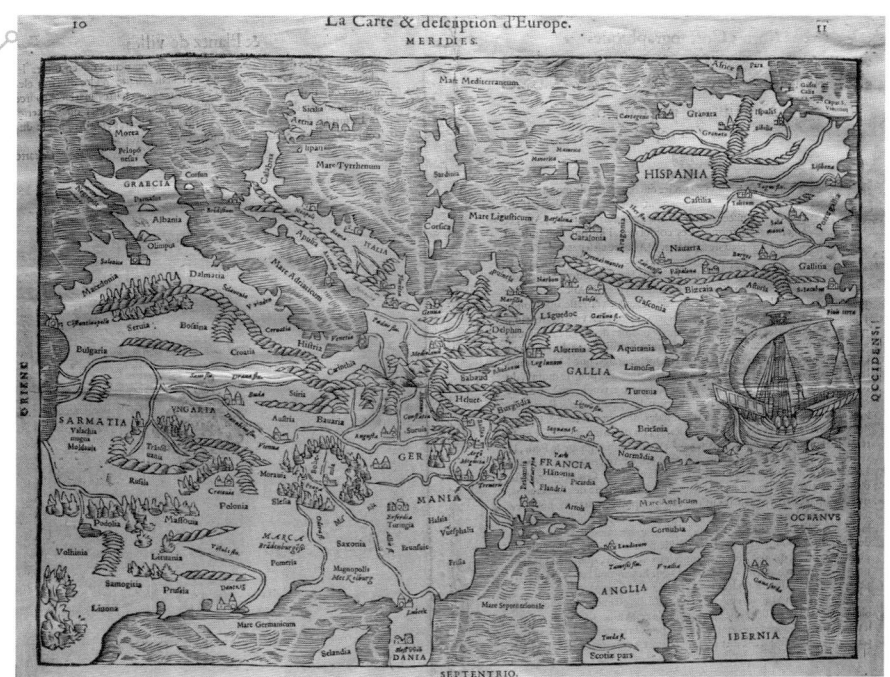

knowing the world and about codifying, categorizing, arranging, and presenting that knowledge in such a way that it can serve as a practical guide to action, for example, navigating ships, accessing natural resources, or "pacifying" a colonial territory. In addition, in our account "mapping" turns out to be a two-way street with some curious twists and turns. Mapping may not only be reciprocal, that is, the mapper gets mapped, but also reflexive, that is, the mapper maps himself in relation to his near and far neighbors. These themes are explored most explicitly in chapter 1, but discovery and mapping also inform the stories in chapters 2, 3, 4, and 5 as well.

The second period is the age of "High Imperialism" spanning the years from 1850 to 1945. The key word here is *exploitation*, several ramifications of which are elaborated in chapters 3, 4, and 5. This period clearly overlaps with the "Ages of Discovery," revealing that discovery, mapping, and exploitation were coeval. In this period Europeans deployed their technologies as part of an agenda of development *and* of domination and exploitation that we commonly call "colonialism." In the course of these hundred years, a panoply of refined European artifacts, practices, and infrastructures were

unleashed in the service of systematic, routine, efficient, bureaucratic colonial exploitation. These superseded earlier more crude
technologies wedded to more violent forms of appropriation like
pillage and outright extortion. In the glory days of colonialism,
Europeans developed elaborate colonial infrastructures to insure
their military control of territories, to transport labor power, and to
get the goods to market.

In this period, advanced military technologies based on mechanization and standardization made European armies and navies seem
invincible; colonial harbors were outfitted with the latest facilities
and became interfaces between colonial production and the markets
of the world (and vice versa); colonial railroads transported people
and goods into and out of the most remote reaches of "the interior";
and telegraph lines enabled colonial governments and enterprises to send almost "real time" messages clear across the globe.
Europeans conceived themselves as masters of the universe, despite
the obvious downsides of their colonial practices. These were manifest not only in conflicts between competing European empires in
the field but even more so in encounters between Europeans and
Non-Europeans. What one writer calls the proliferating "tentacles
of progress" provoked different reactions and strategies from
the colonized—those forced to live with and become part of the
implacable invasion of Europe and its technologies. These reactions
ranged from total and violent rejection to hybridization and wholesale appropriation.

Lastly, we identify a period in which direct colonial exploitation
gives way to more subtle forms of (unequal) *exchange*, a period
lasting roughly from 1920 to 2000. This is addressed in chapters
6 and 7. Again the temporal overlap suggests that processes
coexist in time. In the present case it is clear that the seeds of the
new postcolonial international order were already being sown at
a point in time when imperialism was still very much alive and
kicking. This period begins with the challenges Europe faced after
the First World War, especially in the form of U.S. ambitions to
assume leadership of "the West." After 1918, Europe's hegemonic
aspirations were also put to the test by anti-colonial movements in
China, Korea, India, and Egypt that seized on the Western strategy
of nationalism both as a royal road to social integration and as
a way to mobilize export markets. Even prior to the First World
War, a few farsighted Europeans had argued that in the long run

the colonies (and colonial subjects) would come back to haunt the metropolis. But promises of ready access to raw materials, a cheap colonial workforce and the establishment of global food chains suppressed these qualms and fueled the drive for global expansion throughout most of Europe's twentieth century.

We will examine how imperial technologies survived in the context of now-decolonized territories, both by being adopted and hybridized in novel social and economic frameworks, thus becoming "new technologies of poverty," and by leading a new life as state technologies, although now under new nationalist regimes.[29] But as noted, the main challenge to Europe was the rise of the U.S. with its militantly-capitalist approach towards problems of technical advancement, organization of industrial production, and domestic and global socio-political integration. The debate on whether to copy the American way or to insist on "a European path" divided Europe and fed various attempts by national elites to achieve complete control over the Continent's development. The most radical version was, of course, the Third Reich's occupation of large parts of Europe and the annexation of Eastern European territories as imperial "complementary space" to benefit an economy lacking easy and unimpeded access to both raw materials for its industry and to its overseas territories.

After the failure of the Nazi project to dominate the world, Germany's unconditional surrender, and with rapid decolonization after 1945, Europeans were forced to rethink their position in the world. The Cold War hardly simplified relations between continental Europe and the victorious allies since the Eastern parts of the Continent had disappeared behind the Iron Curtain, and since the U.S. and the territorially rather small European Economic Community were economically bound together and also politically and militarily integrated within NATO. But in this crisis a new interpretation of Europe slowly emerged—first with French Gaullism distancing itself from full integration into the West; after 1969 by a German *Ostpolitik* and the new orientation toward Eastern European societies, economies, and technological infrastructures; and finally with the growing interest (on both sides of the Atlantic—but in different ways and with different consequences) in Asian-Pacific affairs, from Japan to the Asian tigers and China. In the early 1980s a new conception of the global economy began to inform Europe's self-definition, with consequences that became manifest mainly after 1989.

The periodization roughly correlates with the concepts *mapping*, *exploiting*, and *exchanging*, echoing the subtitle of the book. This is emphatically not to suggest that the activities referred to are unique to the period with which we associate them—far from it. In each of the periods Europeans were busy mapping, exploiting, and exchanging—and fighting, hunting, inventing, gambling, farming,

Fig. 0.6   The Cake of Kings and of Emperors: *A French political cartoon from 1898 showed a pastry representing China and kings or emperors wishing to "consume" it: Queen Victoria of the United Kingdom, Wilhelm II of Germany, Nicholas II of Russia, the French Marianne, and a samurai representing Japan, carefully contemplating which pieces to take. A stereotypical Qing official tries to stop the Imperialists, but is powerless.*

bookkeeping, partying, and a million other things. Our character-ization of each period by only one of these elements is intended to suggest only that that particular activity was key to the global dynamics of that period—a practice in respect to which other practices were subservient or just emerging.

A final point is that Europe's globalizing centuries were a world-wide extension of the chronic continental competition among various national powers. This competition was all about developing means of control over an imagined world order. Technology played an important role, inasmuch as it was and still is the source of tools that control flows of goods, people, ideas, and capital from and into Europe. Thus, competition over managing space, people, or goods and for setting standards in a globalized world of technologies is at the center of the following narratives. The aim is to examine how European nations played this competitive worldwide game over a century and a half and how that game changed both Europe *and* the world. Since 1989, the world has become "postmodern," the civilizing machine is no longer propelled by steam, and nobody can definitely tell who is at the controls. As Laurie Anderson's pilot announced over the intercom in the crashing airplane in her song *From the Sky:* "There is no pilot. You are now alone. Stand by." But though there may no longer be a pilot, there is and always has been technology. The question is how it shaped globalization in the past and ultimately what global options it holds for us in the future.

1

Europeans Mapping & being Mapped

Introduction

The conventional story of Europeans and mapping focuses on the so-called Age of Discovery—usually dated between 1400 and 1750. The story is that a number of seafaring European nations developed new techniques of shipbuilding, navigation, recording, cartography, and seafaring that enabled them to discover, map and, eventually, rule the "known world." As the latter increased in size and resolution, an image of the cartographic globe emerged that corresponded ever more closely to the geography we recognize today. The mapping of the world was a one-sided affair, a European success that no other civilization had accomplished before.

In this chapter and in fact throughout this book, this simple version of Europeans mapping will be challenged and deconstructed. In the first place, "mapping" is a much more inclusive concept than simply geodetic tracings. The Age of Discovery was not simply about finding one's way at sea or fixing the positions of landfalls and landmasses by "locating" them at abstract coordinates of latitude and longitude—however monumental this achievement certainly was. It was also about mapping natural resources, geomorphological features, plants and animals, and human cultural diversity. All these, like geodetic mapping, entailed the systematic organization of disparate data into some overarching scheme that—if got "right"—provided both insight and served as a guide to making practices more effective and less risky.

Taking "mapping" in this broader sense also makes it clear that the European "Age of Discovery" did not simply end around 1750. It went on through the nineteenth and twentieth centuries as explorers penetrated into "darkest Africa" or hustled dog teams across the polar ice caps, as climbers conquered the highest mountain peaks and divers descended to the deepest ocean trenches, as petroleum prospectors and railroad engineers slogged through tropical jungles in search of new oil fields and new transportation corridors, or as biologists and anthropologists began to chart the amazing diversity of life on the planet. The point is that Europe could only successfully extend its "tentacles of progress" on the basis of a wide variety of "mappings" and "voyages of discovery" that had to be undertaken again and again as the political and economic challenges and the technological means evolved.

But the really interesting feature of these mappings, certainly those of exotic cultures and their technologies, was that they were both reciprocal and reflexive. Europeans were themselves being mapped by those they mapped and thus being forced to re-examine their self-image in these surprising mirrors—including their ongoing mappings of other Europeans and of Europe as a whole. Hence, the mapping of the world by Europeans and the counter-mapping this incited also abetted European self-discovery and especially the reciprocal typifications by Europeans of their own internal schisms and cultural topologies. By the nineteenth century, European internal divisions had become a metaphor for Europe's generic relationship to the world, with the northern parts of the Continent viewing themselves as by rights the enlightened colonizers, not only of the whole world but of southern Europe as well.

The first part of this chapter describes in a rather straightforward way how "others" mapped Europe and how Europe mapped the globe and its inhabitants. This part begins with an account of exotic mappings of Europe, chiefly by nineteenth-century Chinese observers, who though their country's military defenses had been outclassed by superior European technology, were by no means convinced of European superiority in any broader cultural sense—and not a few European critics voiced similar doubts. This is followed by an account of global mapping by Europeans, both in the classic geodetic sense and more broadly. The second part of the chapter explores the "maps in the mirror" theme, showing that the way Europeans mapped the world ultimately revealed as much about who they themselves (thought they) were as about the world they were mapping.

The Mapping of Europe

> Our emperor and the prime ministers of every one of our dynasties are no less intelligent than the Europeans. Yet none of them racks his brains about how to open the sky or delve into the earth for the purpose of enriching himself contrary to all reason. Our ancestors planned for the future with a sober and clear vision. Be that as it may, they are not like the English who must always increase their profits, come hell or high water.[1]

Ambassador Liu Xihong, one of the first Chinese to visit Europe on an official mission at the end of the nineteenth century, was by no means alone in his pointed criticism of European culture. The dynamic Europeans seemed positively obsessed by the "progress" they had made. According to Liu's observations, this led to some strange behaviors:

> In England everything is entirely different from that which holds true in China. In the political sphere the people carry more weight than the monarch. When a child is born here, girls are preferred to boys. At table it is not the guest but the master of the house who occupies the place of honor. Writing runs from left to right. When reading a book, one begins at the end and finishes at the first page. At meals, the food is served before the wine. All this can be explained by the fact that England is the antipode to China. The sky above this country is in reality located beneath the earth. Social behaviors and institutions are the exact reverse of what we are familiar with in China.[2]

Europeans themselves eagerly devoured observations of this kind; since ancient times Europeans had been curious to discover how outsiders viewed them. The manner in which "barbarians" saw European civilization from without was thought to promote self-knowledge as well as hold up a mirror to Europe's own failings. The hunger for travel accounts about foreign lands, that—from Herodotus to Marco Polo—had always found an avid readership, was now augmented by the desire to discover the familiar in the foreign and increase self-knowledge by making comparisons with the "other." The earliest of these accounts tended to be written by Europeans themselves.[3] From the middle of the nineteenth century, however, the revolution in transport and communication brought a vast increase in actual testimonials by foreigners visiting Europe.[4]

Thus it was that, during a visit to Berlin, Amur bin Nasur, born in Zanzibar in 1868, was deeply impressed by the steam trams that "like palaces" containing food, drink, shops, and mirrors "run on iron rails," and by the broad, clean streets with street lighting that functioned without oil and outshone the moonlight; by the public toilets, the clock towers, the incorruptible justice system, and not least by the salaried rulers that could be deposed and therefore had relatively little power.[5] Of course Africans were just as often affronted by the manners of Europeans; the sight of men and women dancing to music that issued from "coil springs that emitted sounds" was generally considered shameful.

THE GREAT BARBARIAN DRAGON THAT WHAT UP "THE BROTHER OF THE MOON," &c. &c. &c.

Fig. 1.1 Great Barbarian Dragon: *In 1853, British cartoonist John Leech in "The great barbarian dragon that will eat up 'the brother of the moon' & & &" mirrored Asian fears of European technology. The Promethean machine represents "Progress," and it obviously sets out to threaten Chinese traditionalism.*

In the perception of non-Europeans, Europe often seemed to be a collective symbol of the discrepancies in modernity and might obtaining between Europe and their own countries, regardless of whether the objects of scrutiny were Britons, Danes, Poles, or Spaniards.[6] Many non-Europeans were still too inexperienced to make fine distinctions: "Europe," according to one historian "means Great Britain from the vantage point of India, the Netherlands from that of Indonesia, France from that of Algeria." The wealthy, dynamic, and culturally-creative West was generally defined by iconic locations: Paris, London, Amsterdam, Hamburg, Turin, Nice, Spa, and Baden-Baden. In the years around 1900, not only Egyptians, Lebanese, and Ottoman Turks, but also Scandinavians, Spaniards, and Italians situated the center of world development in an area defined by Southern England, the Île-de-France, Northern Italy, and Brandenburg.[7] However, in more remotely-located societies the double-edged "blessings" of Europe appeared with overwhelming and undifferentiated synchronicity: "in many parts of Asia and Africa the cultural package of Western modernization emerged with shocking precipitousness and accompanied by force of arms."[8]

Mature cultures like China, which had been the object of forceful European advances since the early nineteenth century, were able to counter this in a sophisticated manner. A good example is the statement made in the late nineteenth century by the Chinese reformer Zhang Zhidong who opined that Western learning stood for the "utilitarian," whereas Chinese learning stressed the "essential." As late as 1906 the Indian intellectual Har Bilas Sarda stated that his culture was superior to that of the West in every respect and that numerous European inventions and discoveries had actually originated in India.[9]

In general, European accomplishments and culture engendered rather mixed feelings among foreign observers. Non-European societies tended to be rather scrupulous about what they did and did not like about European culture and accordingly what they were willing to adopt and what they felt compelled to reject. Although European cartography had in fact left its traces in China after the visit of the Jesuit Matteo Ricci in the sixteenth century, Chinese maps up to the beginning of the twentieth century continued to depict the geographical locations of regions outside of China in a very vague and summary manner.[10] Islamic countries were hardly more discriminating: "The Islamic urban elite had a stereotyped image of Europeans as brave in battle, but [otherwise] primitive, unhygienic, deceitful and with very loose morals (especially women)."[11] The Japanese in turn felt especially repelled by Europeans' long beards, offensive table manners, animal-like behavior, and particular affinity with dogs. Yet during the fifteenth and sixteenth centuries Japan developed a heightened interest in European techniques and processes of shipbuilding (especially large ships), clock making, printing, and the treatment of illnesses.[12]

Inherent in all of these perceptions was a strong sense of insecurity generated by European vigor and technological superiority, despite the fact that—or perhaps just because—there was such a cultural gap. And indeed it was Europe's scientific and technological achievements that invariably inspired the admiration—but also the cultural jealousy—of non-Europeans. The nineteenth-century Chinese ambassador Zeng Jize managed to salvage Chinese dignity by arguing (not entirely spuriously) that China had long ago had a machine age but had found itself better off courting technological ineptitude and Confucian traditionalism—a cultural stance he considered inevitable for the rest of the world as well:

> There were machines in ancient China. But the material rotted and the people became lazy. The machines were lost forever. One can conjure up an image of ancient China by looking at Europe today. One can guess the future of Europe by looking at today's China. Unquestionably people will one day seek the original state of things once more, will reject slick inventiveness and revert once more to simplicity and abjure complexity. For material resources are not sufficient for all the countries of the world. Therefore it is unavoidable that the original, simple state of things will replace the sophisticated ones.[13]

Mapping the Global World

It would, of course, have been impossible for non-Europeans to map Europeans if the world had not already been in the process of becoming global. This achievement was in fact due in large part to European energy and ingenuity. It is the sober truth that by 1900 Europeans—including "others" claiming this identity—were justified in seeing themselves as the most dynamic and imaginative culture on the planet. Prior to the eighteenth century, European knowledge of the world and of other peoples and cultures was hardly better than anyone else's, but by the turn of the twentieth century Europeans were poised to wipe out the last blank spots on their mental maps of the world, including its human and non-human inhabitants. By then they had such stupendous faith in their own abilities that visionaries of all kinds were advancing radical proposals to improve a less-than-perfect world—often involving wholesale rearrangements of existing techniques and institutions. Technical capabilities and booming international commerce seemed positively to cry out for such an approach. Exponents of this viewpoint were individuals like the Belgian Paul Otlet with his universal classification system for knowledge, or the German chemist and Nobel Prize winner Wilhelm Ostwald, who attempted to achieve a perfect, simultaneous conformation among language, paper formats, and currencies.[14] Ostwald's "energetic imperative" stated that human happiness would result from exploiting energy and matter to the greatest possible extent and without compromise. The corresponding worldview of "Monism" was the expression of an era entirely secure in its knowledge and confident of its ability to solve the last remaining "mysteries of the world" by reason alone and without having recourse to religious convictions or mere faith.[15]

Europeans had tolerated neither knowledge gaps nor religious taboos since the Age of Enlightenment. In that epoch a race to draft a map of the earth began, keeping the world in suspense until by the first half of the twentieth century even the frozen poles had been explored and the last mountain ranges and deepest ocean chasms had given up their secrets.

From the time of Ptolemy, the mathematically-based charting of the earth had been a persistent if often suspended collective undertaking in the Occident.[16] This geodetic project, involving the sustained co-production of new navigational technologies and the centralized accumulation of positional data with new modes of representation in nautical charts, was the *sine qua non* of the Ages of Discovery that launched Europe on its global career. This involved a dramatic change of physical scale, from hundreds to thousands of kilometers, from a few days to several months, from coastal navigation to open ocean passages, and produced a revolution in terms of the conception of space and of how it was represented. The Iberian expansion was decisive in creating the "space of modernity" in which the world came to be represented not in a symbolic way, but as a topographical continuity.[17]

Crossing the oceans demanded new navigational technologies based on mathematical expertise: this included techniques for eliminating errors introduced by the magnetic declination (the difference between the magnetic north indicated by the compass needle and the geographic North Pole). These errors were quite common and harmless when navigating in the Mediterranean, but were inadmissible and frequently deadly when navigating the ocean. The determination of latitudes on the basis of star sightings became standard procedure; astronomical instruments such as quadrants, astrolabes, and nautical rings augmented an already elaborate panoply of aids to navigation; ships increasingly exhibited cutting-edge technology in the realm of mechanics.

During the sixteenth century nautical maps were inspired on the Cantino planisphere, the earliest extant nautical chart on which places (in Africa and parts of Brazil and India) are depicted according to their astronomically-observed latitudes.[18] By the end of the sixteenth century, the Flemish geographer and cartographer Gerardus Mercator was able to publish a cylindrical map projection (the Mercator projection) that, after improvement by Portuguese

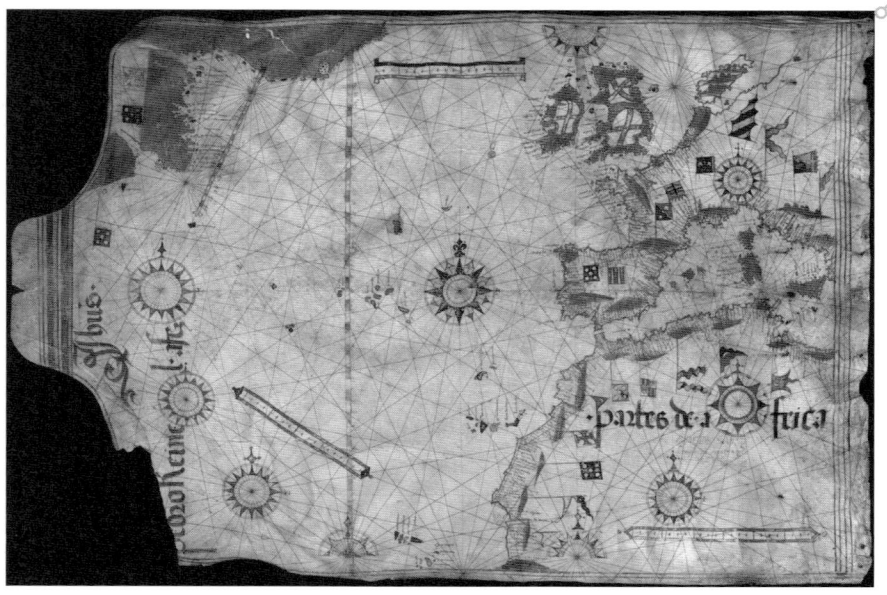

Fig. 1.2   A New Cartographic Standard: *The Cantino planisphere opened a new age for mapping by depicting places according to their astronomically-observed latitudes. Alberto Cantino was an undercover agent in the service of the Duke of Ferrara looking for strategic information brought by the Portuguese from remote lands. The image shows a map (1504) by the Portuguese cartographer Pedro Reinel, one of the earliest to use a graphic scale of latitudes and to introduce a secondary scale to accommodate places which would otherwise be impossible to represent (in this case Newfoundland).*

and British mathematicians, became the standard map projection for nautical purposes. The beauty of the Mercator projection was that it showed constant true bearings as straight lines, a great boon to navigators at sea.

Though mapping the geodetic world was a necessary precondition for establishing stable trade routes embedded in networks of watering and supply stations, and ultimately for the maintenance of colonies, it was by no means sufficient to enable Europeans to exploit the world's rich resources and, later, its willing markets. The same spirit of unremitting classification, standardization, measurement, and calculation that underpinned the geodetic project also inspired a whole range of other European mappings of the physical and cultural world. The Swedish naturalist Carl Linnaeus, who published the first *Systema Naturae* in 1735, also distinguished *Homo Europaeus* from *Homini Asiaticus*, *Africanicus*, and *Americanicus*. Europeans soon prided themselves on seeing the world in a more orderly manner and thus believed themselves capable of profiting from it more than members of other civilizations. Numerous recording techniques gave testimony to this: mapmaking, geography, statistics, mathematical calculations, the standardization of time and space, systematic listings of flora and fauna, field research, and experiments performed under laboratory conditions, and so

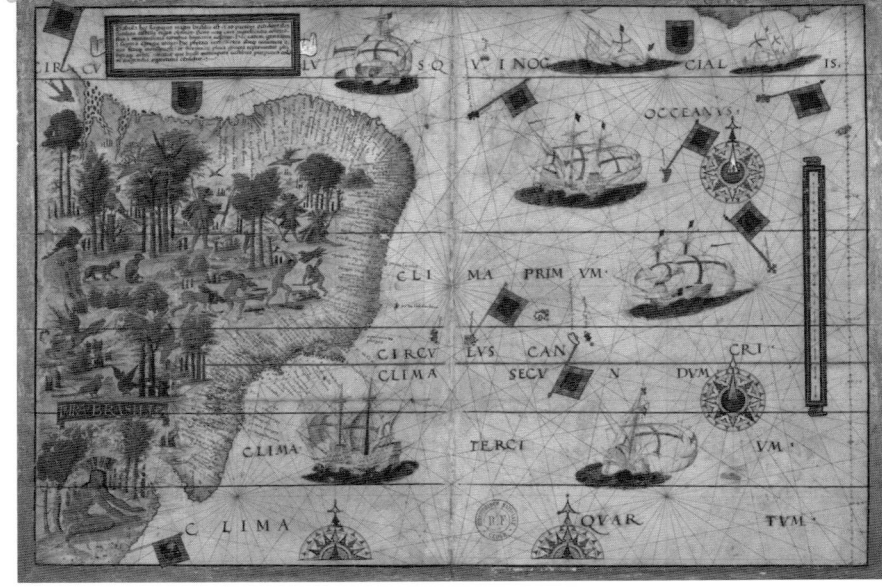

Fig. 1.3 Impressions from Brazil: *This map (ca. 1519) by the Portuguese cartographers Lopo Homem and Pedro and Jorge Reinal shows the Atlantic and the Brazilian coast. The map is very rich in decorative elements, mainly representations of people, flora, and fauna. The legend on the top left is written in Latin and reads "This map depicts Brazil. The inhabitants of this land have dark skin. They are ruthless savages who feed themselves upon human flesh. They handle their bows and arrows with great ability. One can also find in these lands multicolored parrots, lots of other birds and monstrous beasts."*

on. As a concrete expression of the formula "knowledge is power," these disciplines were meant to simultaneously manipulate, adapt, and improve nature in order to facilitate its utilization by humans.

Utilizing and valorizing the world's resources was a laborious enterprise and in the course of the nineteenth century "work" became the central virtue of the modern world as well as an essential criterion for vindicating and implementing Europe's civilizing mission.[19] Given this attitude, it was easy and convenient to view non-Europeans as inherently "lazy."[20] As early as the mid-nineteenth century the German geographer Carl Ritter saw Europe as the "powerhouse of the world" and a "continent for educating the human race." Like most Europeans, he attributed this state of things mainly to the Continent's diversity and its temperate climate.[21] The British writer Rudyard Kipling echoed the same sentiment in his iconic poem *The White Man's Burden* (1899).

Yet it was no secret that for many centuries Europeans had utilized knowledge and methodology gleaned from other parts of the world—Arabia, Egypt, Turkey, China, India. Europe had always been a flamboyantly assimilative civilization and it certainly owed the world.[22] However, after the Enlightenment and its political apotheosis in the French Revolution, Europeans tended to forget their considerable intellectual debts and to lose their respect for the achievements of others. This culminated in an unregenerate sense of superiority, and the nineteenth century became an era of "massive and largely unthinking Euro-assertion."[23] Instead of simply assimilating, the West harnessed already-existing networks and endeavors for its own benefit and simultaneously utilized them in a "parasitic" manner.[24] Now the Europeans saw themselves as being "in the vanguard of the age" and the remaining world—from the vantage point of historical development—as backward or at best as loitering "in the waiting room of history."[25]

Maps in a Mirror

Europe Maps Itself

Nevertheless, this phase of heightened self-confidence, which each imperialistic conquest outside of Europe seemed only to intensify, was short-lived. After 1850, questions of identity began to plague Europeans in at least equal measure. Every encounter with the "other" prompted a comparison that was not always favorable to the European. This encouraged European governments to uncover ever-new adversaries, threats, and perils purportedly lurking outside Europe, ready to take over world leadership. Cases in point were the United States of America, Russia, Japan (or "Asians" in general, for example the "yellow peril," especially threatening because of their ostensible "ant-like industriousness") or, even more generally, the "coloreds." Europeans regarded the rapid growth of these exotic populations with great suspicion, particularly in the twentieth century, fearing catastrophes of Malthusian proportions.

But within Europe itself there were persistent doubts as to the desirability of belonging to a confederation defined by technological primacy and a specific set of common values. This secessionist self-doubt is evident in the *leyenda negra* (portraying Spanish conquests as a barbaric enterprise), in Portugal's quasi-estrangement from Europe, in Greece's Orthodox religious orientation, in England's *splendid isolation* from the Continent, in Iceland and Greenland's relative proximity to America, in Scandinavia's autonomous relationship to the rest of Europe, in Ireland's self-assertiveness as a British "colony," in the role of the Balkans as bridge to the Orient, and later, during the Cold War, in the overarching question of whether the countries east of the Iron Curtain still belonged to the common fold.[26] In addition, from time to time there have also been some hard-to-define examples of isolationism within Europe, such as Albania between the 1960s and 1990.

Despite these eruptions of secessionist uncertainty, Europe's overall image of itself was always grounded in the ideal of an advanced civil society as the precondition for an advanced level of science and technology. Europeans saw their successful efforts to educate themselves and to work productively as essential sources of their authority over others. Once biology and genetics matured into coherent sciences during the nineteenth century, European supremacy was "geneticized" and subsequently spawned the idea of a *Homo Europaeus* and a Caucasian race.[27] In the process, all the qualities ostensibly responsible for scientific, technological, and cultural superiority over "Negroid" and "Mongoloid" races were attributed to the "Europoid."[28]

Literary scholarship and a massive amount of postcolonial analysis and research have shown that the more Europe both viewed and advertised itself as rational, the more it projected its suppressed irrational, mirror-image tendencies onto foreign regions. This was by no means limited to the (Near) East. At various times and places, especially when threatened by non-European economic competition, Europeans viewed Arabian and Sub-Saharan Africa, Japan, China and the Far East, India and the South Seas, as equally "oriental." In so doing Europeans invariably emphasized the exotic and erotic elements of these cultures and portrayed them as rationally and politically inferior to Europe.

But in the same breath they were also often depicted as being in a state of natural harmony and liberated from the burden of rationalizing and instrumentalizing everything. Jean-Jacques Rousseau expressed this widely-shared sentiment in the influential image of a "noble savage." This imagined idyllic life stood in sharp contrast to the clear and present tensions and ugliness of industrial Europe. "Oriental" places and peoples were considered sanctuaries of beauty and contemplation. Yet if there was a conflict or it appeared opportune for various tactical reasons, the "noble savages" could just as easily be made out to be savage barbarians.[29] This schizoid mental mapping of the familiar and the "other" required, of course, a certain ability to ignore facts lest this image dissolve into thin air. And it most assuredly revealed far more about those who were doing the mapping than about who was being depicted.[30]

Two further twists in the relations between Europeans and non-Europeans are worth mentioning: one was the exodus of masses of Europeans fleeing the oppressions of a continent that was being rapidly industrialized. The U.S., seen as a mecca of freedom and tolerance, was the main target of this mass migration enabled by new means of transport, but it also impacted other parts of the world where some Europeans eventually became completely assimilated. As the *Delhi Gazette* reported in 1856:

> Instances have been known of Englishmen coming out to India early in life and becoming in the course of time so thoroughly Indianized, so identified with the natives (usually with the Mohammedan natives) in habits and feelings as to lose all relish

for European society, to select their associates and connections from among the Muslims, to live in every respect in Mussalman fashion, and to openly or tacitly adopt the Mussalman creed, at any rate ceasing to manifest any interest in Christianity ... These have frequently been men of very superior ability ... and their familiarity with the ways of the natives may have paved the way for successes otherwise dubious or impracticable.[31]

The growing mobility afforded by sea travel made it possible literally to leave the map of Europe behind and submerge oneself in other cultures. And in an ironic parallel, from the eighteenth century onwards, undesirable Europeans were exiled to islands that were as remote as possible—a penal strategy that among other things helped to populate the Australian continent. Cities like Sydney and Brisbane were the offspring of British penal colonies.

A second twist was a certain reverence for Asian civilization, which in Europe as a rule meant the ancient cultures of Japan, China, and India. It was often argued that the future of Mankind lay in the fusion of the supposedly rational, enlightened, and technological culture of the West with the purportedly philosophical, deep, and spiritual culture of the East. The most prominent advocate of this idea was the Japanese-Austrian writer Count Richard Nikolaus Coudenhove-Kalergi, born in Tokyo in 1894 and founder of the Pan-European Union. He had written an *Apologia of Technology* in 1922, which he expanded in 1932 into *Revolution Through Technology*. The book expounded a philosophy of history that strove to bring ethics and technology into harmony, equating the former with Asia and the latter with Europe. Ethics was to solve the social question from within, technology from without. Ethics, he wrote, strives to transform man whereas technology strives to transform things. Ethics controls the natural forces within ourselves, technology the natural forces around us. The former strives for harmony, the latter for energy.[32]

Coudenhove-Kalergis' diagnosis was not all that different from that advanced by Liu Xihong a generation earlier. Yet unlike the latter—and in diametrical opposition to the visions of Zeng Jize—he saw these opposites as an opportunity to fashion something insuperable by unifying both. His summation contains all the pathos of which the idea of using technology to develop the world was capable: "Thus for mankind, the second Paradise means a return to nature on a higher plane. With the aid of technology, the entire globe will be transformed into a Garden of Eden with streets and canals, irrigation and drainage, hygiene and building engineering, central heating and cooling."[33]

Along the same lines, but in a rather more pessimistic mode, British scientist A.C. Egerton offered this advice to the decolonized Indian people in a broadcast on *All India Radio*, October 2, 1948:

> May I suggest, do not be too attracted by all the glamour of Western technology…it is wonderful, but we have in some ways industrialized too far and not made the world happier thereby. You have the chance of distilling the best out of the West and fitting it into the age-old civilization of the East.[34]

"Latin" & "Teutonic": European Regionalism

Though "Europe" may often have been viewed as a monolith by outsiders, Europeans themselves subscribed to internal mappings that emphasized an enormous diversity in culture and politics. Take Belgium, for example, whose well-lit and modern highways can still bemuse foreign travelers. Driving the E-40 from Liége to one of the glamorous seaside resorts of De Panne, Oostende, or Knokke is a vivid experience of European regionalism. Again and again historiography has stressed that the European era of nationalism created coherent territories that were both technologically and culturally integrated, turning local peasants into Frenchmen or Flemish farmers and Wallonian workers into Belgians.[35] Yet the evidence tells the tourist in Belgium a different story: as the highway passes through both Walloon and Flemish territory, road signs continuously alternate between French and Flemish as do the names of cities and directions. Lucky indeed is the driver blessed with linguistic skills or a GPS.

Founded in 1830, the Belgian state was quick to integrate and thus insulate itself against foreign "oppressors" from the North, the South, and the West. In his *Histoire de Belgique*, historian Henri Pirenne stressed its fruitful regional dualism, promoting the self-image of Belgium as a "bridge building nation" between Romanic and Germanic cultural and linguistic spheres. In point of fact, rising industrialization, wealth, and infrastructures did indeed keep the mainly agrarian Flemish and the industrialized Walloons together for many decades. Nevertheless, sustained efforts to rally the Belgian populace towards common ends and cultural integration failed. Not even the country's colonial involvement in the Congo was able to foster integration, with the Flemish in particular demanding full partnership in terms of culture and language. Nowadays the once poor but now booming Flemish North grimly supports the de-industrialized and depleted Walloon South.

Fig. 1.4 Inner European Dividing Lines: *Along many cultural dimensions, Europe displays a distinct North–South divide: religions, cultural traditions, landscapes, favored beverages, and weather— among many others. The divides fail to observe national borderlines and can cause tensions as well as inspirations.*

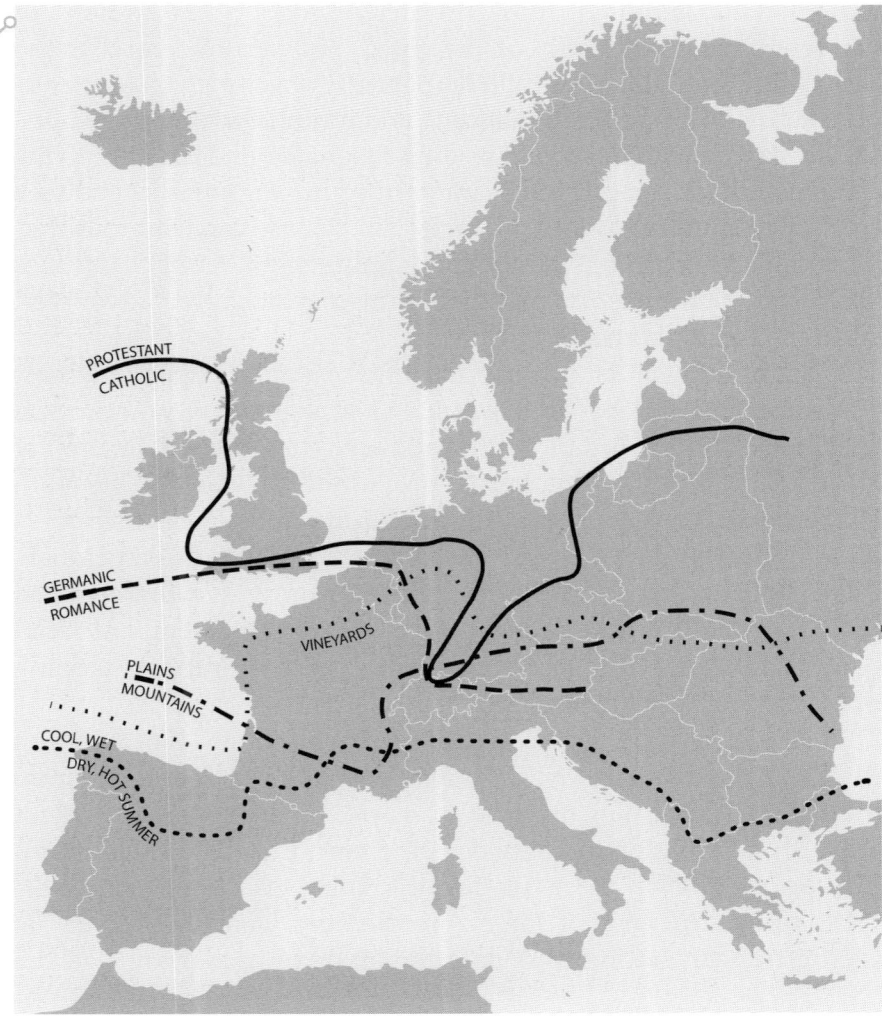

Belgium is far from the only European nation divided against itself, whether politically, culturally, economically, or on the grounds of religion. Regionalism is thus alive and well in the heart of Europe: in an era of expanding mobility and increasing personal contacts made possible by the growth of traffic and communication technologies since 1850, ancestry and provenance have taken on new meanings. Home and birthplace came to be seen as being no more than formative. Coming to terms with the formerly unfamiliar became a lifelong task. Within a mobilized society, cultural "mapping" in the form of stereotypes and clichés performed like currencies of interaction. Regionalism served to make specific areas more recognizable, a tendency augmented by a burgeoning tourist industry that profited from distinctive folklore, regardless of whether it was "authentic" or "invented." But as one can see in the case of Belgium, this cultural heterogeneity was by no means irreconcilable with being an "advanced" and technologically-developed nation state. Regionalism had its own rationales, and contrary to the conventional assumption that modern technology has a tendency toward cultural homogenization, it also has the capacity to differentiate. This is reflected in mental mapping that not only exoticizes "primitive" cultures overseas, but also exoticizes "advanced" cultures close to home.

Such regionalism is an issue not only *within* European nations; it also afflicts Europe as a whole. Most significantly, reciprocal mappings within Europe have produced stereotyped images of a Northern versus a Southern Europe and an Eastern versus a Western one. There are, of course, multitudinous *national* mappings as well, purporting to reveal typical characteristics of Frenchmen, Finns, Italians, and so on. But the culturally most revealing regional divide, certainly from the perspective of the Enlightenment rationale for Europe's conquest of the globe, is the reciprocal mapping of North versus South Europe.

Of the many tensions besetting Europe, the North–South divide is one of the oldest. Whenever Europeans come to draw imaginary landscapes or "mental maps" of their own continent, a deeply-felt distinction between Northern Europe and Southern Europe inevitably rears its head.[36] The very term "Europe," once coined in ancient Greek mythology, was lifted from its Mediterranean setting to denote the lands of barbarians like the

Scythians, who lived to the north. For others, "Europe" refer-
enced the land of the Hyperboreans, a fabulous realm of eternal
spring located in the far north beyond the land of winter. Though
the idea of "Europe" did function as a rallying cry against
common aggressors from abroad—like the Persians or, later on,
the Huns or the Turks—the peninsula remained torn by regional
(and especially north–south) distinctions. This regionalism and
the multitude of closely-connected rival centers of power have
arguably stimulated a peculiar European disposition to cultural
distancing and competition.[37]

North–South antagonism within Europe finds its clearest expression in the opposition between the Mediterranean Sea region
and the Northern Baltic Sea region, since both have been poles of
cultural gravitation for ages. Sometimes "natural" borders like the
Alps were recognized as the dividing line; often political, cultural,
or strategic boundaries served as lines of demarcation. The continental *Limes* and Hadrian's Wall, for instance, attempted to separate
Roman civilization from Celtic "barbarians" like Pictish or Scottish
tribes. These "curtains" were not "iron," and in the long run they
did little more than retard processes of diffusion and interaction.
Nevertheless they were of long-lasting importance for social division and cultural orientation. After the Mediterranean Roman
culture had conquered much of central Europe in late antiquity,
early medieval times saw northern Germanic tribes move south
to subdue the supposedly eternal Roman Empire. These processes
of encounter and entanglement between North and South became
major themes in European historiography during the nineteenth
and twentieth centuries as historians struggled to determine which
culture, in the final analysis, had been the most influential in
Europe.

Though Europe has been a theater of almost constant warfare
waged to acquire territory, extend hegemony, or propagate the
true faith, in the long run seafaring trade and trans-regional traffic
pioneered by merchants and immigrants acting from a variety of
motives turned out to be a stronger force and a medium of regional
integration. Fernand Braudel's classic study on the shaping of a
Mediterranean culture offered a convincing interpretation of how
geography, society, and history interacted in later medieval and
early modern times. Others identified a movement away from

the Mediterranean towards the Atlantic and North Sea as wealth and power shifted from south to north in the pre-modern era. The Baltic Sea area developed as a major commercial region in northwestern Europe, and in the course of several centuries the Hanseatic League established new socio-economic patterns based on long-range business and trading connections among urban regions. Whether the details of these interpretations will stand the test of time or not, it remains evident that after the Middle Ages there was an undisputed reorientation from the South to the more central and northern parts of Europe in respect to trade, art, music, science, and innovation. This was followed by a religious schism that eventually separated a predominantly Catholic South ("ultramontane" and thus closer to the Pope) from a mainly Protestant North (divorced from Rome as a religious center and creating "modern Romes" instead). Max Weber famously argued that this religious divide also influenced cultural habits, mental attitudes, and a specific disposition for science and technology. Some contemporaries have postulated a difference between a "Teutonic Europe" with Great Britain leading and a "Latin Europe" with France as the possible major actor, but there were in fact many more "oppositions" that shaped the character of the Continent.

In fact, cisalpine and transalpine (that is, this side of the Alps versus the other side, as seen from Rome) regions had developed differently since the beginning of European industrialization. Since then, as seen through the lens of technology, Europe has generally been dominated by the North, while the South tended to be less technologically advanced and in most respects viewed as struggling to catch up. This split between North and South also applied to European countries themselves. The southern parts of countries like Spain, Italy, or Greece, but also of Germany, the Netherlands, or Belgium, regarded as lagging behind in terms of economy and technology and sometimes judged to be "less European." In the South, striking contrasts between a presumed "heroic" past and a "deplorable" present dominated national self-imagery. One pattern of argumentation maintained that the South could be "revitalized" only with help from the North. Many ascriptions that were later applied to non-European cultures were in fact first deployed on a smaller scale within Europe itself. Even those elements that form a specific, common European heritage as compared to other world regions—humanistic universities, traveling scholars and artists, separation of church and state, parliaments, the printing press, journals and newspapers, a culture of correspondence and travel, for example[38]—were considered unequally distributed across European regions. These differences were often emphasized and reproduced in a pejorative mode, but sometimes, in celebrating a "Europe of the regions," praised as well.

"Head" & "Body": Imagined Maps of North & South

The European Enlightenment's mapping obsession expressed itself in numerous projects to collect all available data on almost every aspect of life, to convert them to statistics, and to make comparisons between entities with the aim of "organizing the world for Europe."[39] It was important to position oneself, one's culture or nation, and finally Europe as a whole on a map or ranking list, according to yardsticks of performance like the frequency and importance of inventions. Since the eighteenth and nineteenth centuries this ordering obsession has informed European "knowledge cultures," especially when presumed "decline" seemed to be intruding on what was supposed to be continuous "progress." This obsession with decline and progress resonated with the simultaneous emergence of a modern style of historical thinking that was steeped in the ideology of continuous improvement. The new historiography further stimulated comparative and competitive attitudes among European nations and ultimately between Europeans and non-Europeans. "Imagining Europe" became entwined with "emphasizing the centrality of Europe and exaggerating its relative size" on the cartographic as well as on the imagined maps.[40] But it also became interlaced with internal European mappings that were simulacra of those projected onto the world—especially, as we just argued, with the supposed differences between the peoples and nations of Northern versus Southern Europe.

Reciprocal mappings of Europe north and south exhibited a remarkable stability over time, despite the highly unequal tempi of development and modernization in the two regions. In the long run, the northern countries saw themselves as the torchbearers of

Enlightenment progress, whereas the southern countries exhibited deficiencies in dynamism, at least in terms of economy and social welfare. From a truly northern, that is Scandinavian, perspective, the southern parts of Europe were viewed as Catholic, capitalistic, authoritarian, conservative, and colonial. This image has lingered on until the present, though "colonial" has given way to "underdeveloped in terms of equal opportunities."[41] The southern perspective holds fewer distinct stereotypical views about the North, but the harsh climate of the North has frequently been held responsible for many peculiarities like the very different patterns of social interaction and a more robust "civil" society. In addition, the North is often perceived as puritan, disciplinarian, and lacking in human warmth. On Central European maps that became fashionable during the Renaissance and that often depicted Europe as a human body—like Sebastian Muenster's allegory of 1544—Scandinavian countries were marginalized. They appeared as something "foreign" like Africa or Asia.[42] The medieval experiences with Viking raids may have contributed to this purported "otherness." The Roman Catholic Church added fuel to this fire by suggesting that the devil himself dwelt somewhere in the North.[43]

In fact, even up to the present, Scandinavian countries seem to remain notably detached from the rest of Europe. This is not just due to their geographical remoteness, but to a distinct self-reliance that is reflected in a deficiency of common traditions with other European regions. Contributing factors are also the smooth functioning of the Scandinavian nation states and their history of intense, indeed in some cases colonial, interrelationships.[44] Scandinavian solidarity and Nordic self-assurance were fostered by historical accomplishments like the Northmen's adventurous journeys and their settler colonialism that extended as far as North America.[45] In the second half of the nineteenth century a pan-Scandinavian movement even became popular among intellectuals, though this ultimately foundered on the rocks of growing nationalism.

Between North and South there were no ideological tensions like those of the East–West divide during the Cold War, with distant powers like the United States or the Soviet Union influencing Europe in manifold ways. Seen from a Central European perspective, the North and the South in fact became synonyms for "developed" and "undeveloped" regions, a sphere of advanced

Fig. 1.5 *Europa Regina:* *The image of Queen Europe, a young, graceful woman, is an Early Modern allegory of the* Europa triumphans. *Although based on the Holy Roman Empire, the image shows a more general map-like depiction of a unified Christian Europe as the dominant power in the world. Europa Regina was a popular image showing Europe standing upright with the Iberian Peninsula forming her crowned head, Germany her heart, Italy and Denmark her arms (with Sicily representing the orb and the sceptre), and the rest of the European countries her long gown.*

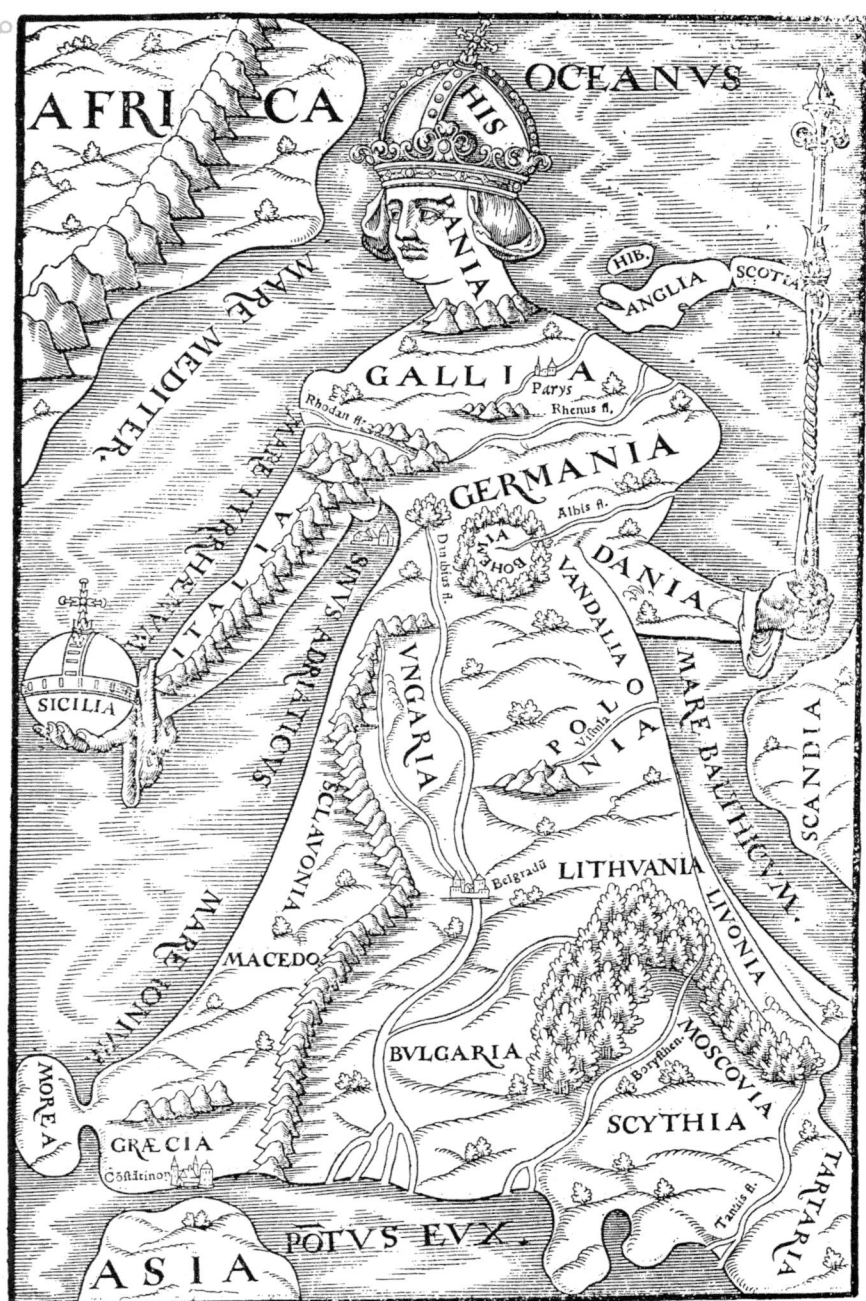

technology contrasted with a sphere waiting to be developed with technology, science, a spirit of inquiry, and knowhow.

This innate European tension was upscaled to global proportions in the nineteenth and twentieth centuries. This was the era of a global North–South dichotomy, with all of Europe being seen as a zone of wealth and technological advancement in contrast to a "Global South" of colonies and later "less developed countries" attempting to catch up. The Cold War distinction drawn between the First and Second World on the one hand and a benighted Third World echoed a saga of economic and technological dominance dating from colonial times. In this new vision of different "worlds," the "North," comprising the industrialized countries of predominantly European origin, came to be seen as a product of historical imbalances and injustices—and as a moral actor with an historical responsibility to undo them.

All these regional stereotypes have persisted throughout European history and remain cogent even to this day. They are components of "mental maps" that, while not precisely reflecting what is "out there," define a reality of attitudes and predispositions that is to some extent self-fulfilling. The mental construction of a North–South dichotomy gives the lie to the illusion of a unified continent. The closer one looks, the less appropriate this illusion of unity appears. Nonetheless, the fantasy has frequently been mobilized to buttress the notion of a radical difference between Europeans and "others." "Historically," a Swedish historian once stated, "the idea of a European civilization project has three Others in particular: the Orient, America and Eastern Europe."[46] We will discuss these in the next chapter. But the seeds of "otherness," as we have seen, were already planted long ago in European soil itself in the form of the regional identity of Northern countries, an identity based on distancing themselves from Europe rather than drawing nearer.[47]

Even more marginal than the Scandinavian countries were those fringe territories that impinged on other hemispheres. Iceland, which like Greenland was arguably not a "European" country because it was not even on the continental shelf, was sometimes defined as the "first America" having been explored and settled by Europeans as well as having established an early version of "democracy."[48] During the Second World War it was occupied by the British who subsequently ceded it to the United States as a military outpost. Supported by the U.S., Iceland proclaimed its independence from Denmark in 1944. In his novel *Atomic Station*, the writer and Nobel Prize laureate Halldor Laxness describes an Iceland torn between Europe and the United States—a position underscored by its geological location and its volcanic disposition. Recently, however, Iceland has gravitated towards Europe: whereas Greenland (*Kalaallit Nunaat*) distinguished itself in 1986 by becoming the only country so far ever to leave the European Community, Iceland applied for membership in 2009.

Finland has been a contested border region between North, East, and West since the nineteenth century, since which time it suffered several wars and periods of occupation by Sweden and Russia. Though it took until 1992 for Finland to join the European Union, it has since became a model member. In Southern Europe similar problems of definition arose. The Iberian Peninsula was often said to have turned its back on Europe, focusing more on Latin America, Africa, and the Spanish and Portuguese colonies. With the end of the autocratic regimes in the 1970s this orientation weakened, but the "Atlantic network" is still very strong. The same holds true for the southeastern Balkan region, which until the nineteenth century was labeled "European Turkey." These problems of definition and orientation still linger.

Filling in the Gaps: Mapping North & South

South to Africa

Southern Europe, as we have seen, was an object of lively cultural mapping—resulting in physical and mental maps that not infrequently sidestepped inconvenient empirical evidence in favor of more tendentious representations.[49] This also held for the more distant South, Africa, save that there more adequate geodetic and topographic mapping was also an important aim. On all modern maps since Mercator's in the sixteenth century, South is down and Africa is accordingly placed below Europe. Moreover, to an "enlightened" mind the continent appeared as a major obstacle on the European routes to India, Indonesia, or China. Consequently,

Fig. 1.6   Port Said
in 1880: *Named after
the Wali of Egypt,
Muhammad Sa'id
Pasha, who granted the
concession for building
the Suez Canal, the city
was founded in 1859.
Right from its beginning
Port Said was a global
hub and—like Tangiers—
an object of "international
colonization," with
passengers and crews
from waiting ships
leaving many local
cultural imprints.*

the opening of the Suez Canal in 1869 was widely acclaimed as an achievement to make the given geography "more effective," and it became as much a symbol of triumphant modern technology as the completion of the transcontinental railroad in the U.S. in the very same year. The canal also facilitated penetration of the "dark" continent, and galvanized the "heroic" age of African exploration. Fostered by newly-established institutions like those in Paris (founded in 1821), in Berlin (1828), or the Royal Geographic Society in London (founded in 1830 as an extension of the 1827 Raleigh Travellers' Club) and often financed by newspaper expeditions like those to the sources of the Nile, exploring the "Dark Continent" became the supreme challenge for ambitious Europeans. Mungo Park, Richard Francis Burton, John Hanning Speke, Gerhard Rohlfs, or David Livingston: they all modified established topographic insights on the basis of local knowledge obtained in person or gleaned from natives. By exploiting useful information they made formerly uncharted territories accessible to European knowledge systems:

> Maps from the earliest times, and in almost all cultures where they existed, have been conceived of as representation either of spaces of real or potential power, control or negotiation for princes, merchants and traders, or of universal and religious space, or yet again as directional aids for travelers, pilgrims and navigators—in

other words as objects which represent spaces beyond the world of the actors concerned.[50]

This is particularly true of Africa: long before Europeans had filled in all the blanks on the map of Africa, the 1884/1885 Berlin Conference, proposed by the Portuguese government and convened by the German Chancellor Otto von Bismarck, negotiated a partition of territorial spheres of interest in Africa which have shaped African borderlines until today. There was only one bearer of first-hand topographic knowledge present: the Englishman Henry Morton Stanley. No native African was asked to participate in the European "scramble" until the mandate system was established shortly after the First World War with the introduction of South Africa as an international "player."

Following the race for the last non-appropriated territories, frontiers were extended in every direction. These were ultimately to be "claimed" by the Europeans in competition with new "imperial" rivals in the East and West. North and South (like "the West" in the United States or Siberia in Russia) became European "frontiers," territories into which to expand, to be cultivated and civilized. The northern arctic regions no less than the southern tropical regions served as imagined "supplementary spaces" for the Europeans, as regions justifying both expansionist and imperial ambitions, which were generally characterized and reinforced by insufficient knowledge and an almost complete disregard for the native populace.

North to *Ultima Thule*

Mapping matters, however imaginary its points of departure. It contributes to cognitive schemes of orientation, stereotyped or not. Perceptions of other places and times may even exist in a sphere completely beyond empirical evidence. This is space for imagination and for utopian or dystopian projections, which can be relatively indifferent to what some consider simple facts. Such mappings become landscapes of the soul. This certainly applied to Africa and indeed no less to the North, which since Tacitus and more particularly since the nineteenth century, non-Scandinavians saw as a pre-industrial idyll providing a "better," more organic, societal environment and a purer, more holistic, relationship between man and nature.[51] Before the boom in tourism to the North Cape began, German Emperor William II had already traveled to Norway whenever possible (and even at inappropriate times—for example at the outbreak of the First World War). He gave the Norwegians a statue of Fritjov looking as he imagined a northern hero should. Much later, one of the most successful European entrepreneurs, Ingvar Kamprad, was able to capitalize on the established images of Nordic living styles with his IKEA chain, Europe's most popular home furnishings retailer.[52]

But, of course, mapping can be and (no doubt fortunately) often is informed by empirical data. In fact this is what the Enlightenment ideal of mapping is all about. The European Ages of Discovery were predicated on mapping practices that bridled unrestrained fantasies and imaginings by confronting them with what were considered cold and irrefutable data. Data informed the nature of the mapping and the mapping in turn contextualized bits of knowledge within new constellations of orientation and meaning. This is how, for example, navigational charts were compiled that actually enabled seafarers to get from some position A to another position B with some chance of success. This is also the route by which fantastic European imaginings of life and nature in exotic regions like Africa gradually succumbed to the weight of volumes of ethnographic and natural historical data.

In many cases acquiring this data demanded strenuous and dangerous forays into "uninhabited" *terra incognita*. European explorers often risked life and limb to acquire geodetic and ethnographic data on the uncharted seas and territories of the world. Voyages and journeys of discovery to relatively-hospitable environments like Africa, Asia, and the Americas had been the bread and butter of the European Ages of Discovery. But the frigid Polar Regions, along with the seafloor, long remained uncharted and prey to more or less fantastic mappings. But these, too, finally succumbed to technological advances that vastly improved chances of survival in such inhospitable environments. The Arctic, *Ultima Thule*, was the first of these truly-hostile environments to disclose its secrets to European curiosity, endurance, technology, and the unceasing search for land and resources.

For technophiles, the Arctic (along with the Marianas Trench) were the last undiscovered inhospitable regions on earth. They were the Moon and Mars of their time: alluring and provocative at the same time. During the interwar years "The Cold" had become an obsession for European intellectuals: filmmakers like Leni Riefenstahl went there, fascinated by the eternal ice that challenged man. "Coolness" became popular in social habit, fashion, and architecture, because it was assumed that modern man would have to adapt to the "effective" environment that technology was about to create. The Austrian engineer Hanns Hoerbiger even promoted a "World Ice Theory" or "Glacial Cosmogony," stating that life and history came from the ice and would eventually return to it.[53]

In the early 1930s the German journalist Colin Ross decided to move to the Arctic Circle with his family and all their belongings. He sought to demonstrate that—with the help of modern technology—it was actually possible for modern Europeans to live there.[54] Ross had been educated as an engineer and became a disciple of the geopolitical writer Karl Haushofer. Ross was at that time already one of the most successful authors and filmmakers in Weimar Germany. In the early 1920s, after a stint at the German Museum of Natural Science and Technology in Munich, he turned to journalism. His assignments took him so far afield that his editors soon advertised him as "the German who has travelled the world most extensively."[55] Ross was one of the popular authors of interwar geopolitical writing, a genre that strove to assess worldwide "potentials" in resources, human population, and technical skills. He was particularly enthusiastic about "young" countries and continents like Japan, Latin America, Canada, or Australia. Not surprisingly, Ross was especially interested in Africa and the Arctic regions since they were closest to Europe and hence served as major spheres of European interest and as "supplementary spaces." Whereas in 1927 he labeled Africa as an "awakening Sphinx" which had already adopted European technology for its own sake and use—however strange this appropriation and adaptation may have appeared to European eyes—the Arctic remained virtually untouched by European civilization. But Ross was convinced that "the whites are moving north."[56]

Ross was a descendant of the British Polar researcher and explorer James Ross, who in 1818 had first encountered and described the Inuit.[57] James Ross' travels marked the beginnings of the systematic exploration of the North. Up to then, Europeans had been motivated to venture north only in pursuit of luxury goods like arctic fox or polar bear furs, or ivory from sea mammals. Towards the close of the nineteenth century, travels to the North were inspired by a set of aims that also underlay European exploration in the tropics: a fundamental curiosity; the quest for routes and passages, for minerals and other resources; the desire to erase the remaining white spots on existing geodetic and cognitive maps; ethnographic endeavors (like the first field trips by ethnographer Franz Boas to the "Eskimos" of Baffin Island in 1883); or the empirical testing of hypotheses about the relative positions of the Magnetic and Geodetic North Poles. One major additional ambition was to test the combination and endurance of man and machine in utterly extreme conditions of climate and remoteness. Decisive preconditions for all these northbound tours were organizational skills and seaworthy vessels capable of navigating icy waters.

This final chapter of European exploration had begun and did not reach its peak until far into the twentieth century. The great era of polar expeditions from 1880 to 1930, including the race to the respective poles was predominantly a European endeavor, though soon taken up by Americans and later also by the Soviets. But Europeans had been knocking on the Arctic door since at least the fourth century BCE when the Greek voyager Pytheas made allusions to a northern island he called Thule. This bit of classical lore was forgotten until the late sixteenth century when explorers like the Dutch Willem Barentsz or the Danish Vitus Bering went searching for a northern passage to the East. They were among the first Europeans to cross the Arctic Circle and in the process they discovered several northern islands including the Spitzbergen (Svalbard) Archipelago. The next pioneer to arouse European interest in the "eternal ice" was John Franklin, who in 1845 organized what turned out to be a dramatic expedition to discover the long-sought Northwest Passage. Years after his dramatic death in the cold, scientists found out that he and his comrades had died from eating canned meat containing toxic levels of lead—on this occasion modern technology had proved fatal. The Norwegian Fridtjof Nansen later traversed the Greenland icecap from east to west on foot, a journey he described in his book *In Night and Ice*. In later years this title became almost a synonym for the Nordic region. Ross noted the extent to which Nansen and his ship, the *Fram*, had contributed to European interest in and visions of the Arctic region.

In explaining his fascination for the circumpolar northern regions, Colin Ross used arguments that were often also applied to the southern deserts: from the vast perspective of geological time the Nordic region had only recently been home to extensive green forests, palm trees, and other tropical plants. Modern-day technology could provide opportunities and means to reopen Arctic territories for human civilization:

> During the last decades the white race has extended its living and governing spaces far beyond the respective temperate zones towards the South and the North. In the South, however, it encounters increasing resistance from the brown, yellow and black people, who demand more living space for themselves. The North, on the other hand, is vacant, apart from those few thousands Eskimos, Tschuks and other primitives.[58]

Because of its austerity, infertility, and the privations exacted by its rigorous climate, the North spawned a specific kind of exalted "heroism." Moreover the Eskimos, masters of ice and cold, could teach Europeans "to balance out machine and magic, mechanics and myth, reason and religion."[59]

Ross and his family, though traveling mainly through Canada, never questioned that it was "Europeans" who had extended their settlements towards the North. The poles were seen as the last remnant of "Okeanos," the ancient border sea separating the inhabited world from the uninhabited: "Today we are about to move these borders, to dissolve them, to open up and develop even the Arctic Sea and the last polar territories."[60] Ironically, however, soon after the Norwegian explorer Roald Amundsen and his ship *Gjöa* finally traversed the Northwest Passage through Canada's arctic archipelago in 1906, this long-cherished trade route almost vanished from European consciousness. It was soon to be replaced by the challenge of opening the fastest airplane routes from Europe to America and Asia.

The races to the poles were staged as media events. Did Robert Edwin Peary first reach the North Pole on April 6, 1909, or was it Frederick Cook on April 21, 1908? What did Robert F. Scott do wrong that Roald Amundsen did right when the latter reached the South Pole faster than his rival in 1911/12? What was the connection between Amundsen and the Italian Umberto Nobile that had them flying over the North Pole in the airship *Norge* in 1926? All these expeditions were framed as triumphant victories of

human will and technological achievement (though hardly devoid of nationalist sentiments), but also as personal dramas of almost heroic dimensions, ideal for mass audiences and the new mass media. Symptomatically, these races for the last unexplored territories were no longer exclusive European enterprises. Americans and Canadians now joined Russians, Scandinavians, and Italians. An Austro-Hungarian expedition was even launched in 1872, culminating in the exploration of the northernmost group of Eurasian islands, subsequently called Kaiser Franz Joseph Land.

But expeditions and discoveries in general slowly lost their nationalistic aura and were increasingly touted as achievements of "Man" over nature. The search for and rescue of lost explorers like Umberto Nobile in 1928 became international enterprises. Nevertheless, national territorial claims on the Arctic region ultimately caused much legal and political strife.[61] With regard to the South Pole, where similar interests in whaling and resources conflicted with each other, the international community sought to proceed "more wisely."[62] The Antarctic Treaty System that was

Fig. 1.7 Airship *Norge* on its 1926 Passage to the North Pole: *The race to the poles was pursued with technology, draft animals, or personnel that were not always appropriate. On May 12, 1926, Italian pilot Umberto Nobile, Norwegian explorer Roald Amundsen, and U.S. explorer Lincoln Elsworth officially succeeded in crossing the North Pole. The picture shows the airship setting off from Svalbard the day before.*

signed on December 1, 1959, and that came into force on June 23, 1961, was the first arms control agreement established during the Cold War.[63]

By the mid-twentieth century, the mapping imperative that had driven Europe out to Africa and Asia in the early sixteenth century came to an end. The final frontier had been reached and all the once unexplored and mysterious corners of the world had now been tamed, "domesticated," and pinpointed. In Antarctica the European Ages of Discovery with all their mental and cartographic mapping, their drawing of distinctions and their delineating of "otherness," for once converged upon a common goal. However, with no people living there, the penguins were the only witnesses.

2

Europe's Significant Others

Introduction

In this chapter we focus on the relationship of Europe to a small number of "significant others," that is, nations and regions that for one reason or another were especially consequential for the course of European and global history, particularly in terms of technological interpenetrations. This allows us to cast our conceptual nets wider than in the previous chapter and to extend our concerns beyond Europeans "mapping," to also include Europeans "exploiting" and "exchanging."

Why do we consider the Ottoman Empire/Turkey, Russia/the Soviet Union, and the Americas to be *"significant* others"? In the first place we are by no means convinced that this is an exhaustive list nor, indeed, where the boundary between significant and nonsignificant should be located. The best we can say is that these three are pre-eminent and durable *examples* of significant others but that through time this torch has been handed over a number of times.

That said, we argue that the Ottoman Empire/Turkey and Russia/the Soviet Union have been durably significant because they form Europe's only overland borders and are in this sense ambiguous mediating zones between what is Europe and what is not Europe—not only in a geographical but especially in cultural and specifically technological senses. The huge differences in the ways Russia on the one hand and Turkey on the other incorporated and modified the European Enlightenment and its technological traditions in their diverse roads to modernization speaks volumes about how cultural and political contexts modulate the material "expression" of science and technology.

The Americas, viewed as a cultural *tabula rasa* by both its *conquistadors* in the south and the Pilgrim Fathers in the north, became Europe's first distant mirror and have remained so to this day. Though even where at first more or less equitable exchanges with the natives had been the rule, relations were ultimately shaped by some version of the doctrine of "manifest destiny," a train of thought that justified the displacement and even extermination of native populations in the name of the God-given advance of (European) civilization. The upshot was the establishment of durable and often prosperous colonies of European settlers that applied their European ingenuity and technological prowess to exploiting the new lands and suppressing their native peoples. Most of these American colonies had achieved nationhood and independence by the mid-nineteenth century, taking their cue from the world's first successful Enlightenment-inspired, democratic, anti-colonial rebellion—the American Revolution of 1776–83. This political situation created an "alternate Europe" in the Western hemisphere. In the north, the new United States took "manifest destiny" to new heights, building military, economic, and technological power of such magnitude that it ultimately became Europe's (first) competitor for global hegemony. In the south, weak and corrupt governments and quasi-feudal social relations retarded modernization and turned South America into a dependent supplier of raw materials and foodstuffs for European markets. European hegemony over South America was formally—though not yet practically—overturned in 1823 with the proclamation by the U.S. of the Monroe Doctrine, in which it declared itself the political overlord of the entire Western hemisphere.

As in the previous chapter, we do not aim at a strict historical chronology. Nor are we striving for any kind of comprehensive account. The point is to show by example how Europe shaped itself and the world through technologically-informed interactions with a select group of "close" neighbors, interactions that transformed those corners of the world even as they changed Europe itself.

From Antithesis to Applicant: Turkey & Europe

Trojan Horses

On November 21, 1898, the German newspaper *Welt am Montag* wrote:

> It is Turkey alone that will become Imperial Germany's India. ... We help the Turks to build railways and to construct harbors. We try to inspire their industry. We sustain them with our loans. We deliver ships and guns as well as officers who will teach them to maneuver these ships and to aim these guns. We lend them German civil servants and German soldiers who will occupy the highest ranks in civil and military administration, first and foremost of course for the benefit of the Turkish Empire. The "sick man" will be cured, cured so thoroughly that, awakening from his convalescence, he will be unrecognizable. One might think him really blond, blue-eyed and Germanic. Our loving embraces will infuse him with so much German vigor that he will be nearly indistinguishable from a German.[1]

The Baghdad Railway was a focus of many lofty German hopes as well as of corresponding British, French, and Russian fears. It was seen as an imperialistic infrastructure: a Trojan Horse-like vehicle of German influence. Writers argued that the railroad would revitalize a culture that had gone somnolescent after the Europeans had begun traveling to India and the Far East by sea. Ultimately this German *pénétration pacifique* was expected to inspire a Renaissance of oriental culture.[2] Prior to the First World War the Berlin–Bagdad project epitomized the German Empire's ambitions to employ technology as a tool of cultural imperialism.[3] Contemporaries saw the construction of the Baghdad Railway as both a means of combating Oriental passivity as well as a harbinger of Europeanization, defined as "the introduction of mental, economic and social ways of life that spread from Western Europe."[4]

In fact, the Germans were granted rights to exploit resources in an area stretching 20 kilometers on either side of the Baghdad

Railway line. They were authorized to build public facilities like railroad stations, streets, hospitals, or schools and also power stations, irrigation systems, and communication facilities. The Baghdad Railway extended the southeastern European railroad lines constructed in Romania by "Railway King" Bethel Henry Strousberg as well as lines built by Baron Maurice de Hirsch in the Balkans and finally the Anatolian Railway that already connected the city of Konya to the famous Orient Express.[5]

A German-Austrian engineer, Wilhelm Pressel, was appointed Chief Executive of the Turkish Railways. Sultan Abdulhamid II himself expected the new infrastructure to stabilize his empire, spark economic growth, and enable the Turkish army to control the country more effectively. Between 1900 and 1908, the Germans (under direction of engineer Heinrich August Meissner) also constructed the Hedjaz Railway between Damascus and Medina to facilitate Muslim pilgrimages to Mecca. During the First World War, partisans led by T.E. Lawrence ("of Arabia") attacked the railroad and partially destroyed it. Remnants remain to this day.

The Baghdad Railway, constructed in several stages, was still unfinished at the outbreak of the First World War. It deeply troubled the British Empire, which viewed it as a competitor to the Suez Canal, their vital "Gateway to India." Another factor was the discovery of oil near Mosul, Mesopotamia. This resource had already become a critical object of international entrepreneurial and political maneuvering. In the event, the British did not complete the railroad until 1940, when air traffic had already begun to marginalize rail transport.[6] Prior to the First World War, however, the project had already engendered the usual international tensions associated with every large-scale project in Europe: submarine cables, telegraphy, the canal projects in Suez and Panama, the Emperor Wilhelm Canal in Northern Germany, transcontinental railroad lines in Africa—all of these were hyper-sensitive issues in this period. These imperial infrastructures served as "tentacles of progress" and as "tools of empire," and held out the promise of peaceful conquests and the myth of placid networks.[7] Later on, leaders of venerable but shaky non-European empires and those of new independence movements alike gladly embraced these European infrastructure projects. They bargained on sharing and eventually acquiring these powerful tools to control their (postcolonial) nations and therefore adopted them eagerly for their own ends.

A European Nation?

On October 30, 1973, exactly 50 years after modern Turkey was founded, the impressive Boğaz Koepruesue Bridge across the Bosporus was opened. It was and remains a symbol of Turkey as a bridge *between* Europe and Asia rather than as a European enclave *within* Asia. The European identity of very few transitional zones—with the possible exception of Russia—has been as hotly disputed during the past few centuries, as has that of the Ottoman Empire and, later on, modern Turkey. Since the emergence of a self-conscious Europe, its inhabitants have persistently defined the Turks as their biggest threat. Europeans have long viewed them as the antithesis of what held Europe together—its Christian and later Enlightenment culture—especially since the Crusades and the final fall of Constantinople in 1453. But next to warfare there was also a great deal of positive interaction—and by no means limited to the adoption of coffee. Islamic scholars, for instance, preserved some of the key texts of ancient Greek scholarship and were thus instrumental in inspiring the European Renaissance.

The ambivalent feelings shared by both Turks and Europeans persisted through time. They reached their apex in the second half of the nineteenth century when "orientalist" portrayals of the culture of the late Ottoman Empire achieved enormous popularity in Europe.[8] On the other hand, Turkey gradually adopted European administrative techniques and European technology in an attempt to stabilize a fragile empire that more than once had been labeled "the sick man on the Bosporus." By the twentieth century Turkey retained only a small remnant of land on the western side of the Bosporus—territory that was universally understood to be geographically "European." This has not allayed perennial and fundamental disputes about modern Turkey's affiliation with the European sphere: should it be associated with the European Union, serve as a bridge to the Muslim world, or rather be excluded from Europe as an inherently foreign culture?

European–Turkish interaction within the last two centuries was not only about religion and culture but about technology as well. The Ottoman Empire, often chastised for its traditionalist attitude towards modern technology, eventually acquired a great deal of European knowhow in military strategy and European administrative techniques as a means of protecting itself against the onslaughts of European empire-building. In the wake of Greek independence in 1830 the Turks reluctantly had to acknowledge the superiority of the united European force mobilized against the Ottoman incursion into Europe. For several decades the Ottoman sultans had tried to capitalize on internal disputes among the Europeans, accepting aid from those European powers that appeared less inclined to penetrate and dominate the Ottoman Empire.

Shortly after its defeat in the Greek cause the Empire engaged the services of European military experts like Prussian general Helmuth Count von Moltke as advisors to the Ottoman army. This adoption of "Europeanism" was gradually intensified, although Sultan Abdulhamid II labored "to fashion Islamist modernity in contrast to the West." On the other hand, a new intellectual elite in which engineers played an important role "anticipated the Darwinian triumph of science over religion."[9] After their rise to power in 1908, the Young Turks took over what had hitherto been a rather cautious and selective process of "modernizing" Ottoman culture and radicalized it. This was the overture to Turkish involvement in the First World War—including the abuse of the Baghdad Railway for the deportation of the Armenians.[10]

The Young Turk General Mustapha Kemal in particular, who after 1934 was known as "Father of the Turks" (Atatürk), initiated one of the twentieth-century's most uncompromising programs of national modernization. He had enjoyed a European-style education and greatly admired European efficiency, dynamism, and creativity—as well as the work of the sociologist and Positivist Auguste Comte. After assuming office as president in 1923, Mustapha Kemal embarked on a radical campaign to transform the shambles of the Ottoman Empire into a Europeanized Turkish nation state. This entailed core elements of European nationalism like clear-cut borders and the concept of ethnic, cultural, and linguistic "homogeneity" in explicit contrast to the multiethnic tolerance that the Ottoman Empire had practiced for centuries.

Building on the modernization efforts of the late Ottoman rulers, Mustapha Kemal engaged in a campaign to forcibly reshape his country, if necessary by autocratic methods. In Atatürk's eyes, the world of the future would almost inevitably be European.[11] He replaced Arab script with the Latin alphabet, Ottoman apparel and customs with Western clothing and civilization, and he set up European-style education and administration systems. Like the other Young Turks, he was convinced that Turkey's future could only

Fig. 2.1 An Armenian View of the Bagdhad Railway: *This 1915 illustration shows Kaiser Wilhelm II promising military and political support to Abdulhamid II's persecutions of the Armenian people, in return for Ottoman cooperation in building the Berlin–Baghdad Railway. One of the first genocides in the twentieth century, the Armenian massacres, facilitated by the railroad, are heavily disputed until the present day.*

be assured if it became a secular nation state modeled on Europe and if it adopted similar institutions, constitutional provisions, and technological infrastructures. But Atatürk was also a soldier, and the fact that he regarded the armed forces as the guarantors of modernization has had consequences down to the present day.

A number of Atatürk's measures served as models for other countries moving towards "European modernity": with Ankara he established a new capital located close to the center of the nation state; he fostered historical research to establish a national identity; he opened Turkish culture to Western music, dancing, and art by inviting European artists to Turkey. He promoted schools, universities, the creation of new infrastructures, and the expansion of existing ones like the famous *Tuenel* in Istanbul, which became the third subway line in the world when it was opened in 1875. Atatürk also established a legal system based on German, Italian, and Swiss law. Other elements of European culture and technology adopted by the Turks included the metric system, the Gregorian calendar, and the nominal equality of women, including their right to education. Additionally, Atatürk drafted plans for an ideal republican village (*Ideal Cumburiyet Koeyue*), which imitated Central European spatial planning schemes. The village included up-to-date facilities like mailboxes and telephone switchboards, radios and paved streets—all centered around a monument to Atatürk, who never ceased to insist on the "folly" of seeking guiding principles other than science and technology.[12]

In 1927 Mustapha Kemal made a public speech lasting exactly 36 hours and 33 minutes in which he explained everything he had imposed upon his people. And in 1931, a tad more briefly, he summed up "Kemalism" in six guiding principles: republicanism, nationalism, laicism, populism, statism, and an oxymoron he called "revolutionary reformism." Almost all the changes Atatürk initiated had a deep effect on Turkish society. He considered himself the supreme teacher of his people; and for several years before his death in 1938, he toured his country to preach the gospel of modernity.

In fact, today—at least in some places like Istanbul or Antalya—Turkey can hardly be distinguished from other European settings. This should not be credited only to the "educational dictator" Atatürk but is also the result of reciprocal interaction. The rise of Fascism, for example, brought an influx of numerous exiles from Europe—paradoxically, mainly from Germany—which certainly

Fig. 2.2 Mustapha Kemal Teaching his People: *During extended travels throughout his country, Mustapha Kemal in 1928 spread his agenda of European modernization by teaching Latin characters, thus replacing Arabic script. The "father of the Turkish people" (Atatürk, as he was known after 1934) ostensibly displayed a "Western" outfit and tried to ban the Ottoman fez, the traditional men's hat.*

proved beneficial for the young Turkish republic. Several hundred renowned scientists went to Istanbul and Ankara at a time when they were badly needed, infusing European knowledge into the Turkish educational system or contributing their expertise to the Turkish economy, infrastructure, and administration.[13]

"Europeanization" versus "Occidentosis"

Turkey's non-participation in the Second World War did little to clarify its ambivalent relationship to the Western hemisphere.[14] The fact that the Soviet Union had strategic designs on Turkey was reason enough to include the country in the Marshall Plan and to invite it into the North Atlantic Treaty Organization in 1952. Responding to the implementation of the NATO infrastructure program, one author recalled the historic mission of the Germans:

> Today American engineers and strategic experts are helping Turkey to build railways, roads, and harbors. Today Truman's policy helps Turkish armed forces by equipping and training them on land, sea and in the air—in order to preserve and invigorate Turkey, but to defend the United States as well. Today Eisenhower's policy seeks to unite Germany, Austria, Yugoslavia, Greece, and Turkey in a line stretching from Berlin to Baghdad, and with greater success than the former Berlin–Baghdad enterprise was able to achieve.[15]

In 1959, Turkey first applied for membership in the European Economic Community (EEC). This sparked an endless discussion on whether or not Turkey was already sufficiently European to be integrated into the Community. Although Turkey had backpedaled somewhat since the days of Atatürk and reverted to a stricter adherence to Muslim culture and religion, it nevertheless endeavored to preserve Mustapha Kemal's heritage. Meanwhile the rest of Europe—or at least the members of the privileged community of EEC, European Community, and European Union—continued to keep an exceedingly sharp eye on the country's development. Yet influences continued to move in both directions: every step Turkey took towards becoming "more European" was accompanied by an influx of Turks coming to Europe as "guest workers," blurring the clearcut distinction between Christian Europe and the Muslim world in both societies. Although since 2002 the Turkish premier Recep Tayyip Erdoğan has turned back laicism to a certain extent, it is possible that this factor has helped to keep Turkey from reverting to fundamentalist attacks on "Western imperialism."

This contrasted sharply with the situation in other Arab countries like Persia/Iran.[16] Under the leadership of two Western-oriented Shahs, Persia had also pursued a modernization program that mimicked what Atatürk had accomplished in Turkey, especially during the White Revolution of the 1960s. However, Iranians had already been angered by foreign interventions, especially by the role of U.S. and British agents in overthrowing Prime Minister Muhammad Mossadegh in 1953 for the sake of retaining Western control of Iranian oil resources.[17] Moreover, Shah Mohammad Reza Pahlavi was less successful in reforming his country than Atatürk. In fact, fundamentalist Muslim criticism was expressed very early and culminated in the Iranian Revolution of 1979, which ousted the Shah from power and from his country. As early as 1962 the Persian writer Jalal Al-e Ahmad had published a book called *Occidentosis: A Plague from the West*, in which he settled scores with the prevailing "Euromania." He likened the intoxication with the West and susceptibility to things Western to an illness transmitted by Western mass culture and technology and compared this to the Trojan Horse. The book can be viewed as epitomizing complaints about Western attempts to infiltrate traditionalist societies, and it contributed to what was later to be labeled the "clash of civilizations."

Jalal Al-I Ahmad identified five conflicts arising from the progress of Western materialist culture, which he called a "machine": 1. In response "to the machine's call to urbanization, we uproot people from the villages and send them to the city, where there is neither work nor housing and shelter for them, while the machine enters the village itself." 2. "Urban life demands security" but cannot provide for it: "Walls have been raised to the sky within each person as well. Each seeks refuge behind ramparts of suspicion, mistrust, and isolation" and seeks refuge in cigarettes, opiates, or cinemas. 3. "The machine entrenches itself in the towns and villages, be it in the form of a mechanized mill or textile plant: it puts the worker in local craft industries out of work" and "renders the spinning wheel useless."[18] 4. Old-fashioned "implements, from plow, *kursi*, canvas shoes and oil lamp, to sickle, spinning wheel, and carpet loom, engender a traditional mode of thought (or vice versa)." By pushing traditional thinking aside, the "machine" turns the individual "into a criminal, a complete cynic, or an outright opportunist." 5. "The emancipation of women that tears the veil from their faces and opens some schools to them," draws "women, the preservers of tradition, family, and future generations, into vacuity, into the streets."[19]

Al-I Achmad further asserted that "90 percent of Iran's population look upon the state as an agent of oppression and usurper of the rightful rule of the Imam of the Age." So one should neither pay taxes that would be invested in arms, nor attend the schools, which would "supply the cultural milieu for the breeding of occidentosis." The machine "demands an end to borders, an end to all gates, and the internationalization of everything and every place…in a time, when boundaries throughout the world serve only to mark off the domains of various corporations, to say that up to here belongs to General Motors, here to Socony Vacuum, here to Shell, British Petroleum, Pan American, or Agip Mineraria."[20]

Machines, Trojan Horses, and the means and modes of "infiltration" have become stereotypes of anti-Western attitudes all over the world. Sometimes their primary focus was anti-Americanism, which gave Europeans the opportunity to distance themselves from the "ugly American." With the benefit of hindsight, this criticism can be seen to mirror European and American attitudes about non-European societies during the age of colonialism. It represents one of four major mindsets non-European elites held about Europe or the West: 1. Attempting to ignore Europeans altogether; 2. Adopting the European colonizers' assumption that one's own society was backward and pre-modern and should be subjected to "Europeanizing" reforms (as Atatürk did); 3. Embarrassing Europeans by contrasting their purportedly enlightened values with quotidian practices of violence, forced labor, racism, and the like; 4. Criticizing Europeans from a local and traditionalist point of view, their consumerism, industrialization, their mechanization of life, overvaluing of the national state, individualism, selfishness, and godlessness. This often went hand-in-hand with the conviction that oriental culture was spiritually if not practically superior to European materialism (as Al-I Achmad and many others assumed).[21]

Throughout the twentieth century, this question was hotly debated in many non-European societies: how can European (or Western) achievements, specifically in science and technology, be adopted without succumbing to European (or Western) mores? Interestingly, almost none of these societies actually questioned the desirability of Western science and technology in general—with the possible exception of Cambodia whose Khmer Rouge government

Fig. 2.3 The Lure of "the West": *Marjane Satrapi's graphic novel* Persepolis, *originally published in French in 2000–2003, and the 2007 film version, inspired debates on the "clash of cultures" in general, the role of women in negotiating Muslim traditions, and "Western" influences like pop music in particular.*

tried forcibly to return the country to a Stone Age existence in the years 1975–79. The "war against the West" was long misunderstood as a war against Western technology and comfort, which it was not; opposition was aimed against what technology purportedly entailed: materialism, liberalism, humanism, capitalism, individualism, socialism, decadence, and moral laxity.[22]

After its victory over Russia in 1905, Japan (besides Turkey) seemed to be one of the major exceptions to this rule, providing what seemed to many a viable alternative to Europe.[23] With the help of *rangaku*, a specialized science for the study of European achievements established by the Bureau for the Examination of Writings of the Barbarians, the Japanese succeeded in creating *wakon yôsai*, a kind of amalgam of Western technology and Japanese spiritualism (see chapter 5).[24] By contrast, the Turkish newspaper *Cumhurriyet* took an extremely generous view of the matter, stating that: "Europe" is neither a continent nor a geography nor an organization. Every country that adopts the culture of modern civilization can be designated "European."[25]

Russia & the Soviet Union: Czarism & Stalinism as "Internal Colonialism"

In 1977 the sociologist Alvin W. Gouldner published an article in which he described Stalinism as a form of "internal colonialism."[26] His idea was that the fledgling Soviet state, run by a small group of Western intellectual Marxists seated in an urban power center and governing in the name of a miniscule proletariat, was in effect a colonial power seeking to dominate a vast peasantry spread out over an equally vast territory. In his words:

> What had been brought into being was an urban-centered power elite that had set out to dominate a largely rural society to which they related as an alien colonial power; it was an internal colonialism mobilizing its state power against colonial tributaries in rural territories.[27]

This vision of the Soviet Union also seems to apply, *mutatis mutandis*, in retrospect to Russia under the Czarist regimes. Of course the Czarist court in no way resembled the Bolshevik party, but they exude the same sense of an island of European Enlightenment culture (especially after Peter the Great) awash in a great sea of rural "backwardness" extending all the way through Siberia to the Pacific Ocean. Both of these regimes had to resort to "colonial" violence in order to maintain their exploitative grasp over their extended territories and populations.

From this perspective, both the Russian and the later Soviet states acted as a kind of proxy for Europe in extending a colonial form of rule and imposing colonial infrastructures and technologies over the vast Northern Eurasian continent. At the same time, of course, both the Czars and the Bolsheviks pursued their own interests, interests which more often than not brought them into conflict with European powers, particularly where Russia bordered Europe in the west and on its southwestern flank.[28] The Crimean War was a "hot" version of this simmering conflict while the Cold War showed its more latent aspect. The ambiguous geographic identity (is it Europe?) of the border regions and indeed of Russia/the Soviet Union itself (which recalls but hardly resembles the ambiguity of Turkey) is mirrored in Russia's ambiguous cultural identity. What seems beyond dispute, however, is that both Russia and the Soviet Union were very much colonial empires in the European style.

Russia & Europe

As in the case of Turkey, the big question about Russia was: is it part of Europe? Both geographically and notionally, Europe's eastern boundary with Russia remains the most ill-defined. In political, economic, and cultural terms, large parts of the Eastern bloc that were subjugated by the Soviet Union after 1945 had by all accounts been integral parts of Europe in the past and have been reorienting themselves towards Western Europe since at least 2004. Whereas Belarus and Ukraine constitute an interstitial and ambiguous space, post-socialist Russia has jealously guarded its political independence in preference to any affiliation with the European Union. Yet because of global economic integration, Central European border regions are more interconnected with the rest of Europe today than ever before. This is above all true for science and technology.

The relatively strong cultural and technological affinity between the satellite states of the former Eastern bloc and the nations of Western Europe, still evident even in the Cold War years between 1945 and 1989/90, justifies the term "East-Central Europe." In the 1980s, Polish, Hungarian, and Czechoslovakian intellectuals and dissidents additionally coined the term "Middle Europe," and would later refer to the period after 1989/90 as a "return to Europe."[29] East-Central Europe has without doubt always been a flexible zone in which the concepts of "Europe" and "the Other" have intersected. In earlier centuries the region was often up for grabs and an object of bargaining both in peace and in war.

From a historical standpoint, what we now call "Eastern Europe" long registered as the "North" on the mental maps of Europeans.[30] Since the Middle Ages, Central European countries endeavored to either push the Slavs, who were regarded as foreign, back towards Asia, proselytize them, or civilize them: that is, to domesticate them to agriculture and urban development.[31] Conversely, Russian politics to the present day has retained a deep ambivalence towards Europe. All the Czars since Peter the Great strove to modernize their empire (or at least its major cities) according to European fashions and techno-scientific models, and to integrate Russia into the Concert of Europe. At the same time, they attempted to maintain Russian autonomy in order to reinforce their privileged position with respect to Asia and to vie with Europe by vaunting the greatness and diversity of the vast Russian domains.[32]

For the Soviet Union, it was possible to approach the question of its affiliation to Europe in a somewhat more relaxed fashion, as the country felt strong and independent enough not to view this particular connection as an existential problem. In a politically symbolic gesture, it was thus able to situate its showpiece city Magnitogorsk (1929) exactly on what was conceived to be the border between Europe and Asia. This followed a pivotal suggestion made amongst others by historian and geographer Vasily N. Tatishchev, who in the first half of the eighteenth century had simply drawn an arbitrary line of demarcation along the Ural River and Mountains. While residential neighborhoods were built in the industrial city's "European" districts, factories were erected in the "Asian" sections of the city, partly for the purpose of showcasing the Soviet Union's self-imposed mission to industrialize its "underdeveloped" areas.

To this day, a granite pillar in the mining city of Nizhny Tagil near the Ural River continues to identify East and West as quintessentially Asian and European respectively. According to unverified legend, the local mines and copper smelters can be credited not only with supplying copper for the construction of New York's Statue of Liberty, but also for the roofing of Britain's Palace of Westminster.[33]

Many of the Russian intellectuals who were forced into exile after 1917 had little truck with this geopolitical conception of the new Soviet Union. In the style of their like-minded predecessors in Czarist Russia, they conceived of a third quasi-continent called "Eurasia," rooted in a Christian-Orthodox mindset and considered independent from Europe rather than simply underdeveloped. Traditionally, this had tacitly enhanced the symbolic function of Moscow by elevating the city to the position of a "third Rome."[34] Another notion prevailed alongside this concept, namely that the border between Europe and Asia ran through the middle of Russia, and that the state comprised a European "mother country" as well as an extra-European "periphery"—a view that transformed

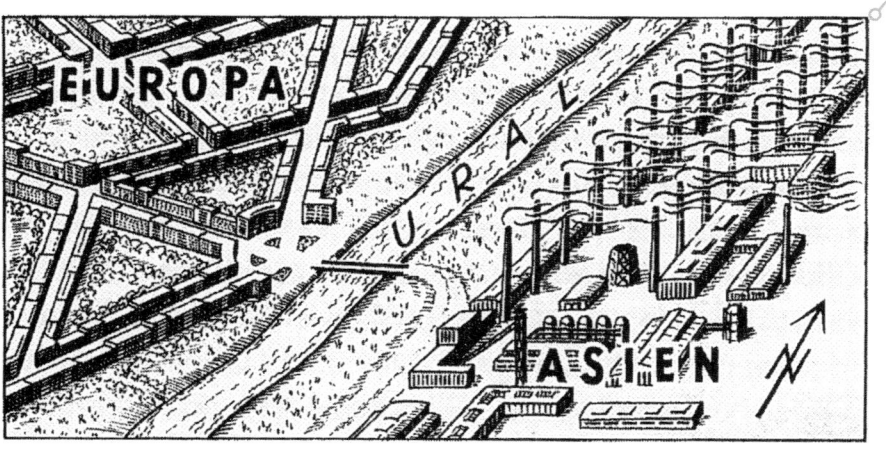

Fig. 2.4 "European" & "Asian" Parts of Magnitogorsk: *The Soviet showpiece city of Magnitogorsk, founded in 1929 as an exemplary industrial city to fulfill the first Five Year Plan, was intentionally constructed at an intersection between Europe and Asia. German urban planner Ernst May devised it as a linear city, copying the model of Gary, Indiana. The Ural River separated residential areas in the west and industrial areas in the east.*

Russia, and by implication the Soviet Union, into a colonial power like other European colonial nations.[35] During the Cold War, this image was to be reversed: while Eastern European regions under Soviet rule now appeared as "colonies" of Moscow, the rhetoric of some Western European politicians seemed to suggest that Asia extended all the way to the Iron Curtain.

Freeing Mankind—Taming Nature

> The construction pace achieved by Soviet people living under the Stalin constitution is unparalleled in the history of the human race. Not a capitalistic country outside the USSR has ever demonstrated such a tremendous surge and comparable speed in building activity. The construction of the Nile dams took sixty-five years, that of the Panama Canal took thirty-five years, and the Kiel Canal was built within a period of fourteen years. While the American hydroelectric power station Boulder Dam was erected within a space of some forty years; the Grand Coulee water power plant took twenty-five years to construct. The Soviet people finished building the White Sea–Baltic Canal, which was named after J.W. Stalin, in less than two years. And in the near future (five to seven years), the great Communist works of architecture—large water power stations and canals will be built.[36]

This quotation, from a book published shortly after Stalin's death in 1953, illustrates the extent to which, in the middle of the twentieth century, technological achievements shaped the substance and the mindset of the Soviet Union, conceived as an Eastern Region of Europe. At its founding in 1917, the Soviet Union had become the world's "first socialist state" and it had now reached the acme of its might. "Communism is soviet power plus the electrification of the entire country," a dictum coined by Lenin, pointed the way to the final consummation of the socialist dream:

> The great Soviet people's accomplishment of electrification under the direction of Lenin's and Stalin's party harnessed a powerful technology which made it possible to transform nature, change the climate, avert droughts, metamorphose deserts into blossoming gardens, and create a life more beautiful than that depicted in fairy tales.[37]

The GOELRO (State Commission for the Electrification of Russia) plan had been announced on December 22, 1920 in the Bolshoi Theater in Moscow for the purpose of implementing the electrification of the Russian Soviet Federal Socialist Republic. At the same time Lenin introduced the concept of a planned economy, which certainly accelerated the industrialization of the country but not without severe collateral damage, including serious ramifications for those involved and ruinous effects on nature. This early example set the stage for the way technology was handled in the Europe's Eastern reaches for much of the twentieth century.

The metaphysics of dialectical materialism propounded by Karl Marx and Friedrich Engels and the consequent Marxist-Leninist obsession with the generative power of new economic arrangements produced an especially unadulterated version of man's instrumental relationship to nature. The socialist aim of subordinating nature to the needs of mankind through technology practically became an end in itself. Ecological concerns about unbalancing nature were voiced very late in the game, if at all. The Chernobyl disaster of 1986 became an instant symbol of the pervasive carelessness and irresponsibility that had characterized previous high-risk technological undertakings. One of the most harrowing examples was the so-called Northern river reversal, a megaproject being carried out since the 1940s. The hydraulic engineer Mitrofan Mikhailovich Davydov based the project on plans that were an easy match for Soergel's earlier hydrological megalomania in the Sahara (see chapter 4). Davydov proposed to revitalize the arid countryside in the Soviet South with water from the Siberian rivers by damming and partly redirecting their flows to the south. These plans were not scuttled until the late 1980s under the leadership of party secretary Mikhail Gorbachev.[38] However, the current state of the Aral Sea, large sections of which are completely dried out due to canal construction, is sobering testimony to what might have happened had the project been fully realized.

Despite the manifest environmental consequences, the fact remains that between the 1920s and 1950s the Soviet Union was widely seen as a paradigm for autonomous industrialization. The staged illusion of a smoothly-functioning planned economy, seconded by equally effective propaganda, ensured that any related difficulties, let alone unrelenting state violence—the palpable form of "internal colonialism" with its holy grail of forced industrialization—were rarely visible outside Soviet borders. Thus occasional visitors, and, even more, those admiring the communist "project" from afar, were under the impression that they were

Fig. 2.5 Soviet Construction Site:
The Soviets portrayed huge construction sites with their hydraulic shovels, tower cranes, and trucks in countless images and films. This shot differs from Western photographs in presenting a woman, who is introduced as a brigade leader and is shown conversing with the engineer in charge on an (almost) equal footing.

seeing something new and utopian, something that was in fact going to change the history of the world.[39] The "West," the capitalist remainder of Europe and the U.S., had a world to lose and understandably felt compelled to rise to the challenge. Nations that considered themselves underdeveloped, as had the Russian empire prior to 1917, were particularly keen to model themselves on the Soviet Union. They established "modernizing regimes" that relied on dirigiste organizations, the leadership of "avant-garde" parties, and extensive social reorganization. They also frequently mimicked the Soviet penchant for mega-infrastructures, showcase cities, and rapid industrialization; understandably so, for the lesson the Soviet Union seemed to convey was that it was possible for agrarian countries at the European periphery—or for that matter in the colonial peripheries—to achieve superpower status within a short period of time. Conversely, the Soviet Union was more than willing to provide technological support to anyone willing to call themselves "friends," including weapons such as the notorious and now ubiquitous Kalashnikov. As with the United States, the Soviet use of the term "friend" became more liberal according to the temperature of the long Cold War.

The example of the Soviet Union is proof positive that twentieth-century European technology had far-reaching political and social connotations: the rhetoric of "developing a new society" by mobilizing technological and industrial resources was immensely alluring and ideologically compelling. Besides legitimating Stalinist "internal colonialism" for the home front, the new industrial and infrastructural "construction sites of Communism" also enjoyed great international resonance. They became icons in the crafting of a new phase of history.[40]

This rosy vision of Utopia has since given way to the reality of an unrelenting dystopia. It is now widely recognized that the Stalinist *modus operandi* not only ruined the environment but exacted untold human sacrifices as well: workers were often forcibly recruited; alleged "saboteurs" (commonly senior technicians, architects, and engineers who were held responsible for any failure to meet overly-ambitious standards) were imprisoned or summarily executed; and countless victims died in famines caused either unintentionally by bad planning or deliberately on the basis of calculated risk.[41] Another revelation is the extent to which Soviet engineering continued to depend on the import of foreign technologies—vehicles and agricultural machines in particular, but also American know-how in the field of hydroelectricity.[42] In a later phase it also drew on European imports, such as the Togliatti/AvtoVaz plants inspired by FIAT. In the period immediately following Stalin's death, the Soviets were able to maintain their developmental momentum and to sustain their preference for quantity over quality a while longer, as the Cold War fanned the flames of a worldwide race for technological superiority. Yet the Soviet Union's subsequent successes were mostly inspired by military feats, such as their arsenal of atomic and hydrogen bombs, their ballistic missiles, and the legendary satellite *Sputnik* in 1957. But of course competing in this armaments race entailed extraordinary expense and this hardly led to improving everyday life for the majority of Soviet citizens. Starting in the late 1960s, the Soviet Union secretly dealt with Western European countries to import technology in exchange for oil or petroleum gas, but it was nonetheless unable to maintain a credible R&D posture in the field of computers and micro-technology. Though the Soviet Union had managed a believable (even if extreme) "first" and a "second" industrial revolution, it certainly proved unable to stage a "third" one. This incapacity, in combination with a dearth of reforms in social and economic spheres, finally ushered in its downfall.[43]

Brotherhood of Cement & Iron

Since the seventeenth century, the project of "Europeanizing" Russia was concerned not only with adopting courtly and urban lifestyles modeled on London, Amsterdam, Berlin, or Paris and establishing administrations and infrastructures, but also with encouraging qualified persons to settle in the "European" regions of eastern Russia. With regard to the nineteenth century, it is appropriate even to speak of a Russian frontier society analogous to the American Wild West. All manner of immigrants, including deportees and political and religious exiles, now populated and cultivated the large expanse of Siberia, principally its southern regions.[44] Even before the First World War, a mere 10 percent of its approximately 10 million inhabitants were indigenous people.[45]

While the Soviet Union's unprecedented Gulag system, which put "enemies" of the regime to forced labor in the harshest regions of the North Siberian Tundra, was basically a copy of methods already employed by the Czarist regime, the latter had also committed itself to developing Siberia's infrastructure. In 1879, for example, St. Petersburg commenced construction of the Trans-Caspian Railway. Its intended purpose was to facilitate troop movements to the southern flank of the Russian empire which, of course, fueled British fears of losing the so-called "Great Game": that is, their fight for influence and supremacy in the transitional zone connecting to the Indian subcontinent. However, it also facilitated the export of large quantities of cotton from Central Asia to Russia and was thus instrumental in transforming the economy, politics, and culture of Central Asia.[46]

The Trans-Siberian Railway was an even more dramatic symbol of the expansion of Russian culture, Russian administration, and Russian technology into East Asia.[47] The Baltic Count Sergei Witte was appointed to manage its construction in the wake of the completion of the First Transcontinental Railway in the U.S. in 1869. The Russian "transcontinental" was built between 1892 and 1904. Even today it remains the longest railroad line in the world, as well as a legendary link between Europe and East Asia. It presently takes seven days to cover the 10,300 kilometers between Riga and Vladivostok. Scholars disagree about whether the railroad actually succeeded in opening Russian territories east of the Ural.[48] However, in respect to the volume of goods transported, the railroad was the busiest in the world for quite some time. In 1932, when its status as Siberia's trunk line caused constant overburdening, it was decided to build a parallel line. But although planning for this Baikal–Amur Mainline commenced in the 1930s, construction was delayed until 1971 when party leader Leonid Brezhnev reignited the torch. The line was subsequently completed in 1984. Once more, Western technology came to the rescue, this time in the form of 9,500 Magirus-Deutz trucks built in West Germany that were used to haul payloads for the construction.[49]

Both the Trans-Siberian and the Baikal–Amur Railway were powerful symbols of Russian and Soviet colonizing ambitions and technological prowess. Conquering the vast space between Europe and the Far East had long occupied the minds of Europeans. Ferdinand von Richthofen, a prominent founding figure of transport geography, had suggested two transcontinental railroad routes between Europe and China as early as 1874, one of which would have led through Russia, the other across India.[50] The great Russian railroads thus partly embodied the contemporary fascination for global transportation networks designed to access the various European "hinterlands." And indeed, geopolitical authors continued to call attention to Siberia's rich mineral resources. Even before the First World War, Fridtjof Nansen had described it as a "land of the future," and a little while later, it was even referred to as "another America."[51] In 1926, a delegate to a Siberian writers' congress exclaimed:

> Let the delicate green breast of Siberia be clad in the cement armor of cities, armed with the stone muzzles of factory chimneys, and girded with iron belts of railroads. Let the Taiga be burned and felled, let the steppes be trampled. Only in cement and iron can the fraternal union of all peoples, the iron brotherhood of mankind be forged.[52]

It is quite possible that these plans or those of Davydov with the Siberian rivers would have been realized had not the Soviet Union so quickly exhausted its economic resources. In the Second World War, Siberia became a safe haven for wartime production plants, distant enough to keep Soviet industry beyond the reach of German occupying forces—though not for lack of trying. In June 1941, the German National Socialists launched an ultimately abortive attempt to achieve what Napoleon had already failed to do one and a half centuries before: to extend their Central-European hegemony deep into Russian territory. The crushing defeat at Stalingrad (now Volgograd), generally considered the turning point of the Second World War, put a definitive end to such illusions.

While Russian plans for Siberia were modified after Stalinism and the Second World War, they lost none of their import. Among other things, sundry "cities of science" were built. Nikita Khrushchev himself championed the city of Akademgorodok, which was erected in 1958 near the Siberian metropolis of Novosibirsk, and constructed in "cement and steel," that is, using the modern building material reinforced concrete in lieu of bricks.[53]

Europeans have always had an ambivalent attitude toward Siberia, not least because it has remained a land of exile for politically-undesirable persons up to the present. But it also seems to be an inexhaustible reservoir of natural resources to secure the future prosperity of Russia, be it oil or gas, diamonds, gold, or other metals, or even wood and coal.[54] The environmental impacts of all this potential extraction have mostly been swept under the rug. A rare exception was Andrei Konchalowsky's epic 1979 film *Siberiade*, which traces the history of Siberian development in the twentieth century and in the process hints at man's responsibility to nature. Despite the fact that parts of Western Europe have meanwhile been connected to Siberian energy reserves via pipelines, one occasionally has the impression that the "eighth continent," 10 million

Fig. 2.6 "**Best Portland Cement**" **in Siberia:** *An advertisement from the late 1920s hints at U.S. imports of building materials. U.S. engineering expertise deployed at large Soviet construction sites such as dams or power stations accompanied them.*

kilometers square and spanning five time zones, is still very much at the beginning of its process of "colonization."[55] New extractive technologies have recently shifted the focus of economic attention towards the colder permafrost regions, previously considered inaccessible.[56] It seems as though once more Russian and European economic (and environmental) interests are facing off in the arctic North. The "Great Game" for power and influence that was played out in South Asia in the nineteenth century is thus being replayed in the North in the twenty-first.

Dependencia—Independencia

Ecological Imperialism

Once in the fifteenth century, and perhaps only this once, the *rabbits* won. When Portuguese settlers landed on the uninhabited island of Porto Santo, one of the ships' captains had a pregnant rabbit on board. This appealing animal and her offspring were delighted with the island; there was an abundance of food, but the rabbits unfortunately did not distinguish between plants that grew wild and those that the settlers were attempting to cultivate. "Defeated by their own ecological ignorance," these immigrants could not prevail against the prolific rabbits. For the time being, at least, they were forced to abandon the island.[57]

This story highlights an element of European expansion that continued to play an important role even in the years after 1850: ecological imperialism. At least since the discovery of America, Europeans had had major impacts on the flora and fauna they encountered in all the new "little Europes" they established throughout the world.[58] Most spectacularly, diseases like smallpox, dysentery, and influenza repeatedly played a role in weakening, decimating, and sometimes even exterminating indigenous populations. But other more mundane forms of ecological imperialism also accrued to the benefit of Europeans, despite occasional defeats like Porto Santo. In fact the success of European imperialist ventures depended to a great extent on biological and ecological factors.[59] This was by no means limited to the worldwide proliferation of certain breeds of domestic animals over the past several centuries; in point of fact, the liveliest objects of exchange were crops and other cultivated plants:

> Coffee, for example, was smuggled out of Yemen by the Dutch, cultivated in Java and Ceylon and then brought to America, where Europeans had already introduced sugar cane, bananas, ginger and

later Asian spices. On the other hand, tobacco, paprika, pineapple, cashew nuts and rubber traveled in the opposite direction. American vanilla found its way to Java and the islands of the Indian Ocean, where cloves and nutmeg that the French had purloined from the Spice Islands were thriving. Cocoa, manioc, and maize crossed the Atlantic from the New World to Africa, whence in turn the oil palm found its way to Southeast Asia. A whole array of overseas crops, vegetables and fruits were suited to cultivation in Europe: Potatoes, maize, tomatoes, various types of beans, gourds or the "Chinese apple": the orange.[60]

Although Europeans for their part contributed wine, olives, wheat, and several other cereal grains to this worldwide traffic, they were mainly concerned with enhancing their own cuisine with new ingredients that were both exotic and nourishing. Many types of foodstuffs that today bear European Union seals designating them as indigenous fare are actually based on raw materials like spices, sugar, chocolate, paprika, potatoes, and fruit that originally came from outside Europe.[61] Cuisine was one domain in which Europe proved to be an especially "assimilating" culture. The marketing of colonial products would become the forerunner and model for modern market economies based on consumption generated by the creation of "needs" and the development of ever-new "tastes." For many decades and in numerous cities and towns, shops selling colonial products were visible proof of the mutual exchange relationships linking Europe and the rest of the world.[62]

New tropical commodities engendered new patterns of consumption and promoted the creation of a variety of new jobs, crafts, and industries. There was a demand for porcelain so people could drink tea, coffee, and cocoa in style; for tins and pipes to store and smoke tobacco; for mechanical looms to weave imported cotton, and for all the crafts necessary for dyeing and processing the finished fabric. Trade in colonial goods expedited the establishment of specialized centers for processing raw materials such as diamonds and ivory, since these facilities were often located far from seaports and commercial centers and scattered all over Europe. Here a paradigmatic process of hybridization can be seen at work, where the "foreign" was creatively integrated, interwoven with local traditions, and soon viewed as indigenous.[63]

The history of specific products like pepper, salt, nutmeg, and sugar demonstrates the extent to which colonial wares have shaped the European economy since the early modern period. Even commodities as seemingly trivial as nutmeg or tulip bulbs could prove to be high-risk objects of trade and speculation on European markets.[64] High prices spurred institutions like the British Royal Botanic Gardens in Kew near London further to systematize their knowledge of botany and agriculture and place it at the service of colonial expansion.[65] "Acclimatization" became a signal word for Europe's utilitarian attempts to appropriate the tropical world: "Attempting to Europeanize the tropics and simultaneously render Europe more exotic and cosmopolitan, acclimatization organizations espoused a practical approach to science, one promising economic prosperity, improved diets and health, and aesthetic enjoyment."[66] The first and largest of these acclimatization societies was the *Société Zoologique d'Acclimatation*, founded in Paris in 1854, which opened a *Jardin d'Acclimatation* in 1860 to exhibit colonial flora, fauna, and people.

But aside from bringing the tropics to Europe, Europeans also transplanted animals and crops across the globe from one colonial setting to the other. France made sustained attempts to establish tropical crops like cane sugar, exotic spices, or fruits in Algeria after losing Haiti, their former source of these crops.[67] Beginning in 1803, the British brought deer, foxes, rabbits, hares, partridges, monkeys, prickly pears, and the Spanish merino sheep to Australia. Here, once again, rabbits became the plague of settler agriculture, but this time the sheep (and the settlers) prevailed: around 1850 sheep already numbered twelve million in New South Wales and in 1891 over one hundred million on the entire Australian continent.[68] Aside from these few striking successes (and unintended consequences), it must be said that successful biological transfers were probably quite rare. As a case in point, the Germans vainly attempted to establish a system of cotton plantations in their model colony of Togo by bringing in Afro-American experts from Alabama.[69]

Starting around 1850, this facet of European-global interaction became transformed by new technologies on several levels: improved transportation systems made it possible to send ever more exotic products to Europe from ever more remote regions. This transformed several commodities from elite luxury goods into mass products and hence provided new culinary bounty in Europe for a broader class of less-affluent consumers. The food industry, geared to mass production, was able to import increasing quantities of raw materials to Europe, where processing was cheapest. Since distance was no longer a serious handicap, global supply chains came into being; refrigeration and canning made it possible to preserve fresh products so that Europe gradually witnessed the emergence of a food supply in which seasonal or environmental fluctuations played a diminishing role.[70]

This process induced new dependencies and was hence anything but apolitical. On the contrary, Europeans (or individual countries within Europe) were becoming painfully aware of their growing dependence on imports and their consequent vulnerability to economic and political disturbances and manipulation. The risks of interrupted deliveries, deliberate price hikes, or politically motivated restraints on shipping caused them to work diligently at finding substitutes for raw materials from the colonies. First the chemical industry and later the biotechnological sector proved capable of increasing domestic production on the one hand and discovering substitutes for extra-European raw materials on the other.

Supply Chains to Latin America

Whereas the eastern limits of Europe remained geographically, politically, and culturally ambiguous, its western borders, although apparently sharply defined by coastlines and territorial waters, reached across the Atlantic, creating new well-defined Euro-American spaces. In the north this ultimately became the cradle of a technologically informed *Western* globalization; in the south it created a stubborn *dependencia* based on unequal exchanges of technologies, foodstuffs, and natural resources. The interplay between Europe and extra-European producers of food, alcohol, tobacco, and other raw materials described above was nowhere as manifest and complex as in Latin America. The story of the tobacco that Christopher Columbus brought back in his baggage is well known; it is also common knowledge that shortly thereafter Europe began importing maize, tomatoes, and potatoes that readily became staple foodstuffs and were soon perceived as being "European." Yet the exchange also moved in the opposite direction: in 1536 the Spaniards brought the first cattle to the Argentine pampa. These animals were to provide a source of food for European settlers during the ensuing centuries.

Until the late nineteenth century meat was seldom exported; it was mainly animal skins that were sent to Europe. The vast quantities of "useless" meat destroyed in the process moved the German chemist Justus von Liebig to suggest using it to make a nourishing meat extract. The engineer Georg Giebert was in fact able to market this process in the Uruguayan town of Fray Bentos beginning in 1863. In addition to canned beef, meat extract was produced there on a large scale, revolutionizing European soup kitchens and becoming an essential ingredient of hospital fare and military rations. A leading German historian even claims that meat extract was no less than an "indispensable ingredient for fueling modern European world conquest," since it was to be found in the knapsack of every European explorer and field researcher.[71] After the end of the nineteenth century, the development of railroads and refrigerated railroad carriages and ships increased the feasibility of transporting meat all the way from South America to Europe. The seasonal difference meant that it arrived in Europe just as the demand for it was greatest.

At the turn of the twentieth century, Argentina had actually achieved the status of a kind of "second U.S." because, like the nation of immigrants to the north, it had succeeded in attracting the creative potential of Europe. Between 1830 and 1950 3.5 million

Fig. 2.7   Liebig Picture: *From 1875 onwards Liebig issued collectible images to market its beef extract. The trading cards showed scenes from history and the present, mostly in an exotic or even orientalized manner. This one depicts the historic paper market in Turkestan. Legend has it that in 751 CE the first paper mill outside of China was founded in Samarkand. Papermaking was one of the few technologies whose origins were credited to non-Europeans.*

immigrants moved to Argentina from Italy alone.[72] Argentina also mirrored North America in that it possessed a highly-industrialized core around Buenos Aires as well as rich agriculture and a cattle industry in its vast hinterlands. Unlike the U.S., however, even Argentina, the most "European" of Latin American nations, did not develop a coherent innovative space that fused its own qualities and genius with that of Europe.[73] On the contrary, following its emancipation from Europe and the declaration of the Monroe Doctrine in 1823, Argentina, like the rest of Latin America, continued to see itself as trapped in a lasting state of dependency. For many years after 1945 the continent's economy relied on exports to the industrialized countries of the North, including dubious products like drugs and tropical hardwood. Soybeans, exported north as animal fodder or a meat substitute, were even cultivated at the expense of primal forests and subsistence crops.

The Mexican author Carlos Fuentes once pointed out that Latin America, as part of the New World, had at one time embodied the utopian dream of helping to regenerate the Old World and to usher in a new Golden Age. But European greed for the purely material worth of gold and other goods became so overwhelming that Portuguese and Hispanic America were made a repository of everything that modern Europe considered insupportable: "Privilege as the norm, a militant Church, indecent pomposity and the private use of resources and of power belonging to the people," so that the colonial period ultimately "turned the Utopia into a leperhouse."[74]

For centuries South America vainly struggled to escape the economic clutches of North America and Europe. In nearly every instance that Latin America managed to establish strong positions or even monopolies on the world market in the course of the eighteenth and nineteenth centuries, the U.S. and Europe were able to respond in a way that changed the game to their own benefit. This was the case with Caribbean sugar cane, for example. By the late eighteenth century anxiety about the continuity and pricing of foreign commodities drove European merchants to limit dependence on imports and to encourage the development of domestic substitutes. Brought to a point by Napoleon's continental blockade, these anxieties prompted the successful cultivation of sugar beets in Silesia in 1806, the yield of which soon equaled that of cane sugar. Experiments with coffee substitutes made from chicory were carried out for the same reasons and a prize competition sponsored by Napoleon III in 1869 stimulated the development of margarine. By the mid-nineteenth century, domestically refined sugar was well on its way to becoming a staple commodity in every European country.

The fate of Brazilian rubber was different and especially dramatic. At the end of the nineteenth century and in defiance of strict export bans, growing demand induced England to purloin rubber seeds (of the *Hevea brasiliensis* tree) from the Amazon region and cultivate them in the Kew Gardens conservatories. Subsequently the seedlings were taken to the East Indies and planted in British Malaya and the Dutch East Indies. At first this did not threaten Brazil's well-established rubber monopoly; in 1910 rubber and coffee still accounted for 80 percent of the country's exports.[75] By 1913, however, rubber production on Asian plantations outstripped that in Brazil and the Congo. In addition, Latin American and African production facilities had by then acquired a reputation for inhumane treatment not only of the workers but of the general populace as well. The "red rubber" from Putamayo and the Congo made it seem an act of downright humanitarianism to establish new plantations.[76] As with the campaigns to abolish slavery in the nineteenth century, it was difficult to determine precisely the extent to which humanitarian considerations were mixed up with commercial rivalries.[77]

Yet increased rubber production did not ease tensions on the world market. On the contrary, with the Stephenson Restriction Act of 1922, Britain, which by this time dominated the rubber market, artificially created a rubber shortage that spurred other nations to action. The U.S. pioneered recycling with the slogan "Use Less Rubber" but at the same time sought new land for rubber cultivation.[78] The tire manufacturer Firestone established plantations in Liberia, but soon exploitative conditions prevailed there as well. The Ford Motor Company established a progressive plantation in Brazil that included a model village called Fordlandia.[79] The Italian firm Pirelli as well as the French Michelin also invested in Asian rubber plantations. England was forced to bow to these united efforts and soon rescinded her restrictive policy.

Germany reacted to the shortage in its own particular way. Initial experiments with artificial latex had been carried out during the First World War in response to the prevailing rubber shortage. Beginning in the 1920s and to an increasing extent after

1934, Germany produced artificial Buna, which Nazi propaganda presented as a triumph over nations that had supposedly banded together in a "raw materials" cartel.[80] But the technological triumph was more apparent than real. Although the synthetic product satisfied Germany's wartime needs, the biggest projected Buna plant, which was to have been built at Auschwitz, was never completed, while at the same time the former British and Dutch plantations in British Malaya and the Dutch East Indies had fallen into Japanese hands. In the long term, moreover, Buna proved inferior to natural rubber. After 1945 only the socialist German Democratic Republic, which lacked natural resources, continued to produce synthetic rubber in appreciable amounts in a plant in Schkopau near Leipzig.

That said, the German chemical industry was certainly also able to boast a few spectacular successes in their search for ersatz tropical raw materials. In 1897 the aniline and soda factories in Baden (BASF) had succeeded in breaking the British monopoly on indigo dyes

Fig. 2.8   Rubber & its Artificial Substitute: *This German school poster of 1940 shows the production path of natural rubber on the left and of artificial rubber (Buna) on the right, on their way to their respective end products. Due to its loss of colonies after 1918, Germany embarked on early research and development programs to create surrogates that could compensate for lost overseas resources.*

by developing a synthetic substitute. The subsequent synthesis of ammonia, for which the chemist Fritz Haber was mainly responsible, turned out to be a rather more ambiguous development from an ethical perspective. On the one hand it served to make first Germany and then the rest of the world independent of both Chilean saltpeter and the guano reserves that until then had been the dominant ingredients of artificial fertilizer, a product that had been invented by Justus von Liebig in the nineteenth century. Because Haber's discovery enabled the conquest of hunger, the most fundamental problem of humanity, it contributed to the survival of roughly half the present world population. Historians therefore justly consider it to be one of the most significant technical developments of the twentieth century.[81] On the other hand, during the First World War chemically synthesized ammonia was used to produce explosives and ever more gruesome varieties of poison gas. But it was another invention made under Fritz Haber's supervision that, on balance, gave his life an especially bitter turn. After his death in 1934 the pesticide hydrogen cyanide (Zyklon B) was used in Nazi concentration camps to kill people, including members of his own family.

The creation of new artificial substances and their substitution for natural ones based on scientific insights and technological development in many cases severed the once dense linkages between Europe and tropical suppliers. In the process entirely new industrial sectors emerged in Europe itself:

> The rapid development of these industries in the second half of the nineteenth century amounted to the creation of an entirely new concept of industrialism. At the heart of this lay the successful replacement of natural ingredients—whether in textile dyes, food, coloring, perfumes, fertilizers for farms or medicines—by synthetic substitutes made, for the most part, from the otherwise worthless and environmentally-harmful by-products of the coal, gas, and coke-producing industry.[82]

In sum, global interconnections via products and supply chains, while they initially enriched the European landscape, also launched a creative search for new ways to circumvent the dependencies thus created. Many consumer goods and objects of daily use no longer display their extra-European origins. Europe nowadays celebrates biotechnology for having emancipated the Continent from the political and climatic vagaries of its southern markets. Today our taste buds can hardly distinguish whether the taste of vanilla has its origin in a test tube or in vanilla beans (pods) harvested in Madagascar, Reunion, Indonesia, or the Comoro Islands.

The New Europe: The United States of America

Atlantic Communication

While South America settled into the mode of neocolonialism that came to be called *dependencia*, North America—and in particular the United States—developed into Europe's closest mirror and ultimately into a technological powerhouse that rivaled, even as it continually interacted with, the inventiveness of the "Old World." The following anecdote exemplifies the heroic technological mythification of Atlantic history since 1850: in 1858, the American industrialist Cyrus W. Field and his Atlantic Telegraph Company succeeded in laying the first transatlantic submarine cable between Newfoundland and England. In so doing, Field, who had formerly acquired great wealth in paper manufacturing owing to the booming American newspaper market, had done much more than merely demonstrate the promise of a new technology. His copper wires insulated with gutta-percha from Malaysia also adumbrated a new Atlantic sphere of communication. He was said to have thought up the idea while studying a globe in his library—the suggestive power emanating from the map had made the project, which he had initially considered utopian, suddenly seem possible.[83]

However, a reliably functioning transatlantic telegraph link was easier said than done. It was not achieved until July 1866, and then only thanks to the aid of the steamer *Great Eastern*, by far the largest vessel afloat at that time. Field's first pioneering cable worked for less than a month and finally exhausted its communicative capacity after transmitting the request: "Repeat, please!" Soon after the successful laying of the second cable, the Americans laid another that connected Greenland, Iceland, Norway, and Denmark. Since the first cable landed in Britain, Americans feared that it would not be available to them in case of conflict.[84] Yet the heroic narrative surrounding visionary inventor and "system builder" Cyrus Field is so deeply embedded in the history of technology that it continues to be told in the above fashion. In his anthology of *Decisive Moments in History*, Austrian writer Stefan Zweig, for instance, went so far as to consider the event of "the first word to cross the ocean" a unique technological feat and concluded that from that moment on, "the world has but a single heartbeat."[85] A contemporary poem conveys a similar notion:

> 'Tis done! The angry sea consents,
> The nations stand no more apart;
> With clasped hands the continents,
> Feel the throbbing of each other's hearts.
> Speed, speed the cable, let it run,
> A loving girdle round the earth,
> Till all the nations 'neath the sun
> Shall be as brothers of one hearth.[86]

Fig. 2.9 Linking the World through Cables: *The image shows the Atlantic telegraph cable network and the Atlantic Strategic Triangle, spanned by stations on the Portuguese mainland (Lisbon), the Cape Verde islands, and the Azores archipelago (late 19th century/early 20th century). The Atlantic telegraph cables exemplify how a transnational infrastructure may empower a peripheral country (in this case Portugal) by turning it into a decisive gateway for Europe.*

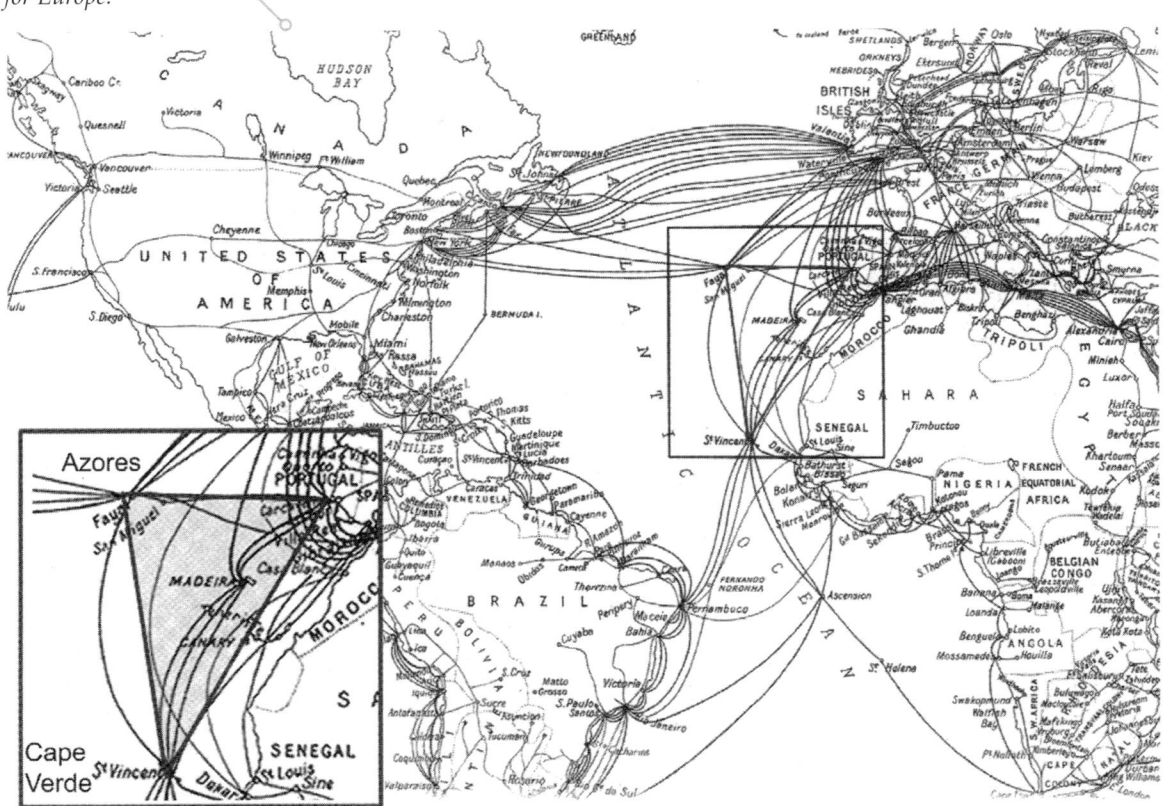

As common and clichéd as the image may seem in its presentation of new technology as a unifying force among nations, there is a kernel of truth to it. The laying of the cable between Europe and the U.S. did inaugurate a transatlantic "era of synergy" that lasted from around 1867 to approximately 1914 and that has not been equaled since. Indeed, the transatlantic collaboration between European and American inventors, scientists, and practitioners laid the foundation for modern technology in four central areas: 1. The generation and utilization of electricity—in fact, the project of laying undersea cables and then getting them to work had already embroiled an international elite of technicians and natural scientists such as William Thomson (Lord Kelvin), Michael Faraday, and Werner von Siemens and given electrical engineering an enormous boost. 2. The invention and commercialization of combustion engines, which laid the foundation for the twentieth century's individualized mass motorization. 3. The development of particularly durable synthetic materials that enabled the creation of entirely new facilities, as well as housing and the built environment. 4. The advancement of modern communication and information networks which, much like the first submarine cable, fostered the perception that the world was gradually coalescing into a single entity.

The decades prior to the First World War saw the transatlantic development of all manner of electrical appliances, engine-driven vehicles, steel-frame skyscrapers, huge suspension bridges, reinforced concrete, urban infrastructure facilities, dynamite, aluminum, the refrigerator and air conditioning, modern print technology and the typewriter, the telephone, (video) cameras and audio devices, as well as the ballpoint pen, Coca Cola, zippers, cornflakes, and the drinking straw. Although these inventions were developed and optimized in the course of the twentieth century, making them more accessible as well as democratizing them, it has been argued that this was only an aftermath during which few radically new innovations appeared, with the (admittedly notable) exceptions of the gas turbine, the computer, and nuclear power.[87]

The "era of synergy" that preceded the twentieth century demonstrated once again that in an important sense the western boundaries of Europe were not defined by the Atlantic Ocean. Rather, Europe has always been closely linked to its former colonies beyond that seeming divide.[88] For one Swiss historian, America was rooted in Europe and shared its global status, though it was also Europe's mirror. "It was simultaneously the duplication and antithesis of Europe."[89] The course of the nineteenth and early twentieth centuries saw not only the development of a common image of the "West" as a community of shared values but also an "Atlantic" sphere of technological innovation in which any attempt to differentiate between European and American contributions utterly missed the point.[90]

The European/North American innovative space established foundations for technical modernity that are still in place today[91]: intellectual property should enjoy the same protection and rights as technological patents and the conviction that technological development profits from a division of labor and a spirit of lively competition. In this climate, specifically national achievements were all but indistinguishable from one another, especially in revolutionary fields like electronics engineering and chemistry, the leading technologies of the "second industrial revolution." The contributions made by people like Werner von Siemens, Nikola Tesla, Thomas Alva Edison, Charles P. Steinmetz, and Michael Faraday, were always closely intertwined. Their research was embedded in intense networks of exchange and communication and constituted the emergence of an international scientific community.[92]

In terms of "mirroring" or differences, the U.S. soon began to challenge staid European conceptions about marketing and consumerism. It excelled in aggressive advertising and the creation of needs, the sale of low-priced consumer goods whose dependence on oil or electricity would then entail high running costs, production based on so-called "Fordism" and "Taylorism," the formation of ideals of mobility and individual tourism, and the readiness to purchase standardized products whose planned obsolescence required willingness to replace them. This stood in stark contrast to still-influential "European" traditions of craftsmanship and one-of-a-kind production and it gave rise to a chronic culture of European distancing from American "mass-culture" that has persisted over time and periodically resurfaces in various manifestations of anti-Americanism.

Europe being Colonized

Whence comes this terrible wave that threatens to wash away everything colorful and unique from our lives? Everyone who has been over there knows it: from America. Historians of the future will record, on the very next page after the great European war, that the conquest of Europe by America has begun in our time. More: it is already in full spate, and we simply don't notice it (all conquered peoples invariably think too slowly). Every country here, and all its newspapers and statesmen, still rejoices when it gets a loan of dollars. We still flatter ourselves with illusions about the philanthropic and economic goals of America: In reality we are becoming colonies of her life and lifestyle; servants of an idea that is foreign to Europe in the deepest sense: the mechanical one.[93]

This statement by Stefan Zweig from his *The Monotonization of the World* (1925) reflects a fear rampant in Europe throughout the twentieth century that the United States of America and the American way of life were coming to dominate the modern world. The very same Zweig who had lauded the laying of the first transatlantic cable as one of mankind's greatest hours, was dismayed at what came flooding back to Europe from the U.S. two generations later. The European "outpost" in the West that had achieved independence by the end of the eighteenth century and in 1823 proclaimed itself de-facto hegemon of the entire Western hemisphere with the promulgation of the Monroe Doctrine, developed so impressively during the nineteenth century that it threatened to push the Europeans to the sidelines. After the First World War it was clear that the prediction made in 1840 by the astute French observer of America Alexis de Tocqueville could no longer be ignored: the future world would be ruled by the two "superpowers"—the U.S. and Russia—and Europe would be marginalized. Like Zweig, many Europeans already considered themselves vanquished, reduced to finding a kind of bitter solace in Oswald Spengler's idea of the "decline of the West."

Of course the new great powers in both West and East were founded on European preconditions: for the Soviet Union this came down to Marxist ideology as well as a materialistic orientation towards the reshaping of both nature and society. For the U.S. it meant an unceasing influx of "huddled masses," victims of oppression, famine, or simply those who found Europe too confining and hoped to realize their stifled aspirations in America's wide open frontier society.

Between 1800 and 1914 alone roughly 65 million people emigrated to the United States, many of them destined for the frontier.[94] The term "frontier" originally referred to the western regions of North

America, the opening and violent "pacification" of which was justified by the missionary notion of the "manifest destiny" of the young U.S. In the twentieth century, however, with the closing of the physical frontier, the concept has come to refer to more generic challenges like economic intensification and improving technological efficiency. For the Atlantic cultural and economic sphere, which consists of the U.S., Canada, and parts of Europe, this common, albeit strongly competitive, orientation towards science and technology formed the nucleus of their identity as well as the cornerstone of their well-nigh religious notion that their civilization was the acme of human development.[95]

From the beginning of the twentieth century this "West" was seen in the U.S. as being the unique source of the economic and technical modernization that would alleviate scarcity and bring prosperity to the whole world. This "American Dream" had been launched on its global career at the Columbian Exposition of 1893:

> In Chicago, Americans flaunted the cheap mass products, the dazzling technology, and the alluring mass culture that, in the coming century, they would spread throughout the world. This American dream of high technology and mass consumption was both promoted and accompanied by an ideology that I shall call "liberal-developmentalism", which can be broken down into five major features: (1) belief that other nations could and should replicate America's own developmental experience; (2) faith in private free enterprise; (3) support for free or open access for trade and investment; (4) promotion of free flow of information and culture; and (5) growing acceptance of governmental activity to protect private enterprise and to stimulate and regulate American participation in international economic and cultural exchange.[96]

To many Americans, their country had become a universal model:

> In order to become a modern society, a nation needed extensive capital investment generated by foreign borrowing and by exports; development of educational, transportation, communication, and banking institutions; a steady supply of cheap labor; maximization of individual initiative for people deemed most efficient; wide-open land use and freewheeling environmental practices; and a robust private business sector solidly linked to capital-intensive, labor-saving technology. This blueprint, drawn from America's experience, became the creed of most Americans who dealt with foreign nations.[97]

These basic convictions and their influence on what has since been termed "the American century" (Henry R. Luce) have frequently been described.[98] For Europeans, the term "Americanization" suggested that Europe had become subject to outside influences it had been able to ignore during the nineteenth century. As early as 1902, in a well-received book, the British journalist William Thomas Stead spoke of "The Americanization of the World."[99] He pointed out the extent to which the U.S. had become a global player with huge economic resources that enabled her to penetrate markets in which Europeans had until then been unchallenged. Stead was, of course, mainly venting the British fear of a "decline of Empire," but in the final analysis his conclusion represented the national variant of a fear that gripped all of Europe.[100] Perhaps ironically, Stead was to perish in the Atlantic when the *Titanic* sank on April 15, 1912—not far from the spot where the first maritime cable between England and the U.S. was joined by the American steamer *Niagara* and the British *Agamemnon*.

In the final analysis it would make more sense to speak of "Westernization" rather than "Americanization," given the continuous transatlantic reciprocity underlying the process.[101] For in the twentieth century, too, the U.S. remained the destination of choice

THE GAP IN THE BRIDGE.

Fig. 2.10 The Gap in the Bridge: *According to Woodrow Wilson's 14 points, the League of Nations would have to be a cornerstone of a new world order. Following an "isolationist" disposition of the U.S. public after the military intervention in Europe, a Republican majority in Congress blocked the U.S. entry to the League. In December 1919,* Punch *ironically illustrated how post-First World War road and bridge building remained incomplete.*

for European emigrants and refugees, and this transatlantic migration of persons and ideas vitiates the notion of simple American dominance, making it too one-dimensional. For example: after both the First and Second World War, the victorious U.S. secured European patents and European scientists for the purpose of enlarging her technological lead.[102] After 1945 the most prominent representative was the German rocket scientist Wernher von Braun who contributed significantly to the U.S.'s greatest technological and organizational triumph up to that time: the 1969 moon landing that had been proclaimed by President John F. Kennedy in 1961 as one of the goals of his "New Frontier."

In Europe the term "Americanization" generally referred (and refers) to certain elements that are seen particularly by the educated classes as being in opposition to supposedly deep-rooted European traditions, to wit: the elements of "massification."[103] But during the interwar yerars of the twentieth century, the Fascist and autocratically ruled states of Europe also believed that they had to wage a "war against the West."[104] "The West" here was a cluster of images and ideas in the minds of its sworn enemies: cities and urban civilization (commerce, mixed populations, artistic freedom, sexual license, scientific pursuits, leisure, personal safety, wealth, and so on), the bourgeoisie, reason and, finally, feminism.[105] Seen from today's perspective, this cultural and religious war on "the West," including Europe and in varying ways its "significant others," is more than ever being waged within Europe itself.

3

Wars & Peace at Home & Abroad

Introduction

The previous two chapters have well documented the fact that the ways Europeans mapped, exploited, and even exchanged depended on their mastery of violence and their readiness to use it if their material interests were at stake. Explorers lived by their rifles, colonial regimes depended on elite foreign troops, and the maintenance of favorable terms of global exchange often depended on the threat of or the actual use of force. Offers that could not be refused were often agreed to amidst the clatter of guns.

Europe's wealth, small political scale, population density, linguistic and religious differences, feudal structure, nationalisms, and revolutionary fervor had historically fostered an inordinate level of violence. This was particularly true for the more organized forms of warfare associated with princes and national states. Like any practice, fighting has an important technological component and it is safe to say that since the Middle Ages Europeans have been at the forefront of efforts to devise ever more ingenious technologies for making and breaking sieges, building "impregnable" lines of defense, and devising new offensive technologies like firearms, gunboats, bombers, and ultimately thermonuclear devices designed to wound and kill as many enemy soldiers (and later civilians) as was humanly possible in a given span of time. This story of how Europeans revolutionized war through new military technologies has been told often enough and we will not take pains to repeat it here.[1]

Less prominent in the literature are accounts of efforts to curb modern war's destructiveness, especially the enormous increase in rates of mortality that went along with "improved" weaponry. This became a public issue in one of Europe's chronic internecine wars, the Crimean War (1853–56), though the American Civil War (1861–65) was in this respect a worthy successor. New hygienic measures such as those advocated by Florence Nightingale and her followers had little impact, of course, on the new technologies and scale of violence, but they did improve soldiers' chances of surviving it.

The new technologies of violence and survival were almost immediately exported to colonial settings—if they were not in fact tried out there in the first place. There, European technological superiority made victories on the battlefield almost embarrassingly easy and enabled small numbers of Europeans to "pacify" vast territories and populations. But the colonial setting also stimulated the development of new strategies and dedicated weaponry. At the same time, colonial fighting was becoming an increasingly "public" affair. This was in the first place due to the general availability of the field telegraph—pioneered in the Crimean War—that in principle enabled the general staff in the home country to exercise day-to-day command over colonial armies in the field—often to the chagrin of formerly autonomous field commanders. The field telegraph had its civilian counterpart in the increasing flood of telegraphed reports (and later wired illustrations) by journalists stationed at the front. Now publics and parliaments, subscribers to imperialist jingoism and those concerned about excessive violence in the colonies, could closely monitor the progress of colonial wars. In an ironic counterpoint, Catherine and William Booth's Salvation Army, founded in 1878, also adopted the imperialist military metaphor. The Booths' moral crusade equated English slum dwellers with benighted colonial subjects and the Army's spiritual logistics utilized the imperial "tentacles of progress" that had by then tied large portions of the world to London's apron strings.

A separate section in this chapter on European domestic and colonial warfare is reserved for the advent of aircraft and air power. This utterly transformed both European and imperial topographies and threatened the traditional conception of national borders. It also completely revolutionized war, as the extensive use of the airplane in the First World War clearly showed. During the interwar period it became evident that the employment of aircraft in a subsequent European war would put civilian populations at risk as never before. But many of these ideas were tried out first in the colonies, where the use of aircraft for reconnaissance and the bombing of civilians was honed to a fine practice—a harbinger of things to come.

The Alchemy of War

War Machines

The scrapping of professional armies and the introduction of mandatory conscription in European states early in the nineteenth century entailed new entanglements between the military sector and civil society as well as the mobilization of greater national strike capacity. Because of enhanced demands on material and human deployment, national armies began to embrace standardization and pursue efficiency in regard to both men and machines. But the military brass could not simply ordain this matrix of mechanization; it had to be accepted by the soldiers themselves. Hence, since the middle of the nineteenth century servicemen had become the objects of "biopolitical" research aimed at creating a powerful amalgam of man and machine. This process fostered military techniques that not only functioned in more precise and disciplined ways, but that were also composed of interchangeable and replaceable parts—again including both man and machine. Isolated machines were arranged into complex systems that ultimately congealed into a destructive universe of precision and standardization.

In fact, nineteenth-century engineers gathered bits and pieces of knowledge from disparate domains like ballistics, kinematics, geodetics, telegraphy, chronography, and cinematography, and rearranged them in standardized technologies capable of unleashing stored energies precisely and instantaneously. Exemplary artifacts for mechanical engineering in those days were the precision lock and the automatic gun. Manufacturing based on high-precision interchangeable parts was in the air. Around the turn of the nineteenth century the chief controller of the French military industry, Honoré Blanc, had developed a kind of "modular concept." It was based on the interchangeability of all parts in advanced military technology, so as to enable rapid assembly as well as replacement in case of breakdown or malfunction. The first major gun of this type was the "Springfield Standard" in 1823; and it was the former U.S. ambassador to Paris, Thomas Jefferson, who in this case was the medium of trans-Atlantic technology transfer.[2]

Here military technology was pioneering new engineering concepts like "permissible values" and "narrow tolerances," aimed at achieving maximum efficiency. These seemed applicable as well to the human bodies of soldiers and workers—and ultimately to societies in general whose "system effectiveness" was to be enhanced.[3] Cinematographers of the nineteenth century were the precursors of those technocrats who covertly strove to run the machinery of state like an imperturbable dynamo.[4] The military pendant was the transformation of foot soldiers into engineer/workers. Loading a gun as quickly as possible and increasing the number of bullets that could be fired per minute became the bread and butter of military drill until Hiram Stevens Maxim's machine gun filled the breach. Its ammunition belt foreshadowed the assembly lines of mass production but also the mass death caused by hails of bullets during the First World War.[5] One military historian concluded:

> In the half century before World War I, military technology changed far more quickly than military organization and doctrine, setting the stage for stalemate. Contrary to most contemporary military thinking, improved weapons and logistics greatly favored the defense. Repeating rifles, smokeless powder, quick-firing long-range field artillery, and machine guns multiplied firepower and extended the killing zone. Doffing gaudy color in favor of field gray or khaki, soldiers left firing lines and maneuver for ground cover and trenches. Runners began giving way to telegraph and wireless, muscle to steam and petrol. Staffs burgeoned to direct vast armies as nations prepared to put millions of men under arms. With the new giving way to the newer ever more quickly, almost every aspect of military life was altered if not transformed.[6]

Fig. 3.1 Krupp Gun Factory in Essen: *One of the most important steelworks in the Ruhr Area, Krupp became a model of an expanding company. Since the nineteenth century, it had maintained strong ties to national politics, kept a global eye on developing markets, and displayed an ability to adapt to short-term demands. In 1917 Essen was a European center for the mass production of arms, and Krupp was notorious for its large-scale artillery. Krupp's range of goods also comprised machinery for colonial exploitation.*

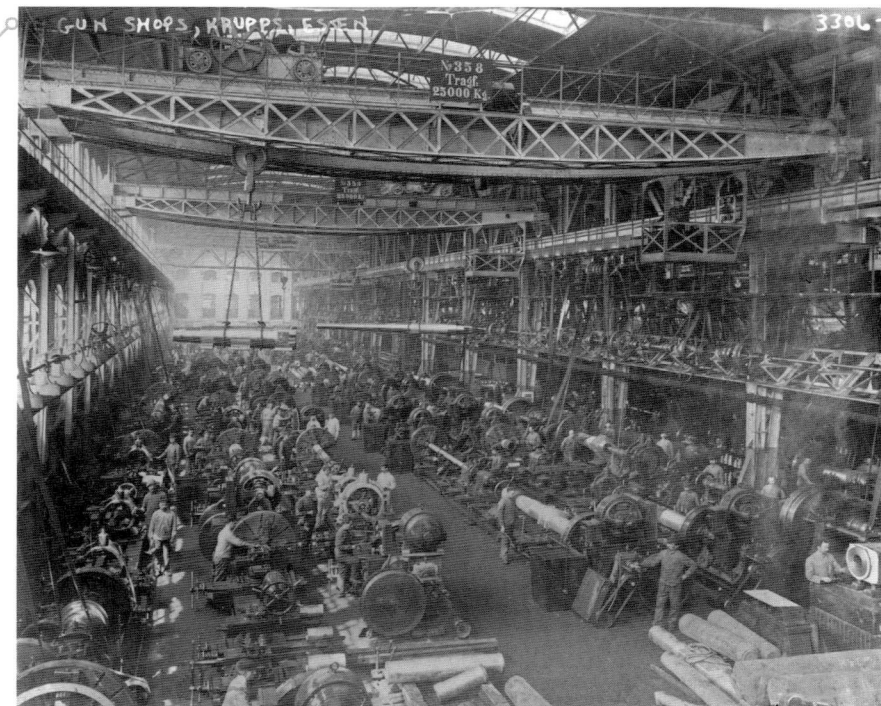

During the twentieth century, training soldiers and improving their fitness became subservient to the international race to acquire superior arms, reliable military resources, and to promote the further industrialization of combat. Soldiers became just another resource, but despite this conflation of man and machine, a nuance remained: during the First and Second World Wars European soldiers were replaced by non-Europeans as "cannon fodder" wherever possible—a substitution that generated a great deal of debate in the interwar period, with Europe's reputation for cultural and technological superiority at stake.[7]

The Crimean Experience

Fighting and war have always been major spurs to technological innovation, and in the age of European globalization wars of all kinds helped to spread the destructive force of modern technology even to the remotest territories. But ironically, by the nineteenth century new military technology was being developed not only to gain the advantage in fatal arms races but also to contain and "civilize" warfare. While by then Europe had begun to share its monopoly on destructive military technologies with the United States, its ever-warring nations remained at the forefront of the search for ways to make wars "safer" and more humane. The new styles of warfare developed on European fronts ultimately found their way to wars abroad, particularly various forms of colonial wars. However, as we shall see, the rules of engagement, the morals, and proprieties that came to govern domestic European wars rarely applied to colonial wars in which tolerance for inhumane tactics and technologies were justified by racism and made possible by distance and ignorance.

This section takes its point of departure from an iconic conflict that caused conspicuous changes in European warfare after 1850, changes intertwined with new technologies. It suggests a composite interplay between technological innovations and media coverage on the one hand and political and social reactions on the other, between the emergence of a technology-driven military-industrial complex and attempts to constrain or even curtail the fatal effects of "modern" warfare. Finally, it traces the export of military violence to non-European settings and the blurring of lines between European and colonial "fronts" of action. Using the example of the courageous Florence Nightingale, it argues that from 1850 onwards several salient tendencies could be perceived. First, there were attempts to limit or control the destructive effects of wars, effects that had been amplified thanks to new technologies. Second, we see the emergence of a regime based on recurrent—and successful—efforts to surpass the enemy's technological capacities, despite the fact that more "advanced" technologies never managed to stabilize military superiority in the long run. Instead they either provoked enemies into creating more "intelligent" (for example, guerrilla) types of warfare or ushered in an arms race. And third, European military establishments began to draw complex lessons from their campaigns outside Europe, based on the exploitation of colonial wars as test sites for new European technologies. Such conflicts were less subject to humanitarian concerns, concerns that had become more salient in European theaters of war, with the notable exceptions of the two World Wars.

The Crimean War (1853–56) stood at the cusp of these ambivalent developments in European military technology. Waged at the European periphery, the war had numerous repercussions throughout Europe, not the least of which followed from the work of the "Lady with the lamp," the redoubtable and heroic English nurse, statistician, and publicist Florence Nightingale. Few persons were so conversant with contemporary European knowledge and could so eloquently articulate Victorian concerns about a society seemingly growing more and more anonymous and calamitous. Single-handedly she developed a new discourse about war and peace, based on the social knowledge and experience garnered within an "imperial" culture. On September 20, 1858, one year after her return as a nurse from Crimean war hospitals, she wrote to her father:

> I was so intensely impressed with the state of the *souls* of the army that I have given up my life for sanitary reform. No moral and intellectual progress can be, by any means of ours, effected without sanitary progress. ... While the dwellings of the soldiers were in a state which debarred them from the common decencies of life—in a state which drove them to drink as the general and only refuge from foul air. ... We have devised a system of statistical registration, which, if it is carried out, will have nothing equal to it in Europe. ... For in the nature of an army and its discipline, the soldier *must be and remain* entirely dependent upon the authorities for life and death, for morality and immorality, for health and disease. ... If I were asked, what is the sin of this generation? I should say *vagueness.* ... It is vagueness, which prevents people from blaming what is blamable and which makes them love what is *not* lovable. ... I am going to send you in a day or two my report to the war secretary, which is to be perfectly confidential, of course. Ever dear Pa, your loving child, F.[8]

The Crimean War was the first major struggle among European powers since the Napoleonic Wars in the early nineteenth century. For a number of reasons, it had an enormous impact on how war and peace have been reflected upon and experienced ever since. It became a focus for debates on the many consequences of warfare in general and the treatment of humanitarian "collateral damage" in particular. Finally, it helped many people overcome that "vagueness" which Florence Nightingale blamed for the social grievances that so plagued her century.

Together with the American Civil War (1861–65), the Crimean War is held to be one of the first "modern" wars and also one of the first with the potential to become a "total" and "global" war.[9] During the campaign a first telegraph cable across the Black Sea

The Daily Mirror

THE MORNING JOURNAL WITH THE SECOND LARGEST NET SALE

No. 2,122. | Registered at the G.P.O. as a Newspaper | MONDAY, AUGUST 15, 1910 | One Halfpenny.

MISS FLORENCE NIGHTINGALE, "THE ANGEL OF THE CRIMEA," WHOSE DEATH IS ANNOUNCED, WITH SOME OF THE NURSES WHOSE PROFESSION SHE CREATED.

Fig. 3.2   A Woman Center Stage: *Shortly after her death in August 1910, "The Angel of the Crimea" and "The Lady with the Lamp" was portrayed in* The Daily Mirror *among nurses "whose profession she created." Today she is also commemorated as a social and sanitary reformer, a statistician, and a feminist.*

was laid, thus establishing a direct line of communication between army headquarters in Crimea and the Allied governments in London and Paris. This resulted not only in greatly expedited news transmission but also in increased opportunities for politicians and governments in Europe to intervene in the campaigns. "The telegraph provided the state with its nervous system and its intelligence-gathering capacity."[10] This was not always to the liking of field commanders, used to being able to make decisions on their own. In one instance a French general ordered a telegraph line cut in order to prevent himself from being "schoolmasterly graded" by his government.[11] Other innovations tested in the Crimean War included static warfare from trench systems and blind artillery barrages and new weapons like the Minié guns, sea mines, and ironclads. The war also generated many secondary theaters and had repercussions reaching as far as Australia.

The combatants in the Crimean War were unequally endowed with weaponry and in particular with the newest military technologies. British industry in the 1850s was about ten times as powerful as that of the Russians, with almost 1,300 steam engines in operation compared to presumably none on the Russian side. The British and French armies were both well equipped with modern ships and arms. At the same time they were poorly administered and in a deplorable state as far as the soldiers were concerned. The troops had not seen combat since the Napoleonic Wars and had to endure frightful hygienic and sanitary conditions. Nursing and treatment of wounded soldiers was at a level comparable to that of the Thirty Years' War. But while this might have left a seventeenth-century European home front utterly cold, new technologies like cheaper and more efficient printing presses, the telegraph and nascent news agencies, made the grim details of the Crimean theater readily accessible to the European public.

Throughout the course of the Crimean campaign European newspapers provided day-to-day coverage of events and conditions on the far-away battlefields. An outstanding figure among the new professional group of "war correspondents" was the illustrious Irish-born William Howell Russell. He regularly telegraphed detailed accounts of his battlefield experiences and findings to *The Times*, which could thus keep the British public informed about Crimean events. Russell's reports focused particularly on the horrendous conditions for sick and wounded soldiers; and he raised uncomfortable questions about the British army's state of preparedness. Russell unfavorably compared British field hospital practices to the superior conditions enjoyed by wounded French soldiers, who were nursed by the Sisters of Charity. Bowing to pressure from a dismayed and outraged public, the British ministry of war asked Florence Nightingale to intervene. For the first time women were officially allowed to travel to a theater of war.[12]

Nightingale, a well-educated, multilingual, and widely-traveled woman with a talent for mathematics, had developed a particular interest in nursing.[13] Arriving at the Crimean war zone on November 5, 1854, she immediately began working in the Barracks Hospital in Scutari. Together with her fellow nurses she had to cope not only with overcrowding, but also with inadequate supplies of basic necessities and facilities and—worst of all—with substandard sanitation. Rats and flies abounded, and only tiny amounts of fresh water were available. While the French had demonstrated that female nursing staffs could do much to improve conditions, British Army doctors had been wary of women's presence on the scene because they feared both sexual harassment by the soldiers and alcoholism among the nurses themselves—an ill to which many untrained nurses in the past had been prone. Nightingale proved them wrong: she reorganized much of the traditional treatment in hospitals and enticed substantial funding from a concerned British public to establish clean and orderly conditions. Death rates soon plummeted. Despite the fact that she was a fervent proponent of the ultimately discredited miasma theory of disease, she succeeded in establishing improved hygiene as a major precondition for successful convalescence.[14]

Nightingale also availed herself of new findings and the wave of innovations that marked her time. In recommending "meat tea" to the British soldiers, for instance, she based herself on German chemist Justus Liebig's ideas that the nutritious elements of meat could be extracted.[15] As a trained mathematician, Nightingale collected systematic data to bolster her statistical arguments for medical reforms. Statistics as a new tool of knowledge and persuasion stemmed from efforts by European nation states to assess their available resources, pioneered by mathematicians like the French Pierre Simon de Laplace, the German Karl Friedrich Gauss, the Belgian Adolph Quetelet, or the British Francis Galton.[16] Statistical organizations had been established in the early nineteenth century, a development that culminated in a first statistical congress at Brussels in 1853 with 26 countries in attendance—officials and non-governmental participants alike. Their aim was to standardize weights and measures on a transnational level, but also to discuss social reforms and alleviate poverty. An International Statistical Institute was founded in The Hague in 1885 with the aim of providing statistics to describe human social behavior for the

Fig. 3.3 Pictorial Efforts at Persuasion: *Florence Nightingale was one of the first women to implement rose diagrams based on statistical evidence. This one, showing "the causes of mortality in the army in the East," was sent to Queen Victoria in 1858. It indicates the number of deaths that occurred from preventable diseases, those that were the results of wounds, and those due to other causes.*

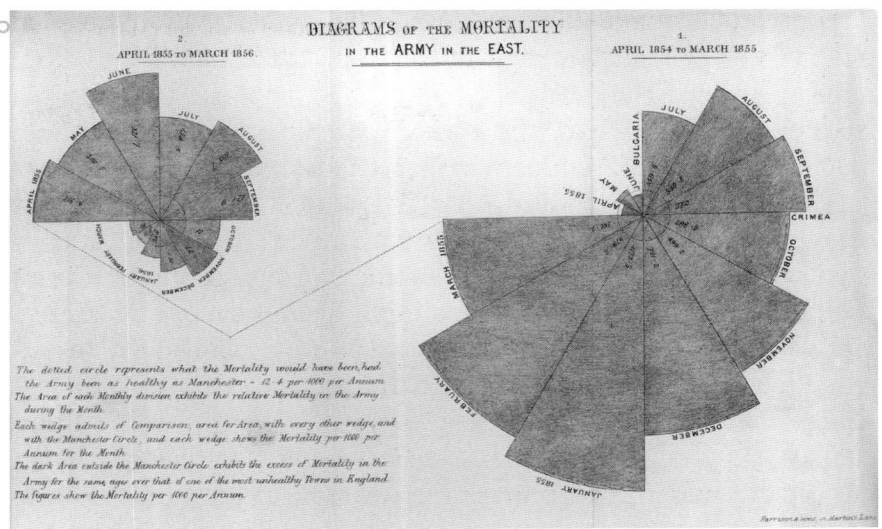

purpose of facilitating social reform, promoting public health, and improving society. After cartography and natural history, statistics became the next step in the systematic "mapping" of nations, of Europe, and ultimately of the world.

Nightingale made frequent use of statistics and developed diagrams better to visualize purportedly "neutral" and "objective" facts and figures. In her report of 1857, which was published later on as *Notes on Matters Affecting the Health, Efficiency, and Hospital Administration of the British Army* and submitted to Queen Victoria, Nightingale employed so-called "polar area diagrams," sometimes referred to as "Nightingale Rose Diagrams." They dramatically displayed the radical reduction of death rates in Scutari hospitals from 42 to 2.3 percent, within six months after her arrival. In 1859, Nightingale became the first female member of the Royal Statistical Society, and like other activists she continued to use graphs to make social problems transparent and immediately apprehensible.

Like Henri Dunant, Elsa Brändström, Albert Schweitzer, and later Mother Theresa, Nightingale became an international role model for European humanitarian engagement.[17] Her interventionist campaigns appealed to a public conscience that had already been massaged by the news media. She also profited from the fact that censorship of war reports was quite lax at that time—at least in Great Britain. While the French public was kept in the dark about death rates among French forces, the British knew by way of their

war correspondents that seven times more soldiers died in hospital
from illness and neglect than in fighting on the battlefield. After
returning home, Nightingale devoted her energy to improving
conditions within the British Army regarding matters of health,
food, and living conditions in general. With Queen Victoria's
support, she managed to establish the "Royal Commission on the
Health of the Army" and an "Army Medical School" in 1857.

Following the Indian Mutiny of the same year Nightingale's
interests came to include India, although she never went there.[18]
Her aims were to improve army sanitation, public health and
irrigation, to combat famine, and to reform land tenancy, nursing,
female medical education, and village sanitation by again influencing government and public opinion. Finally, she facilitated
the establishment of a Sanitary Department in the India Office.[19]
In 1860 she opened a School for Nurses at St. Thomas Hospital in
London and participated in reforming workhouses. A recent record
summarizes:

> Miss Nightingale's Training School, one of the first formal educa-
> tional practices to appear in Europe, was based on a hospital-
> centered model. It stressed the need for *vocation* among young
> candidates and for systematic training in the wards. This dimen-
> sion of professionalism soon crossed Atlantic boundaries and went
> as far as colonial Australia and Canada in the West, and Korea and
> Japan in the East, on the wings of Nightingale nursing superinten-
> dents for schools and hospitals. Around the world, the new profes-
> sional nurses challenged the patriarchal structure of the recently
> medicalized hospitals.[20]

Nightingale pursued her numerous activities right up to her death
in 1910, although she largely remained at home. It is said that for
the sake of fully concentrating upon her duties, she at least partly
pretended invalidism.[21]

Taming the War?

Much has been written about Florence Nightingale. Like many who
have advocated and lived "reform" or "revolution," she has been
both revered and excoriated. Henri Dunant, a Swiss businessman
instrumental in founding the Red Cross, suffered a similar fate. His
experiences at the Battle of Solférino in 1859 are said to have initiated the Red Cross movement as well as the Geneva Convention

of 1864. It is clear that Dunant, like Nightingale, was not simply a heroic figure of humanitarianism but also had the fortune and wisdom to recognize and exploit a window of opportunity. Thanks to the new communication technologies that expedited reporting from the battlefields, European governments could now ill afford to ignore either the fates of conscripted soldiers or the opinions of a well-informed home front.[22] Dunant's "opportunity" was not only the awakening and articulation of a European moral conscience— that had occurred earlier with respect to Negro slaves, abandoned children, and penitent prostitutes. Rather, it was the simple fact that with the establishment of a civilian volunteer organization that could assume responsibility for the care of casualties, European governments and their military advisors could focus in a more single-minded way on waging and winning wars. Inspired by religious sentiments, the Red Cross developed a disciplined and efficient regime of relief: "Murderous refinements of war should have correlative refinements of mercy"—as Gustave Moynier and Louis Appia, two of the leading apostles of the Red Cross movement, stated in their treatise *La Guerre et la charité* in 1867.[23]

It was clear that the champions of humanitarian relief were not pacifists who advocated the abolition of war as such. Pacifist efforts stretched from the International Peace Conference in 1849 in Paris to the establishment of international organizations like the *Institut de Droit international* (founded in 1873) and associated efforts at the international abolition of war in general. Individuals like the French Frédéric Passy or the Austrian Bertha von Suttner tirelessly campaigned to ban war completely from politics.[24] Suttner edited the international pacifist journal *Die Waffen nieder!* (named after her famous autobiographic novel of the same name, published in English in 1892 as *Lay Down Your Arms!*). She also influenced Swedish industrialist Alfred Nobel to institute his Peace Prize.

Pace the pacifists, public opinion still held that waging war was the legitimate pursuit of politics by other means, as Carl von Clausewitz, the Prussian theorist of warfare, claimed in his epoch-making book *On War* (1832). As the *Palgrave Dictionary of Transnational History* recently put it: "The notion of war as a permanent feature of human life prevailed until the end of the 19th century, despite the efforts of many people—notably adherents of various pacifist religions as well as secular thinkers of the European Enlightenment—to devise ways to avoid or abolish it."[25]

Along these patently non-idealistic but perhaps intelligently pragmatic lines, the International Humanitarian Law (or Geneva Law)
sacrificed the best to the better by defining the responsibilities of
parties engaged in armed conflicts.[26] These efforts were at least
partially reflected in emergent international law that applied to
more countries than just a few European great powers and their
ius publicum europaeum. Viewed through the lens of technology,
however, neither the story of war nor that of efforts to mitigate its
destructiveness can be captured in a narrative that focuses only on
legal developments. A recent summary stated that war historians
agree on a paradox governing the long history of war:

> Over time warfare has grown more destructive, and has put the
> lives of ever greater numbers of people at risk; and yet the frequency
> of war has declined steadily so that fewer people, especially in the
> developed world, have experienced war directly. The more trans-
> national war itself has become, the greater has been its potential for
> human and environmental destruction, but the less likely has been
> its active manifestation at any level above the sub national. ... The
> single factor most responsible for the changing nature of war is
> technology.[27]

The development and deployment of new technologies, unpredictable and often unforeseeable in terms of quality and quantity, has
affected warfare fundamentally and gradually rendered military
conflicts unpredictable in terms of their repercussions, their dura-
tion, and their outcomes. Parallel to new technologies in war, old
technologies—sometimes in new guises—also continued to play a
critical role.[28]

Modern wars, which have increasingly been waged by machines
rather than by men, developed a tendency to expand uncontrollably
and to blur the boundaries between military and civilian spheres. The
threat of warfare impacting on civilians impelled peace movements
and international organizations to look for ways to eliminate excuses
for declaring war. At the same time the Western "era of synergy" in
the field of technological innovations from around 1865 to the First
World War ushered in a period of technology transfer that could
hardly be controlled even in peacetime (for example by patent laws).[29]
During wartime it was virtually impossible for any nation to keep the
lid on new technological advantages:

> Such technologies travel across borders as quickly, or even more
> quickly, than armies or navies, especially during wartime with

its immense pressures to innovate. The competitive proliferation of technology, otherwise known as the arms race, has been held responsible for starting its share of wars.... The proliferation of military technology almost always continues during periods of peace.... What came to be called "defense industries" assumed responsibility for entire aspects of postwar economies.... By the mid-20th century there came to exist something of a global military community through the exchange—both explicit and illicit—of standards, technologies, ideas and personnel, so that even some soldiers on opposing sides of the Iron Curtain found that they had more in common with one another than with their respective civilian populations.[30]

In countries with big military R&D sectors and more or less permanently in thrall to the unrelenting arms races of the nineteenth to the twentieth century, responsible engineers assuaged their guilty consciences with a legitimizing mantra: they were merely busy devising a weapon that would ultimately help end all wars. This "consummation devoutly to be wished" had already been associated with the introduction of gunpowder in the early modern period. Now it was just as wishfully applied to submarines, torpedoes, Zeppelins, machine guns, toxic gas, land mines, missiles, and laser weapons. In 1917, the American aviation pioneer Orville Wright blithely asserted that in the near future airplanes would make wars impossible.[31] Until the atomic bomb, engineers believed in—or at least affirmed—this correlation between technological quantum leaps and the "taming," if not termination, of war. But it was not until the establishment of a regime of Mutual Assured Destruction during the Cold War that the initial purpose of "defensive" technology as a means of protecting people finally came into its own—though the stakes of failure or error would have been unimaginably high.

Exporting War

Colonial Wars

In 1898 a British colonial army under the command of Horatio Herbert Lord Kitchener succeeded in suppressing an African uprising led by the Mahdi (the Expected One or the Redeemer), despite being heavily outnumbered. At the crucial battle of Omdurman (in Sudan) the Mahdi's Dervishes, although numerically far superior, were simply mowed down by superior British firepower, including machine guns. Over against 11,000 casualties on the Sudanese side stood 20 casualties among the British and another 20 casualties for the allied Egyptians (see also chapter 4).[32] Experiences like this encouraged twentieth-century military strategists in comparable situations to pursue a policy of ending campaigns as quickly and as victoriously as possible by applying superior technology in an intensive and overwhelming manner.

Although much of the conquering momentum of European expansion derived from the introduction of diseases and "ecological imperialism," military technology had been at least as crucial in conquering colonial space and its indigenous inhabitants. Together with steamboats and tropical medicine, improved firearms were keystones of the New Imperialism unleashed by Europeans in the second half of the nineteenth century:

> Especially from the middle of the nineteenth century, the technologies of firearms, and what historians sometimes call the arms gap, decided outcomes. ... New oblong bullets were invented in France in 1848, and were tested in the colonies: the French used these bullets first in Algeria, and the British against the Xhosa in the Kaffir War of 1851–1852. Around the 1860s there was a crucial technological shift from muzzleloaders to breechloaders. It was the breech-loading gun that created a major discrepancy in power between those with and those without. The American-developed "repeating rifle" and the Maxim, invented by Hiram S. Maxim in 1884, only increased this discrepancy in power.[33]

Victories like those at Omdurman were widely seen as spectacular confirmations of the superiority of European military technology. This could hardly be said of contemporaneous efforts by European colonial powers to "pacify" conquered territories and colonial subjects. "Pacification" referred to military actions and campaigns aimed at achieving fine-grained and routinized control, aims for which superior military technology could not play out its specific advantages, inasmuch as local knowledge was often more decisive and the "periphery" adept at closing relevant military gaps.[34] These

Fig. 3.4 Western Weapons in Colonial Settings: *Henry Morton Stanley is depicted during the Emin Pasha Relief Expedition (1886–89), one of the last major European expeditions into the interior of Africa in the nineteenth century. Emin Pasha (born Isaak Eduard Schnitzer), an Ottoman-German physician, naturalist, and governor of the Egyptian province of Equatoria on the upper Nile, was isolated and surrounded by Mahdists. Stanley, shown here equipped with a Maxim machine gun, was sent to relieve him. Hiram Maxim developed the first fully automatic machine gun (1884), which he offered to the United States War and Navy departments. They refused him and the British War Office adopted it instead. By 1891 it was standard issue for the British Army.*

MR. H. M. STANLEY WITH THE MAXIM AUTOMATIC MACHINE-GUN.
THE EMIN PASHA RELIEF EXPEDITION.

kinds of protracted colonial struggles offered numerous opportunities to reflect on the limits of technological superiority and on the kinds of situations in which imperialists were pretty much "helpless."[35] Favorable conditions, especially pitched battles like Omdurman, were in fact rather rare. Territories were as unfamiliar to European colonial armies as were the tactics of enemy combatants. Neither could be compared to European settings and therefore colonial campaigns were perceived as "unconventional wars." On many occasions, European technology proved to be ineffective on the ground.[36]

If nothing else, colonial wars tended to blur clear-cut battle lines. Tellingly, they were not incorporated into the Hague Conventions of 1899 and 1907, which for the first time made strict distinctions between civilians and combatants. Its Article 22 enshrined an almost revolutionary statement: "The right of belligerents to adopt means of injuring the enemy is not unlimited." Yet since rebellious colonials often adopted guerrilla tactics, colonial wars almost inevitably impacted civilians and their livelihoods.[37] Worse yet, efforts to "civilize" wars often had exactly inverse effects outside of Europe. From war to war and depending on available knowledge and resources, new military technology (poison gas and Dumdum bullets, for example) and new tactics (air policing, for example) were researched and tested on indigenous people. This also applies to "innovations" like incarcerating civilians in camps, ethnic cleansing, and ultimately genocide.[38]

In Leo Tolstoy's *Sevastopol Sketches* (1855) the Russian artillery officer, reflecting on the Crimean War, defined warfare as a kind of handicraft. The same characterization later resurfaced in Tolstoy's masterpiece *War and Peace* (1869). This take on warfare was inspired by the *ius publicum europaeum*, a body of jurisprudence created to regulate conflicts among the great European powers, waged as "cabinet wars." In colonial settings European "manners" were usually circumvented by labeling incidents that led to war as "uprisings" or "crises," or by talking of "disturbances" or "expeditions." As colonial "enemies" were not acknowledged as combatants with equal rights, the rules of symmetrical warfare did not seem to apply. Almost all colonial wars were accompanied by propagandistic efforts to dehumanize the opponent and to apply more or less self-fulfilling clichés. As in the Crimean War, "information warfare" was widely used to massage the home fronts.[39]

For European colonialists, successful campaigns often turned into Pyrrhic victories. They involved disproportionate amounts of bloodshed that eventually damaged a nation's image, even if the wars in their initial phases could be largely concealed from the public. The colonized peoples, on the other hand, understandably perceived Europeans as unpredictable to say the least. In fact, the colonizers often walked a very thin and quite irrational line between the "civilizing mission" on the one hand and the crushing of indigenous opposition on the other. This was exacerbated by a sense of injured honor and the perceived "ingratitude" of the natives, as well as the chronic fear lurking in the wings of a frontier society in foreign terrain. Colonial tactics alternated in erratic ways between the carrot and the stick. Benevolent development and brutal racism were frequently two faces of the same coin. On several occasions, minor incidents sufficed to trigger an escalating spiral of violence.[40]

Some years ago a Swedish author described the rise of genocidal concepts as a gradually developing style of thought among European "conquistadors." Basing themselves on the quasi-racist doctrine developed by Georges Cuvier, Charles Lyell, Charles Darwin, and their worshipful disciples, a doctrine that placed Europe at the top of the evolutionary ladder, European colonizers regularly espoused the idea that inferior races had to be sacrificed to the inevitability of human progress. In the minds of European colonizers this logic clearly justified subjugating the natives and, where necessary, even annihilating them.[41] Everywhere Europeans prowled the wilderness, leaving trails of destruction with their superior weaponry and infrastructural prowess. And they justified their work by arguing that people who refused to monetize their territories and generate profits had simply forfeited the right to keep their lands or even to survive.

Wars in the Ascendant "Information Society"

Extreme attitudes like those justifying genocide provoked growing opposition among a concerned European public. Critics of colonialism made their voices heard and many European colonial nations constructed a system of checks and balances rooted in a strong parliament and a multifaceted public sphere.[42] An exemplary instance of anti-colonialism in the early twentieth century was the launching of a media campaign against African "blood rubber" and the "Congo horrors."[43] This extraordinary campaign not only uncovered a pattern of grievous colonial scandals—it also brought them into the limelight and into the international arena.[44] In the two decades preceding the Congo revelations, the European press had praised Belgian King Leopold II as an altruistic benefactor of mankind and as a patron of public works in the heart of Africa. European politicians and the public both assumed that the funding had come from Leopold's private assets for the benefit of the Congolese people.

In fact, it gradually became clear that quite the opposite had been the case: Leopold had been driven by personal greed and self-aggrandizement. Through unparalleled scheming, he had acquired a huge colony *à titre personnel* and had brutally exploited it, first for ivory and then for rubber to serve the demands of an expanding automobile industry. The king had mercilessly instituted forced labor and tolerated the inhuman treatment of native people in an unprecedented way. When, shortly before his death in 1908, international pressure finally forced Leopold to hand the colony over to the Belgian state, the Congolese population had been reduced by nearly half. The international media and one of the first transnational human rights initiatives had achieved a remarkable success.[45] Both institutions were expressions of a nascent international civil society based upon the idea of globalized circulation of information. The humanitarian campaign for the people of the

Fig. 3.5 **Drill Tower in Kroen Barée on Northern Sumatra:** *From 1892 onwards the* Koninklijke Nederlandsche Maatschappij voor de Exploitatie van Petroleumbronnen *(later Royal Dutch Shell) produced oil in Dutch Indonesia. Exploitation technologies in European colonies were mostly kept simple, and often implemented without humanitarian or environmental considerations. Due to its available oil supply, during the Second World War the Japanese army temporarily conquered Sumatra.*

Congo was one of the first examples of media intervention: "The mobilization of emotions became the alibi for imperial intervention of a new, non-colonial kind."[46]

Fifty years before, the Crimean War had spawned the concept of modern "media wars." Jean-Charles Langlois and Robert Fenton were among the first photographers to transmit pictures from battlefields (because of necessarily-long exposure times, these could be taken only in the quiet after the battles), initiating a "picture war" that transformed wars into "media events."[47] The Crimean War also was one of the first that demonstrated the strategic advantages of news control. The complex interaction between media-influenced public opinion and military aggression was summarized in a popular quip of uncertain origin (possibly by the author H.G. Wells): "The first casualty in a war is truth." Media presence on the battlefield coupled with rapid transmission of reports by telegraph changed the nature of war dramatically because it opened yet another front in which the warring parties had to engage and from which they had to emerge victorious.[48] The media were both chroniclers and participants in warfare, often contributing to morale by depicting the current enemy as "barbaric," as "savage," and "ruthless" in order to stir up soldiers and the home front and incite a nation under arms to wage an unconditional war. War propaganda, which first peaked during the two European World Wars, later became more subtle in an effort to "conquer hearts and minds" by using psychologically-informed tactics.[49] Since the demise of the Cold War, for example, labeling military interventions as "humanitarian aid" has increasingly legitimized participation in "hot" wars. Yet the opponents of war also used the media's influence on the public and appeals to a "world conscience" in a similar way. When in 1948 Norbert Wiener hailed the advent of the "information society" as a bulwark against the barbarism that had characterized the Second World War, he was exploiting just one link in a chain of assumptions leading to the conclusion that "publicity" would help to render crude international aggression impossible.[50] In a longer view,

> the invention of communication as an ideal occurred at a time when the prevailing ideas were those of modernity and the perfectibility of human societies. It was thus the product of a belief in the future... Engineers from the *Ponts et Chaussées* of the ancien régime were among the first to formalize a theory of communication

associated with the organization of a national space and the construction of a domestic market, by applying it to roads and canals. ... Long-distance communication technology was promoted as a guarantee of the revival of democracy. ... Communication means standardizing and doing away with chance.[51]

Global dissemination of information was the common goal of professional communities and international organizations, whose numbers grew exponentially during the second half of the nineteenth century. This transnational sphere of knowledge can be seen as paving the way for what later became "global governance" and its major rationale was the avoidance or at least de-escalation of war and the preservation of peace by fostering communication between parties.[52] Like the community of statisticians, the advocates of international cooperation, though they were also arguing from a European perspective, were nonetheless true pioneers of globalization.[53]

"Darkest Africa" in the East End

The new abundance of public information on extra-European events and conditions upset established stereotypes and prejudices. At the same time, evolving European knowledges were transforming both the material foundations of European and colonial societies as well as the way Europeans perceived the world and themselves.[54] Florence Nightingale had been one of the pioneers in brokering such cosmopolitan knowledge and perceptions; cognitive maps that were produced and applied in metropolitan and peripheral regions alike. Generally, interchanges of this sort were ordered so as to underscore European "superiority," yet they sometimes also seemed to reveal stunning similarities between "inferior" races and the European lower classes. Eminent British social reformers of the Victorian Age, for instance, who in their own language waged "wars" against poverty and moral decline, displayed a penchant for drawing parallels between metropolitan and colonial knowledge systems. When missionary William Booth campaigned for the moral betterment of the British proletariat, he equated European with non-European conditions:

> Darkest England like Darkest Africa reeks with malaria. The foul and fetid breath of our slums is almost as poisonous as that of the African Swamp. ... Just as in Darkest Africa ... much of the misery

of those whose lot we are considering arises from their own habits. Drunkenness and all manner of uncleanness, moral and physical abound. ... A population sodden with drink, steeped in vice eaten up by every social and physical malady, these are the denizens of Darkest England among whom my life has been spent and to whose rescue I would now summon all that is best in the manhood and womanhood of our land.[55]

In 1878 Booth and his wife founded one of the most successful global philanthropic movements. At the time of his death in 1910, the Salvation Army was at work in more than 30 countries where it maintained a flourishing network of schools, hospitals, reformatories, factories, publishing houses, and other institutions. Its publicity continued to rely on colonial metaphors as they presented the Army's global project for moral redemption:

> Britain's urban poor, the prime targets of the organization's redemptive activities, were construed as "heathens" or "savages" in a rhetoric borrowing heavily from imperial travel writing. It is evident that extra-European points of reference had become commonplace in late Victorian public debates. ... The Army's program of "Social Salvation" was conceived on a global scale, as one of its features was the emigration of unemployed plebeian elements of British society to "overseas colonies". Thus, not only the epistemological tools of the Empire but also its infrastructure and the practical possibilities it offered were very much present in public debates and shaped what can, from that point on, no longer merely be called a "metropolitan" discourse.[56]

Booth's description of "Darkest England" was a characteristic example of the impact that imperial technologies of knowledge production had on the emerging discipline of urban social studies.[57] A peculiar dialectic of the familiar and the unfamiliar emerged. Familiarity with the metropolis helped to shape experiences in colonial contact zones and this in turn generated knowledge that later was re-imported to the metropolis. Booth's "tropicalization" of London's East End contained cartographic mappings of spaces of disease and disorder as well as the classification of human types, later on called "ethnology" or "race science." Ultimately, the "Sunken Millions" of the urban poor were put on a par with the "savages," "natives," "pygmies," and "baboons" out there in the colonies.[58] Booth went so far as to propose a colony called "New Britain" for these "Sunken Millions," a plan that was widely discussed in the British public sphere and even partly realized through "assisted emigration." The Salvation Army's General argued:

> The world has grown much smaller since the electric telegraph was discovered and side by side with the shrinkage of this planet, under the influence of steam and electricity there has come a sense of brotherhood and a consciousness of community of interest and nationality on the part of the English-speaking peoples throughout the world. The change from Devon to Australia is not such a change in many respects as merely to cross over from Devon to Normandy.[59]

Salvation Army officers operating in the colonies, on the other hand, often "went native" in order to approach indigenous people as closely as possible and to successfully evangelize non-European "races." Frederick St. George de Lautour-Tucker, Booth's son-in-law, changed his name to "Fakir Singh" and adopted Indian dress in order to pave the way for Salvationist work in India. He also published a book called *Criminocurology or: The Indian Criminal and What to Do with Him.* This took the taxonomic techniques for identifying deviant behavior (in order to "heal" it with Salvationist techniques) to a new global level. The "civilizing mission" had become a global endeavor that targeted classes and races within an "imperial social formation" to transform individuals into what were considered productive and submissive bodies.

Like no one else in his time, Francis Galton showed what a prolific command of overlaps of imperial knowledge taken from both European and non-European settings could accomplish. A half-cousin of Charles Darwin, Galton was a Victorian polymath. He left his mark on the history of "imperial" social expertise and bio-politics, influenced anthropology, criminology, statistics, invention, and exploration, but also contributed to eugenics and finally racist ideology. During the Crimean War, the British army utilized his book *The Art of Travel*, subtitled *Shifts and Contrivances Available in Wild Countries*, for it explained how to maintain hygienic conditions in the wilderness. In 1875, he published the first weather map in *The Times*, later on he developed questionnaires and surveys to obtain comprehensive social data and analyzed fingerprints as a means to unambiguously identify deviant individuals. This new scientific tool of social control was tested first in India during the 1860s. After Galton had spent ten years analyzing pattern types, a

Fig. 3.6 Early Forensic Science: *In his 1892 book on* Finger Prints, *polymath Francis Galton offered methods of indexing and decrypting the "delicate lineations," including plates with principal varieties of patterns like arches, loops, and whorls. He also discussed their "value to honest man." The first Fingerprint Bureau to classify criminal records was established in Calcutta in 1897.*

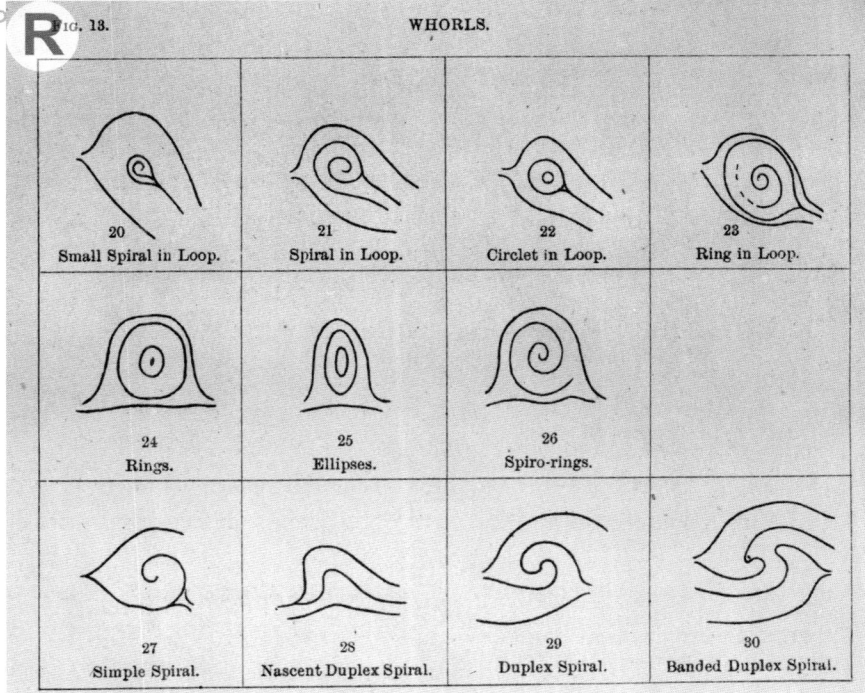

"Fingerprint Bureau" was established in Calcutta in 1897, several years before the London police adopted it in 1901.[60] Like Florence Nightingale (and the "Empress" of the Jewel in the British Crown, Queen Victoria herself), Francis Galton never went to India.

In conclusion, we can say the following about the use of organized violence as a tool of imperial domination: Thanks to technological innovations and the democratization of conscription, warfare moved towards totalization, with as much material and as many men being deployed as possible. Greater participation by the home front was embedded in a media-influenced public sphere. Efforts to contain or control wars and attempts to institutionalize peacekeeping within Europe had their counterpart in an increase of violence exported outside Europe. Unless European colonizers were kept under close scrutiny, "peacekeeping" could easily spiral out of control. Achieving technological superiority in destructive potential usually proved elusive; it either provoked an arms race or more "intelligent" strategies to compensate for deficiencies in competitive technology.[61] Experiences of wars and attempts to control people outside Europe contributed to "imperial" cognitive mappings that could be (and were) transferred and applied at home and abroad. War and peace revealed the ambivalence of technology: asymmetric command of technologies did not guarantee superior power; and technologies of warfare were constantly in dialogue with technologies of peace.

Different Europes from Above

Empires of the Sky

The years between the First and Second World Wars were pregnant with portentous events signaling tectonic shifts in Europe and the world at large. These events were cause for widespread concern about new technologies and their implementation, and for hopes and fears regarding the future of globalization. Concerns about new technologies achieved their apotheosis with the opening up of the so-called "third dimension":

> From its invention in 1903, the heavier-than-air flying machine moved relatively slowly to the center stage of technological wonders, but once it did reach that point, around 1909, it retained its symbolic hegemony until the Second World War. ... The airplane did not simply relativize space and reduce travel time and distance: flying was a new visual representation and appropriation of space.[62]

The airplane, the rocket, and the spaceship gripped the imagination of many Europeans and spawned new mental maps of a globalized world. Aviation fascinated the entire globe and also sparked European dreams of conquering new dimensions of space.[63] European perspectives now shifted away from two-dimensional, primarily static geographies, which had been thoroughly explored during the era of European imperialism. Seen "from above," mapping became three-dimensional, and cartography increasingly embraced dynamic change. An emerging geopolitical knowledge system gauged the potentials of modern communication technologies as well as the availability of industrial resources or "living spaces."

The interwar years were marked by a continuous expansion of communications and traffic infrastructures within Europe and abroad. Aviation took pride of place and airlines became pre-eminent symbols of national progress and prowess.[64] But there was also cause for concern, particularly about the potential role of aviation in wars, both domestic and colonial. "The ease with which airplanes could cross national borders rendered every nation supremely vulnerable to any country with mastery of new flight technologies. Hence the airplane became a vessel of both hope and fear and the object of substantial national investment by European powers."[65] In the aftermath of the First World War, airlines providing mail and passenger service popped up like mushrooms all over Europe. But airlines also quickly proved themselves to be extensions of colonial power by other means. In fact the world's second still-operative airline was established in Colombia in 1919, the *Sociedad Colombo Alemana de Transporte Aéreo* (SCADTA). German financing and the use of Junkers seaplanes moved the United States government to support and invest in Juan Trippe's new international airline, Pan American World Airways. Pan Am, providing regular service since 1927, became a pioneer of international air transport and an instrument of choice of the United States government to secure non-American spaces; its name also alluded to pan-Americanism (as the German "Lufthansa" alluded to the medieval "Hanseatic League").[66]

But other "empires of the sky" had preceded Pan Am's by nearly a decade.[67] On October 7, 1919, the *Koninklijke Luchtvaartmaatschappij NV* (KLM) was founded by banks and business interests at the behest of a former Dutch pilot, Albert Plesman, who headed the company until his death. KLM's first scheduled service between Amsterdam and London was opened on May 17, 1920. It was followed the same year by an Amsterdam–Copenhagen route via Hamburg and in 1923 by a route to Brussels. KLM opened the world's first airline reservations and ticket office in Amsterdam as early as 1921 and is the world's oldest continuously operating airline. In 1928 Plesman also founded the *Koninklijke Nederlandsch-Indische Luchtvaart Maatschappij* (KNILM), the Royal Dutch East-Indies Airlines, which in 1930 inaugurated regular flights from Amsterdam to Batavia (now Jakarta) in the Dutch East Indies, a distance of 13,900 kilometers and until 1940 the world's longest scheduled air route. KNILM merged with KLM in 1945.

Other countries followed in rapid tempo. *Det Danske Luftfartselskab* commenced operation in 1920 as well, followed by *Società Italiana*

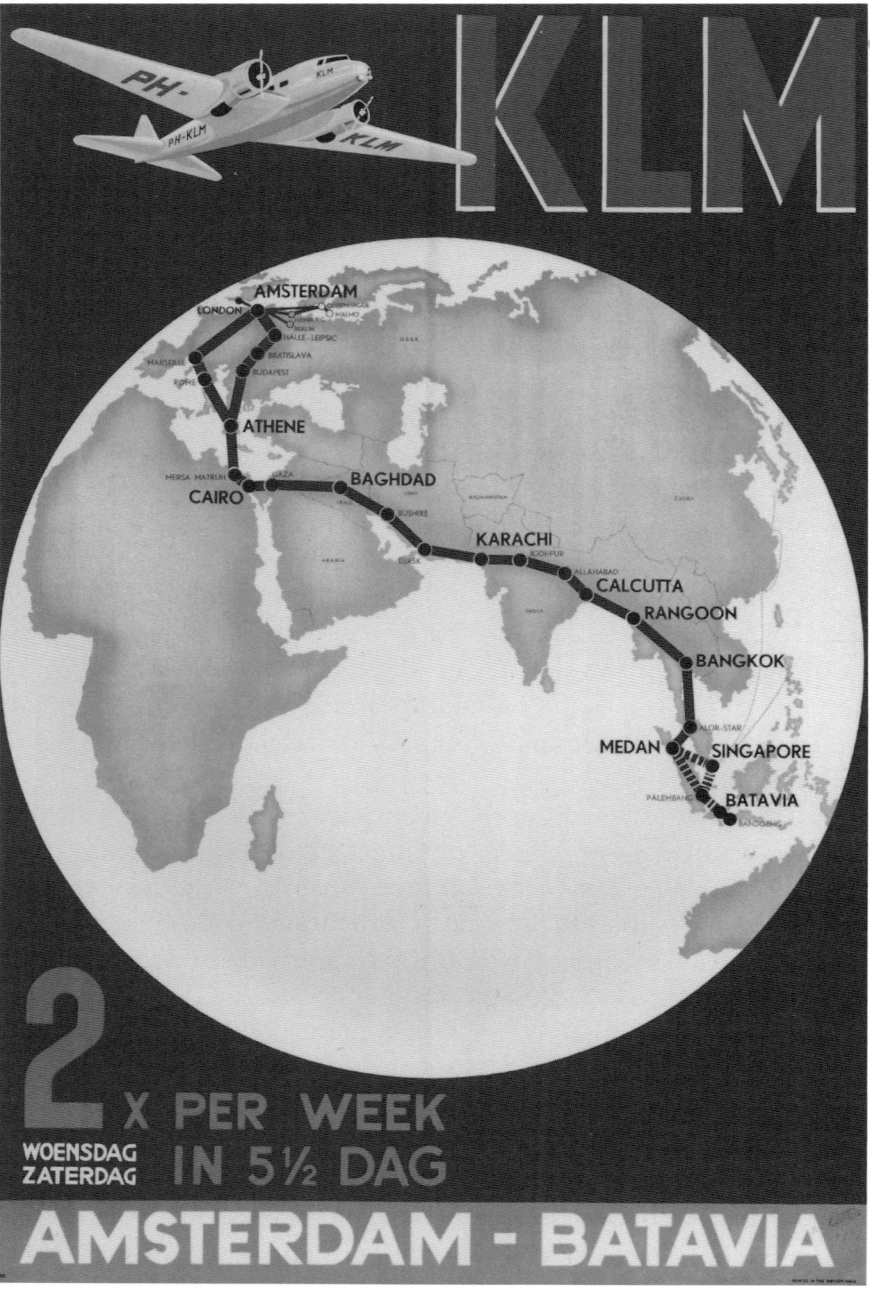

Fig. 3.7   Weekly Service to Batavia: *Like most of the early airlines, KLM Royal Dutch Air Lines boasted about its remotest destinations to illustrate the range of (national) influence. In the beginning, it still took five days to get from Amsterdam to the Dutch Indies. Today direct flights to Jakarta are available, arriving in just 14 hours.*

Servizi Aerei (1922), the Belgian Sabena in 1923, Swissair (1931), and Air France (1933). These aviation empires began to go global by the mid-1920s. By that time British Imperial Airways had established long-distance routes to Cairo, Basra, Karachi, Delhi, and Jodhpur, and by the early 1930s to Cape Town, Calcutta, Rangoon, and Singapore, followed by Australia in 1934, Hong Kong in 1935, and Khartoum and Kano in 1936; the Portuguese Imperial Line to Luanda (Angola) and Lourenço Marques (now Maputo, Mozambique), took off in 1946, just three months after the first TAP flight from Lisbon to Madrid.[68] Even the most remote places on the planet were now accessible to imperial power in a matter of at most a couple of days, and extending the range of possible destinations was just as important for airlines and nations as were the expected profits. Financing these ventures remained touch and go because the metropolitan monopolies were chronically challenged by locally-created services (like Australia's Quantas Empire Airways) or European governments more willing to subsidize their air flight companies.[69]

From Aviation to Air Power

Although in the interwar period Europeans of all stripes were uniformly enthusiastic about the future of aviation, its pragmatic political significance differed from country to country. As far as the still remaining colonial powers were concerned, airplanes implied a major improvement in communication with the colonial peripheries. This was especially true for Great Britain: "The rapidity and efficiency of our imperial communications," Air Vice-Marshall Seften Brancker remarked in 1924, "may well be the measure of our success—or our failure—to hold our empire together in the future."[70] For the colonial administrations at home, increasingly under scrutiny by their parliaments and held to account by a concerned public, air routes brought the colonies dramatically closer. This put a premium on improved administration and investment in colonial development under the motto *mise en valeur*. At the same time the new closeness brought about by aviation made such improved real time administration and strategic investment feasible. The immediacy and elaboration of lines of communication brought about tighter control, enhanced the presence of colonial administrators on the colonial periphery, and produced intense dialog between politicians at home and diplomats abroad. Inky "hearts of darkness" became a thing of the past. Understandably,

Fig. 3.8 Conquering the Air: *During the* Guerra de Melilla *in 1909, one of the so-called* Rif Wars, *Spanish troops launched a balloon better to control their Moroccan adversaries. It took just another four years to discover that "from above" it seemed likely to also be able to drop bombs. After 1914, aerial warfare had conquered the "third dimension."*

of course, colonial officials on the spot, fearing constraints on their autonomy, often bucked at such closer scrutiny by the decision makers in the capital.

While everyone believed that the facile transgression of borders in the third dimension would improve contacts among nations and between colonial powers and their peripheral colonies, aviation was also perceived as a threat to national integrity. Medical experts, for example, feared that increasing air traffic would contribute to the spread of disease. Every time an airplane made a stopover there was the danger that bacteria or infected animals and people might be let off or taken aboard. This led to the establishment of international sanitation standards in the 1930s, but fears of an influx of epidemics from abroad still haunt Europe today.[71] Immigration authorities also suspected that colonial subjects would use airplanes for one-way travel to Europe. To prevent labor migration, British officials therefore implemented legal and regulatory policies that dissuaded British Africans and Indians from flying the national airline.[72]

Aviation also assumed a new role in minimizing the exercise of force in colonial settings.[73] Incidents like the outrageous massacre by British troops of unarmed demonstrators in Amritsar, India, in 1919 led to widespread repudiation of excessive force in maintaining colonial order. The deployment of air power was expected to reduce overt force to a minimum; air policing appeared swifter, more effective, and also cheaper than conventional punitive actions. Views from above made landscapes look like topographical maps, and pilots learned to scan them with the eyes of a field commander.[74] As a corollary, aerial-mapping techniques also opened a new chapter in the saga of the European ages of discovery, improving the accuracy, efficiency, and immediacy of geodetic mapping and extending it to even the remotest territories.

Nonetheless, the terror incited by air policing should not be exaggerated. As colonial rule from the air became more and more familiar to indigenous peoples, its fearsomeness gradually dissipated.[75] Nonetheless, effective or not, the new doctrine of "control without occupation," developed and applied in Iraq, did at least guarantee a future for the Royal Air Force in the early 1920s: "The air control scheme offered the British government the magic formula of continued control with reduced expenditure, and this unorthodox method offered one solution to a central dilemma of post-war imperialisms."[76] In consequence, the balance of power between the central state and the tribute societies on its imperial peripheries was irreversibly altered. But in the end this shift involved not only improved communications and surveillance, but the active exercise of air power as well:

> The control of the Empire and repression of insurrection was secured by new technology without which the British might have been forced to abandon several areas, including part of the Middle East. Foremost was the use of air power for reconnaissance, supply and attack. The bombing of dissidents, at times ruthless, ensured British authority at immense savings of manpower.[77]

During the early postwar years, Winston Churchill, then British Secretary for War and Air, saw Mesopotamia (that is, Iraq) as a laboratory for high-tech colonial control in line with the new colonial obligations stipulated in the Versailles Treaty and by the League of Nations mandate system.[78] Churchill and Hugh Trenchard, Chief of the newly created Royal Air Force Air Staff, supported deployment of air power against Iraqi nationalist uprisings "to police colonial territories more cheaply than by employing large numbers of troops on the ground."[79] At the same time, the British relied on alliances and cooperation with the Iraqi people until the country gained independence in 1932: "The fact that the RAF was put in control of security in 1922 symbolized that this was a new situation in the colonial world, for aeroplanes could intimidate and punish, not rule."[80]

Nevertheless, the fact that indigenous people did not enjoy the same international standards of human rights as whites encouraged the British to employ air strikes with bombs and machine guns to selectively-target rebels in Afghanistan, Egypt, Nigeria, Southwest Africa, and elsewhere. The Italians and Spaniards also relied on the demoralizing effect of air power, which seemed to be both cost-effective and efficient. Spain used poison gas to put police bombing to the test in the Moroccan village of Xauen in 1925.[81] All this presaged a terror from the skies that reached its pre-World War Two apotheosis in the bombing of Ethiopia by the Fascist Italian government.

A Fateful Solidarity

On June 30, 1936, the Ethiopian emperor Haile Selassie addressed the General Assembly of the League of Nations:

> It is my duty to inform the Governments assembled in Geneva, responsible as they are for the lives of millions of men, women and children, of the deadly peril, which threatens them, by describing to them the fate which has been suffered by Ethiopia. It is not only upon warriors that the Italian Government has made war. It has above all attacked populations far removed from hostilities, in order to terrorize and exterminate them. Towards the end of 1935, Italian aircraft first dropped tear-gas bombs upon my armies. ... The Italian aircraft then resorted to mustard gas. ... Men and animals succumbed. The deadly rain that fell from the aircraft made all those whom it touched flee shrieking with pain. ... That is why I decided to come myself to bear witness against the crime perpetrated against my people and give Europe a warning of the doom that awaits it, if it should bow to the *fait accompli.*[82]

Until it was invaded by the Italian army under orders from Fascist dictator Benito Mussolini on October 3, 1935, the East African country of Ethiopia (Abyssinia) had been—together with Liberia—the last African region not under control by Europeans. The Italian attack was in one sense a delayed retribution for the Battle of Adwa in 1896, when the Ethiopian army had successfully repulsed an Italian attempt at reducing much of Ethopia to a protectorate.[83] The conquest of 1935 went hand in hand with the massive deployment of modern technology and was swiftly followed by efforts to exploit and develop Ethiopia through civilian infrastructures. Hard on the heels of the occupation, the Italians announced a 2,800 kilometer road-building plan resembling what had been done in Libya after the transient 1911–12 Italian invasion. High-modern cities like Asmara were planned, based on strict segregation, to provide a "forgotten region" with civilized facilities and to open up Africa for an expected six million Italian settlers.[84]

While many Europeans were taken aback by Haile Selassie's report, it also gave certain European advocates of colonialism new hope. They could hardly have failed to note the feeble response by the League of Nations: that half-hearted imposition of a few economic sanctions on Italy only to rescind them some months later. To them, the Abyssinian example confirmed that a revival of European colonialism under Fascist rule was indeed possible.

Although largely overshadowed by the Second World War, Italy's Ethiopian campaign ultimately became a turning point in twentieth-century history. Six factors make this campaign a watershed in colonial rule and warfare in general: 1. Once again a technology that had been heralded as contributing to peace—in this case aviation—became a deadly military weapon. 2. Ethiopia's struggle

Fig. 3.9 Lined up at Bos Nassibu's Camp, the Ogaden: *Italian lorries loaded with well-armed infantrymen during the war between Ethiopia and Italy (1935– 36). As in most colonial wars, Italians delegated direct fighting to the natives, but deployed modern technology (and mustard gas) and justified their attacks as executing "civilizing missions."*

conflated colonial warfare with modern concepts of "engineered wars" that terrorized civilians and developed into total wars. 3. The war was a test run for the massive air attacks of the German Blitzkrieg. 4. It documented the willingness of Fascist countries to instigate violence to acquire colonies. At the same time it moved beyond traditional colonialism by establishing new kinds of Fascist empires linked by infrastructures in the way the Roman Empire had once been. 5. The Ethiopian War also demonstrated how ill-prepared the League of Nations was to contain these expansionist ambitions, unsuccessfully embarking on "appeasement" efforts instead. In consequence, the first attempt to install a global level of government lost all credibility as colonized peoples realized that no effective help could be expected from that quarter. 6. This experience eventually sparked independence movements. The Japanese conquest of Manchuria in 1931 and the Italian–Abyssinian War were precursors of the Second World War, truly a technicians' war. This war ended in August 1945 with the deployment of another innovative technology that ushered the world into a completely new era: the atomic bomb.

One additional reason why the League of Nations did not intervene on the side of Ethiopia may have been members' ambivalent feelings about technology transfer to non-European peoples. Since the First World War, the unique powers endowed by the command of modern science and technology had begun to evaporate. It was rumored that the "colored" peoples had now become a threat to the "white" race. In 1931 cultural critic Oswald Spengler stated:

> Then, at the end of the [nineteenth] century, the blind will for power committed a crucial blunder. Instead of keeping technical knowledge, the most valuable asset of the white race, secret, they boastfully peddled it at every university in both spoken and written forms, and they were proud of how the Indians and Japanese admired it. ... But instead of just exporting products, they shared confidential information, techniques, methods, engineers, and organizers. ... All the colored races discovered the mysteries of our strength, seized upon them, and utilized them.[85]

Because of their low-wage policies, Spengler added, these countries would become "deadly competitors" and in the long run render white labor superfluous. Concerns like these may have contributed to the ambivalent attitude of Europeans to the Ethiopians' cause. Another reason might have been their guilty conscience.

The failure of the League of Nations to respond to the plight of colonized peoples turned out to be a fateful error of judgement for the Europeans themselves. Encouraged by the League's weak stance, Adolf Hitler embarked on an arms build-up and repeatedly tested the limits of tolerance as he systematically provoked neighboring powers. During the Spanish Civil War, a special unit of the German air force, the Condor Legion, was deployed in several air raids, the most notorious of which was the attack on the Basque city of Guernica on April 27, 1937. Pablo Picasso's famous painting gave powerful symbolic expression to exactly what Emperor Haile Selassie had revealed at Geneva less than a year before.

Axis of Aviation

Those countries that had failed to build up or maintain a colonial empire and that exhibited a proclivity for authoritarian regimes like Spain, Germany, and Italy, had a different idea of Europe than that espoused by the League of Nations.[86] Fascist geopolitical mappings and ideologies turned away from global concepts and reverted to the principle of preserving interrelated, coherent, and manageable territories connected by infrastructure.[87] The potential of aviation to transcend borders seduced them into giving the idea of colonial empires a second thought. The airships of Count Zeppelin, for instance, were seen as enabling their possessor to rule the skies.[88] In 1928, author Ernst Juenger wrote:

> A modern age of discovery has dawned, and all the islands and countries formerly reached by land and sea are being discovered anew; our era also finds itself facing challenges comparable to the discovery of the passage to India or the safeguarding of the first trade routes. It is of paramount importance for Germany in particular, after being placed under such restraints by the peace treaties, to secure for herself in the spatial orders of tomorrow the prestige lost to her in those of yesterday.[89]

The geography of transportation was thus transformed, supported by the construction of new infrastructures, especially airports. Previously neglected regions and islands gained new significance as airbases and imperial outposts.[90] "To a much greater degree than any other type of transportation, air traffic is a tool of politics. Paths in the skies are also paths of political energy and political will. Streams of air traffic are also streams of political power," geopolitical

Fig. 3.10 Ancient Encounter with Modernity: *The original caption says: "The Graf Zeppelin's rendezvous with the eternal desert and the more than 4,000 year-old Pyramids of Gizeh, Egypt (1931)." Since Napoleon's Egyptian campaign in 1798, Europe had been intrigued by the idea of revitalizing ancient grandeur by modern means, and pictures contrasting folkways with modern technology were abundant.*

publicist Walther Pahl wrote in 1938.[91] Africa in particular seemed to "have become a continent *newly*-developed and much more closely affiliated with Europe than only a few years ago."[92]

Italy enjoyed a comparable love affair with modern technology, especially aviation. It was no coincidence that Futurism established itself in the wake of Italy's attempt to conquer Libya. In November 1911, Italian Lieutenant Giulio Gavotti initiated the practice of aerial bombing by actually dropping four 4.5 lb Cipelli grenades on the enemy from his German-built Rumpler airplane *Taube*.[93] Impressed by technological novelties like these, Futurism became one of the precursors of Fascist technological euphoria. It delighted in the unchecked speed, altitude, distance, and force made possible by machines. Italo Balbo, one of the Blackshirt leaders of the Fascist era, subsequently became a national hero for building up an ultramodern fleet of airplanes and aviation training centers for pilots.[94] Balbo succeeded in expanding Italy's network of flight routes in and beyond Europe, also stirring up enthusiasm with his spectacular flights to Odessa, Rio, and Chicago. He trained a class of committed pilots that seemed to come straight out of a H.G. Wells novel in which a brotherhood of such flyers ruled the world.[95]

In the 1920s, Italian air power theorist Giulio Douhet in his book *Command of the Air* foresaw wars of the future being waged and won by bombarding the enemy's factories, power stations, laboratories, and infrastructures from the air, demoralizing the foe and undermining his morale.[96] Referring to Douhet, whose book was read by Army Air Corps all over the world, British Prime Minister Stanley Baldwin predicted that European civilization would be "wiped out" in the future because "the bomber will always get through." Fears like these encouraged a number of countries to undertake the construction of precautionary measures against air raids like bunkers and bomb shelters in the years leading up to the Second World War.[97]

Like the First World War, the Second was also a war of technologies, of arms races, and the struggle for access to energy resources. Aerial bombardment of cities, for instance, expedited the development of radio and ultimately radar.[98] Nevertheless, this war also remained a man-to-man conflict between soldiers wielding guns.[99] In every respect it was more comprehensive and costly than the First World War, and even more truly an actual *world* war. Depending on when its beginning and end are set, hostilities lasted from between six and fourteen years, with most of the casualties occurring outside Europe, as the theaters of this war were scattered everywhere and a significant number of its combatants were no longer Europeans.[100]

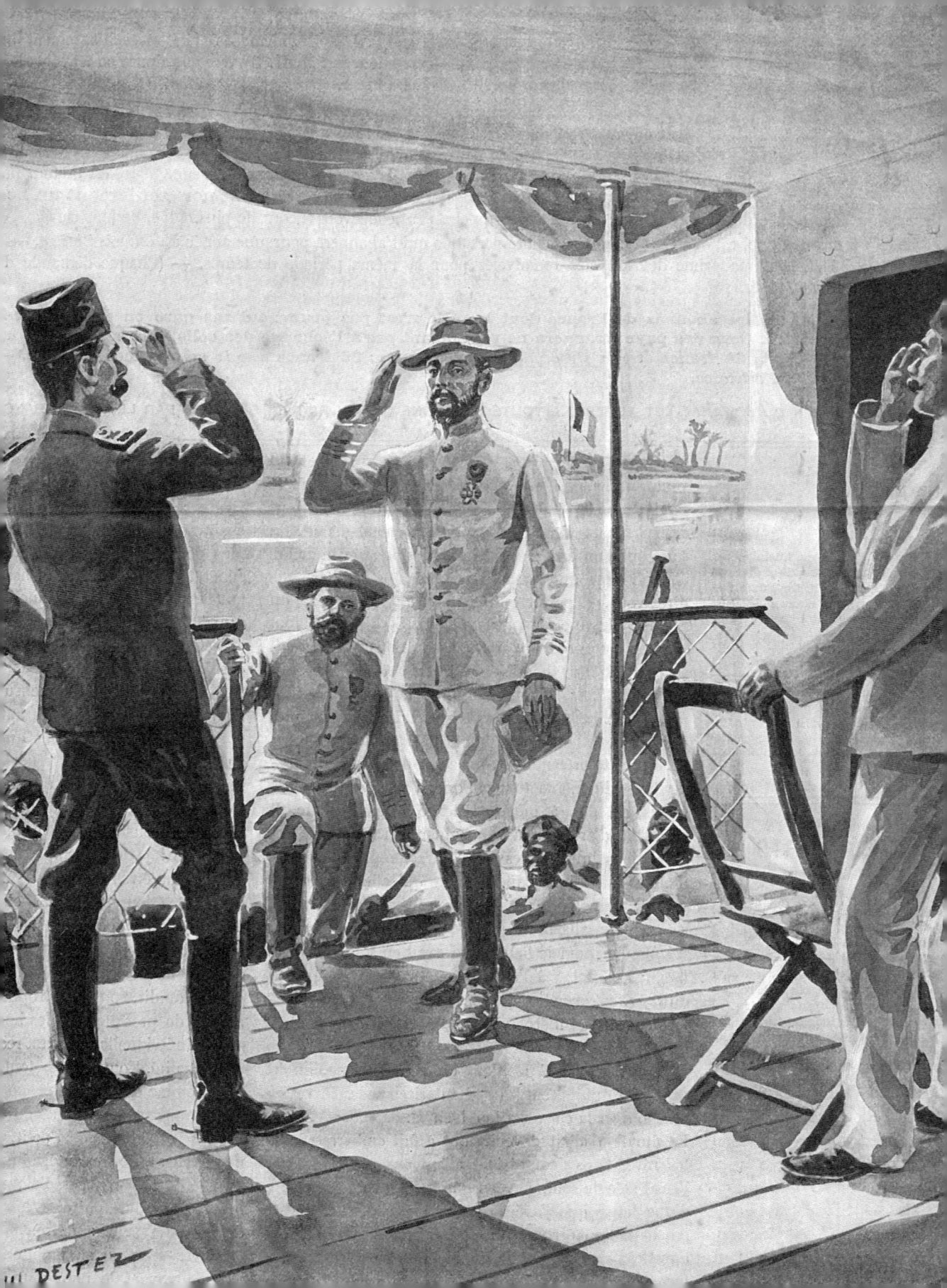

4

Scrambling for Eurafrica: Resources & Axes of Infrastructure

Introduction

Though geodetic and other kinds of mappings and technological superiority in warfare were certainly keystones of Europe's globalizing *dynamic*, they were means rather than ends. The aim of imperialism was not simply to chart and subjugate far-off territories and their native populations, but rather to incorporate them into a global economy in which European nations pulled the strings. Effort was directed at establishing systematic and routinized structures of exploitation based on on-site primary processing of mineral and agricultural resources, high-volume "penetrating" networks of transportation and communication, and on labor contracts that would ensure plentiful supplies of cheap native labor. Though colonial armies remained painfully present, mature imperialism tended to eschew overt violence and instead institutionalize it in the form of discriminatory laws, unequal labor contracts, and orderly if unjust penal systems. In this chapter we examine several aspects of this stage of "high imperialism" as it applied to Africa.

The Berlin Conference (1884–85) rewove the fabric of the traditional colonial order. The principle of the "effective occupation" of colonial territories replaced the traditional historical rights. This new vision of "legitimate" colonization harmonized perfectly with the agendas of Benjamin Disraeli and Cecil Rhodes for the British Empire, of Leopold II of Belgium for the Congo, of France for its African possessions, and of Otto von Bismarck for Germany's colonial ambitions. Africa was an irresistible prize for European industrial powers eager to secure new sources of raw materials and markets.[1]

Great Britain was the key player on Africa's colonial chessboard. Its determined attempts at controlling Africa unleashed the most intense conflicts among European nations: the ultimatum to Portugal in 1890, and the Fashoda conflict with the French in 1898. Despite apparent differences, these were essentially similar conflicts. Both were confrontations between British imperial interests and those of another European nation. In both cases, Britain prevailed. Finally, both of the diplomatic and military conflicts were embedded in tensions between rival projects of territorial domination by means of railroads.

British colonialism in Africa ultimately spawned Cecil Rhodes' project of the Cape to Cairo Railway. Although he was mainly interested in the southern states of the continent, he envisioned a continuous "red line" of British dominions from north to south. The railroad was to be a critical element in the unification of British colonial territory: it facilitated governance, enabled troops to move quickly, abetted white settlement, and fostered trade and mining.

Cecil Rhodes' railroad project and his concept of an "all-red" railroad line (and an "all-red" telegraph line) have become classics of colonial historiography.[2] By contrast, the conflicting Portuguese and French railroad projects have been largely ignored. Portugal proposed the so-called Pink Map project that aimed at linking its west coast colony of Angola with its east coast colony of Mozambique, more specifically, connecting Luanda and Lourenço Marques. France also pursued a plan for a railroad link across its colonies from west to east across the northern part of the continent, from Senegal to Djibouti.

Fig. 4.1 The Rhodes Colossus: *Cecil Rhodes' concept of "all-red" infrastructures, that is, railroads, roads, and telegraphs crossing or using exclusively British colonial territories in Africa, was the centerpiece of the British imperial strategy. Rhodes considered that the success of imperial rule depended on the efficiency of governance, which, in turn, depended on the efficiency of circulation of information, instructions, men, and commodities. In this cartoon, signed by Edward Linley Sambourne and published in* Punch *in December 1892, Rhodes is depicted as a colossus ruling all over Africa and holding telegraph lines that would link Cape Town to Cairo. The Cape to Cairo red infrastructure projects were never completed but they stand for the ferocious rivalry among European imperial nations.*

In 1887, Portugal mounted four expeditions to reinforce or to establish Portuguese sovereignty in the inland territories between Angola and Mozambique. These territories were a kind of "no man's land" (*res nullius*), never claimed by any of the European powers despite various allegiances between local rulers and European countries. The hidden agenda behind the Portuguese move was, of course, the pursuit of the Pink Map plan by assuring control of a right of way through regions claimed by both Portugal and Britain. Armed clashes between a Portuguese expedition led by Alexandre de Serpa Pinto and a British expedition under command of consul Henry Hamilton Johnston near the Ruo River eventuated in a British ultimatum to Portugal in 1890. Its consequences deeply affected Portuguese internal politics, as the ultimatum became one of the main weapons used by the Republicans to overthrow the monarchy.

France pursued the same strategy in 1897. The Third Republic sent an expedition led by Jean-Baptiste Marchand to take possession of the area around Fashoda and make it a French protectorate. This would secure a direct line from Dakar to French Somaliland (now Djibouti) by the Red Sea in the Horn. Marchand's men

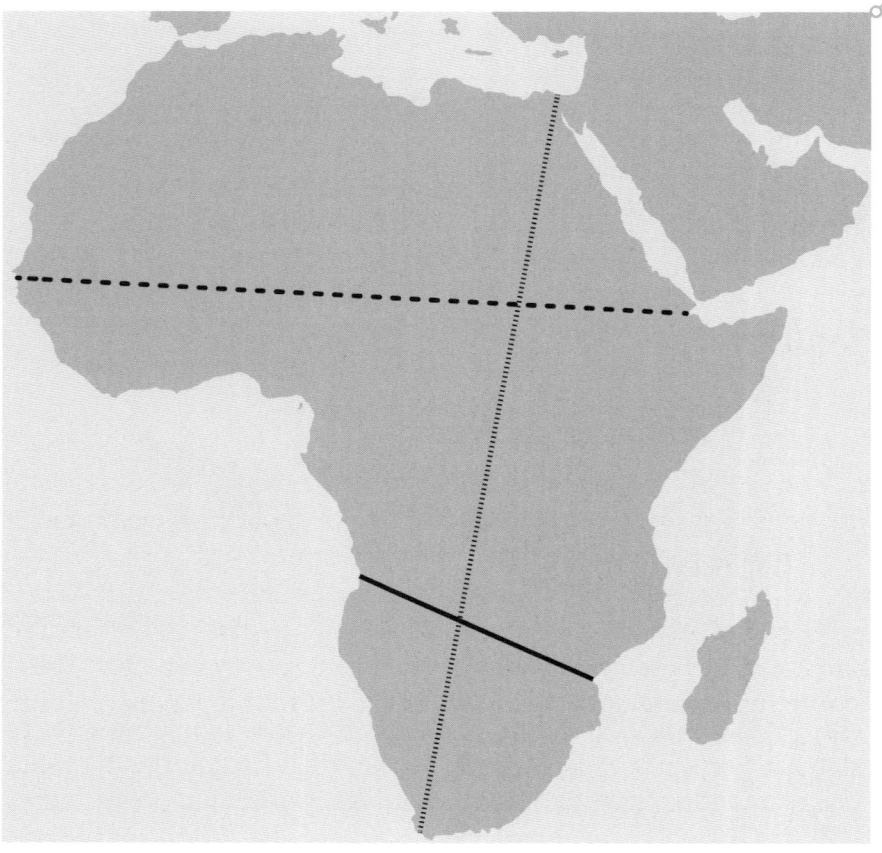

Fig. 4.2 Diplomatic Derailment: *Cecil Rhodes' plan for building the Cape to Cairo railroad line met the fierce opposition of other imperial powers that also sought to dominate the African hinterland and secure trans-continental railroad lines. The clash of interests between Great Britain and Portugal and Great Britain and France were particularly harsh and had severe political implications. In this map, the vertical line stands for Rhodes' Cairo to Cape Town Railway, the northern horizontal line for the French line linking Senegal to Djibouti, and the southern horizontal for the Portuguese project to link the Western and the Eastern coasts of Africa using Angola and Mozambique as the main pillars for imperial transport and communications.*

reached Fashoda in 1898, claiming this territory for France using the *res nullius* argument, again following in Portugal's footsteps in southern Africa. However, a British flotilla led by Horatio Herbert Lord Kitchener soon confronted the French expedition, eventuating in the Fashoda Incident and eventual diplomatic defeat for France.

In the context of the Scramble for Africa, the British ultimatum to Portugal (1890) and the Fashoda Incident (1898) are critical events that once again underscore the importance of technologies and infrastructures in the occupation and exploitation of Africa. To underscore the fact that the scramble for Africa was not only about building railroads and establishing rights of way, we begin by describing the often-brutal and racist exploitation of African mineral resources, of which Cecil Rhodes has become an icon. We also discuss plans to green the Sahara, an infrastructural imaginary aimed not simply at making Africa's riches more accessible, but at producing even more riches. In their most ambitious form, these plans included partial drainage and reclamation of the Mediterranean Sea and the production of hydroelectricity utilizing the new fall lines in the major tributary rivers. These plans implicated more than a "scramble" for Africa; they envisioned a "Eurafrica," a fusion of the two continents—with the former Europe clearly in the driver's seat.

South Africa: "Racializing" Resources

In the spring of 1867, close to the South African city of Hopetown and near the Orange River, a 15-year-old boy called Stephanus Erasmus Jacob discovered a very peculiar stone. He presented it to his mother, who was intrigued by the bauble's radiance. It later turned out to be a 21.25 carat diamond, and after being cut into a jewel of 10.73 carat, was named "Eureka." In 1967, one hundred years after its discovery, the diamond monopolist De Beers acquired the Eureka stone and dedicated it to the people of South Africa. Although by this time the country had become a wealthy and industrialized society, its apartheid policy had also caused it to become internationally ostracized. The Mine Museum of Kimberley, where the diamond was put on display, is located at the site that marks the beginning of the South African diamond rush, one of the many "rushes" for resources in the nineteenth and twentieth centuries. These took place at various locations, such as Pennsylvania, California, Australia, and Canada, leaving deep imprints on their respective landscapes. While nature bore visual evidence of these depredations, local societies were even more deeply affected by the gold, oil, and diamond fevers.[3]

Rarely has the complexity and perhaps savagery of Europe's interrelations with its overseas outposts been better exemplified than in the case of South Africa. Both the country's specific resources and the technologies used to extract them, as well as subsequent conflicts over their appropriation, profoundly shaped the "rainbow nation" by impacting the dynamics of social coexistence within the country, and subsequently foreign relations between South Africa and Europe. The power structure of South African imperialism was more than elsewhere deeply materialistic: governed by expected profit and questions of ownership, the rule of land titles and concessions, and efforts to guarantee availability of cheap human and material resources.[4] Discoveries of precious mineral resources only added fuel to this fire, and finally came to dominate and further distort relationships between colonizers and indigenous peoples.

Such matters were not always negotiated openly. Quite the contrary: they were often discussed secretly and in a way that constantly confused material aspects with a symbolic level of meaning, usually leading to conflicts in which antagonists accused each other of entertaining "dishonest" or "hypocritical" motives. But Europe (and much of the rest of the world) was indeed dominated by material values and greed for possessions, and had the power to quickly and fundamentally change geographical as well as mental maps.

Despite being almost useless for practical purposes, gold and diamonds were of exceeding monetary and cultural value in the nineteenth century; the former especially drawing a flood of immigrating adventurers, fortune hunters, settlers, and workers— demographic shifts that profoundly affected the terms of on-site labor. In South Africa, exploitable resources, especially gold and diamonds, fostered a racism of segregation, which repeatedly turned violent when property rights were at issue. The developing apartheid policy made for a tension-ridden relationship between people of "European" and "African" descent, tensions that were further aggravated by rivalries among European settler societies themselves.[5]

Since the seventeenth century, the European bridgehead at Africa's southern tip had changed from a strategic nautical supply base on the route to the East into a stable Dutch colonial settlement. In 1814 the Congress of Vienna ceded the Dutch colony to Britain, and although the so-called Great Trek of the Boers into what would later become the Oranje Vrijstaat, Natal, and Transvaal was largely able to defuse the resultant conflict, it incited often-violent clashes with native African tribes, particularly the Zulu.[6] With the expansion of British dominance, the pragmatic coexistence of African kingdoms, white Afrikaner republics, and European colonies was threatened by a growing lack of space. Consequently, territorial segregation, which was first tried in the Boer Republic of Natal, was more strictly enforced. While the dwellers of many colonies had previously been free to decamp, migrate, or otherwise maneuver themselves out of clear and present dangers, these were no longer viable alternatives for native tribes or the Afrikaner peasants (or *Boers*) of Dutch origin. Moreover, the discovery of valuable resources had instantly radicalized the political geography of coexistence.[7] The diamond finds exercised an almost magnetic attraction on adventurers, fortune hunters, and workers, mainly from Europe, the United States, and Australia. As a general rule either the technologically most advanced or the best-funded Europeans succeeded in staking claims.[8]

This is the story of famous diamond diggers such as Barney Barnato or Cecil Rhodes, founder of the De Beers Mining Company. After years of free competition for claims in Kimberley, De Beers quickly managed to monopolize diamond production by introducing costly mining machinery and building on extensive previous investments. The first steam engines were introduced to the mines as early as 1875. Technical advantages enabled De Beers to dominate the mining business by buying concessions and subsequently extending its cartel across Southern African territories. Shortly after the last native African diamond digger, Reverend Gwayi Tyamzashe, quit the scene in 1893,[9] De Beers Consolidated Mining Company Ltd. effectively took control of the global diamond market.[10]

Fig. 4.3 Compassion for "the Wretched of the Earth": *In her photo essay on South African diamond miners for* Life Magazine, *September 18, 1950, U.S. reporter, legendary photographer, and former war correspondent Margret Bourke-White concluded: "Now about the mines. One thing that's happened to me ... from now on I just hate gold and diamonds." Four years earlier, Bourke-White had taken another photograph that became iconic, Mohandas K. Gandhi at the spinning wheel.*

This process was re-enacted at an even larger scale during the 1880s, when gold was found near Witwatersrand. Previously a modest town in the middle of nowhere, Johannesburg now exploded into one of the biggest cities in Africa. The "mineral revolution" with its overwhelming returns soon enabled Cecil Rhodes to virtually buy his way to becoming Prime Minister of the Cape Colony, in addition to which he established another "private" colony called Rhodesia. A role model for European colonizers,

Rhodes pursued ambitious plans to create a coherent East African sphere of interest, which was to be controlled and whose infrastructure was to be developed by the British Empire. This Cape to Cairo plan represented a high imperialist scheme that imagined continuous railroad lines as the backbone of a unified European-African continent—a scheme that, however, conflicted with the interests of other European nations.[11] At the turn of the twentieth century, such perspectives made South Africa a destination of choice for European investments.

South Africa seemed rewarding for yet another reason, albeit one that it shared with all other European colonies: in no case had European-style industrial labor relations, established in Europe following a string of collective actions, been implemented in European settler colonies. It seems that European employers were doing their best to realize Karl Marx's diagnosis in practice:

> The discovery of gold and silver in America, the extirpation, enslavement and entombment in mines of the aboriginal population, the beginning of the conquest and looting of the East Indies, the turning of Africa into a warren for the commercial hunting of black-skins, signalized the rosy dawn of the era of capitalist production.[12]

In fact, most of the human casualties of European colonization occurred in territories that were rich in resources, either due to expulsion or forced labor regimes.[13] European entrepreneurs often sought out the most favorable investment conditions, and in the case of South Africa the terms of trade were strictly racialized. Starting in 1872, the governments introduced pass laws, enabling comprehensive control of native African workers and eventually contributing to the creation of segregated "homelands." The Diamond Trade Act was passed that same year, invalidating the "presumption of innocence" which had previously protected workers who had been caught by the police with a rough diamond and had been unable to identify its origin. Whereas colored workers were housed in closed camps, white employees were usually provided with better-quality lodging as well as higher-paying jobs. The Act also allowed companies to set up "searching-houses" in a system of routine surveillance managed by company police.[14] This legislation had far-reaching effects on South African history:

> Segregation policies affected the rights of Africans to own land, to live or travel where they chose, and to enjoy job security or the

freedom to switch jobs, leading finally to a limit on black power in southern Africa. In the twentieth century, segregation restricted Africans to dangerous, unskilled, low-paid jobs in mining and industry or to laboring on white-owned farms, while supervisory jobs in all economic sectors were held by highly paid whites. ... Throughout southern Africa, whites maintained attitudes of superiority, paternalistic benevolence, and social distance toward Africans. Segregationist policies, legitimated by scientific claims from biologists, anthropologists, and other experts provided whites with higher social status and enabled them to maintain economic and political advantages.[15]

But even though the diamond industry became a vehicle of political, social, and economic control, a paradoxical situation developed in which many native Africans who were attracted by the Kimberley mines brought home weapons from South African cities. Their access to guns redefined African–European relations, prolonging the series of wars waged between whites and native Africans seeking to curb white expansion into the interior of southern Africa.[16] The Boer Wars, which took place between 1880 and 1881 and again from 1899 to 1902, were partly rooted in a competition for labor between the "industrialized" British and the "agrarian" Boers in Transvaal. Being one of the few violent clashes among European settlers abroad, these wars attracted great attention in Europe. The fact that they mirrored different concepts of settling and developing and were fought to legitimize a dubious "racial order" designed to secure access to valuable resources further heightened European public interest. Oddly enough, at the exact same time European powers were making a great show of unity by cooperating in quelling the Chinese Boxer Rebellion.

The Hidden History: Politics, Diplomacy, & Technology

In the age of colonialism and imperialism, European countries profited from their colonies in several ways. In addition to the immediate returns on exploitation, colonial powers used their colonies as assets in the European struggle for power and influence. As European countries consolidated and developed their overseas assets they also engineered ways to translate this dominion into power in the European arena.

The European domestication of the newly-conquered territories was simultaneously a matter of *utilization* and *imposition* of Europe's techno-scientific vision of progress. In other words, infrastructural technologies were at the core of imperialism not just as tools to achieve strategic ends, but as actors expressive of and enacting different colonial strategies. More often than not, episodes framed in the media or in political fora as military or diplomatic events had a hidden history in which technology was the true protagonist. The following accounts of African imperial railroads and of plans to reclaim the Sahara and the Mediterranean show that, beneath what appears to be pure power politics and diplomacy, European powers were in fact pursuing conflicting technological endeavors. The real reason why Britain, Portugal, and France were balancing at the verge of war by the end of the nineteenth century was because their African railroad projects, symbolizing but also enacting the core of their colonial strategy, were topographically incompatible. The Portuguese east–west axis in the south, the French east–west axis in the north and British north–south axis through east Africa simply could not co-exist in the same colonial space. Hence, the African fever, the "scramble" for Africa, was clearly a techno-political and a techno-economic phenomenon characterized by the use of infrastructural technology to pursue national political and economic goals. Hidden beneath the complex diplomatic and political games played out in the European and African arenas was the domestication of the African landscape by means of geographically extensive and therefore conflictual technical infrastructures.

From the nineteenth-century European standpoint, African culture was "invisible," and its technology backward, barbaric, and primitive. The invasion and pacification of new territories, indeed their very exploitation, was legitimated by the understanding that the culture and values of the colonial powers were universally valid and hence the rightful heritage of all mankind. In this sense, imperial rhetoric did not acknowledge the "other" as distinct, but rather as suffering a deficit of civilization, which could only be overcome thanks to the civilizing influence of European colonization. Even if this perception applied above all to the Sub-Saharan regions, it was also considered valid for Northern Africa. European powers regarded all of Africa as their playground, a large territory to be used and abused as their interests and needs might dictate. Therefore European technologies in Africa remained mostly a European affair, designed and implemented to serve imperial strategies.

In the Scramble for Africa, Africans played either the role of warriors ultimately defeated due to technological backwardness or of workmen serving European infrastructural and commercial endeavors. The concept of the civilizing mission was unidirectional; African knowledge and skills had nothing to offer the Europeans— excepting possibly survival skills in the bush for hunters and explorers. Ironically enough, the natives, though clearly dazzled by the whistle of a locomotive, did not appear to share the European enthusiasm for, and faith in, tracks and wires. The lack of written testimony makes it difficult to discern exactly how European infrastructural technologies impacted local cultural and technical traditions. However, the scarce evidence based both on European statements and on anthropological studies, suggests that railroads and telegraphs were viewed as alien objects which were not welcome and which often disrupted local communities and econo-mies. The most obvious examples are the artificial barriers created by railroad lines, severing centers of population from their food supplies and other resources and compelling aboriginal populations to move their cattle in search of new pastures and water.

Transportation infrastructure was crucial to African colonization because the European drive to develop its African colonies in pursuit of national economic growth and a thriving global economy was pragmatically restricted to a few areas with rich resources that might be—and increasingly were—hundreds of miles from the nearest seaport. Hence exploitation was essentially dependent on inland transportation and information infrastructures that consolidated and opened large territories by creating a hierarchy of spaces. The successful exploitation of the African hinterland depended heavily on the efficiency of transfers between railroads, harbors, and ships. And until the 1930s this specific entanglement between politics, economy, and technology on African soil remained entirely a European affair.

Pink Map versus All-Red Railroad Line

On January 16, 1890, one of the big Portuguese newspapers, *O Século*, posted on its front page a rousing call to arms urging the people of Lisbon to rise against the British navy, which was supposedly sailing from Gibraltar to shell Lisbon:

> Fellow Compatriots! The English navy is sailing from Gibraltar to the river Tagus, where it is expected today. After having aimed its cannons at Portugal to steal the lands we have discovered, England attacks us once again to prevent the public manifestation of our outraged national pride! Disdain the English and England by showing our love to the motherland and our brotherly ties to friendly countries. Fellow Compatriots! Let our angry cries be louder than ever! Let us shout: Long live Portugal! Long live France! Long live Spain! Long live Italy! Long live all honored nations!

This was one response to the British ultimatum, delivered five days before by the British government to Portugal, demanding the retreat of the Portuguese military forces from all the lands between the Portuguese colonies of Angola and Mozambique. The broadside urged that the British fleet, the "cruel" and "savage" John Bull navy, should be received by patriotic artillery shots, and stressed the support of other "honored" nations for the Portuguese cause, namely France, Spain, and Italy.

The British ultimatum has been all but ignored by World History and in the context of Portuguese history, it has been viewed as mostly a diplomatic and political incident. In contrast, we argue that the ultimatum was basically driven by conflicting technological ambitions arising from two clashing railroad projects projected as the infrastructural backbone of both British and Portuguese imperial strategies to dominate the African hinterland.

The general concept of the Pink Map—a continuous strip of Portuguese sovereign territory between the western and eastern coast of Africa—dated back to the eighteenth century. Portugal saw its advance into Africa's inland regions as the only way to check British and French intrusions and to deflect the Boers' growing appetite for Portugal's colonies. As "informal imperialism" based on historical rights gave place to the doctrine of the "effective occupation" of African territories, rivalries among European powers became increasingly intense. Portugal responded to this new international framework by establishing "civilizing outposts" to secure Portuguese sovereignty over the Angola and Mozambique hinterlands. In postulating the Pink Map, the Portuguese government aimed to recover its status as a major colonial power, a status it lost after the Berlin Conference.[17]

This active policy was reinforced by the 1877 expedition to Africa, headed by Hermenegildo de Brito Capelo and Roberto Ivens, in order to explore the inland territories of Angola and to pave the way for the official implementation of the Pink Map (1886). At the same time, Portugal signed a treaty with France (1886) and another with Germany (1887), in both of which the Pink Map was recognized.

In 1887, under duress from the Berlin Conference, Portugal launched four expeditions to consolidate the Portuguese South Africa project. Their aim was to reinforce or to impose Portuguese sovereignty across the territories of Zambezi, Bié (Angola), Niassa (Mozambique), and the area today known as Shire River. The expedition to the Shire, headed by Alexandre de Serpa Pinto, triggered the British Ultimatum. The British government claimed that the Portuguese explorer had lowered the British flag and taken over by force the territory of a pro-English tribe, the Makololos.

However, it is clear that the so-called Serpa Pinto Incident was a response to the new position Cecil Rhodes' British South Africa Company was claiming on the African chessboard. Queen Victoria had conceded royal privileges and autonomy to the British South Africa Company, thus feeding Rhodes' imperialist and monopolistic ambitions. The Portuguese plan collided with Cecil Rhodes' Cape to Cairo railroad project. The Portuguese west–east axis and the British north–south axis simply could not co-exist.

Portugal's international status, however, depended heavily on its colonial authority and agenda, of which railroads, roads, harbors, and telegraphs were critical features. From the Portuguese point of view, these material signs embodied different messages and expectations. On the international stage, Portugal consolidated its presence in Africa and kept its colonies. In the national context, the profits of this strategy were quite clear: national pride was secured and Portugal could once again, as it had in the sixteenth and seventeenth centuries, fly the flag of the "civilizing mission."

Concurrently with Capelo and Ivens' 1877 expedition mentioned above, two other expeditions to southern Africa were mounted by the Portuguese state. Of a rather more subtle imperialistic bent, they were headed by the civil engineers Rafael Gorjão (for Angola) and Joaquim Machado (for Mozambique) and charged with assessing the viability of a plan to construct two railroad lines: one in Angola,

Fig. 4.4 Behold Civilization!: *Alexandre de Serpa Pinto was Portugal's leading protagonist in the diplomatic incident that opposed Cecil Rhodes' plan to link Cape Town to Cairo by railroad and the Portuguese Pink Map (railroad linking the coasts of Angola and Mozambique) and that led to the British Ultimatum to Portugal. This cartoon, authored by the famous Portuguese caricaturist Raphael Bordalo Pinheiro and published by Pontos nos ii, shows a portrait of Serpa Pinto on the upper left side and, below, African natives in awe looking at a document that reads "Civilization: Railways," presented by Serpa Pinto.*

linking Luanda to Ambaca, and another in Mozambique, linking Lourenço Marques to Transvaal.

These expeditions were the first surveys preparatory to the construction of railroad networks in both Angola and Mozambique. Modern machine technology was clearly viewed as essential to enforce efficient colonial exploitation. In this context, the "civilizing mission" sought to reshape the races and cultures living in Africa according to the dictates of modern colonialism. It was held to be in the nature of things that Europeans would lead this cooperative process and that native culture would evolve in a perfect symbiosis with the construction of the railroad.

The wealth of Africa—coal, copper, iron, cobalt, gold, diamonds, timber, and rubber—and its agricultural potentials—coffee, cocoa— necessitated careful zoning of spaces of exploitation and this in turn placed stringent demands on the quantity and quality of means of communication. Railroads were at the heart of these infrastructural strategies and played a key role in promoting commerce, industry, and agriculture, opening Africa up to the nineteenth–century world economy by extracting its natural wealth and creating overseas markets for European textiles and machinery. Decisions about where railroads should be projected were extremely complex and contentious because they had to maintain a balance between serving the economic status quo and servicing expectations about the leading economic areas of the future. Discussions about railroad lines were therefore often quarrelsome, entailing conflicts between local, regional, and national economic interests and differing engi- neering perspectives, as was the case in the fierce discussion about part of the line linking Luanda to Ambaca. Where international economic interests were involved, these tensions could threaten open armed conflict, as with the British ultimatum to Portugal and the Fashoda Incident.

To combat the growing English and German penetration of its African colonies, the Portuguese government was forced to reassess its priorities. The main objective now became to occupy the inland regions of Angola and Mozambique and link the two colonies by railroads and roads. The absence of these basic colonial infrastructures in the Portuguese colonies had worried both politicians and engineers for decades. There were clearly doubts about the wisdom of investing in such expensive infra- structure. The sketch for the first 178 kilometers of the railroad from Mossamedes to Bié sought legitimation by framing itself as the project cherished by "the illustrious explorers Hermenegildo Capelo and Roberto Ivens."[18] Joaquim Machado, the author, used information collected by these Portuguese explorers to justify the route proposed: to cross the Plain of the Mossamedes district to the interior, a region which had a climate more congenial to Europeans, as well as abundant resources. Last, but not least,

Machado unveiled the real aim of the project, also quoting Capelo and Ivens:[19]

> To emphasize the colonization in the African inner land and to block any attempts from any country to occupy the territories between our possessions and the inner regions, thus preventing us from building our African empire, are part of an enterprise, which we should not give up. … To link those markets [from Angola] to the port of Mossamedes through a railway would bring us considerable immediate results, as well as far-reaching economic and political benefits. … It would incite Portuguese settlers; and the Portuguese, thus gathered, would become a barrier against the German invasion, and would perpetuate our domain up to the heart of the Austral inter-tropical Africa, linking Angola to the province of Mozambique.

There it was, the Pink Map: an invitation to build up an African empire that would actually occupy and exploit the vast territory between Angola and Mozambique, ultimately a dream that European realpolitik would make impossible to fulfill.

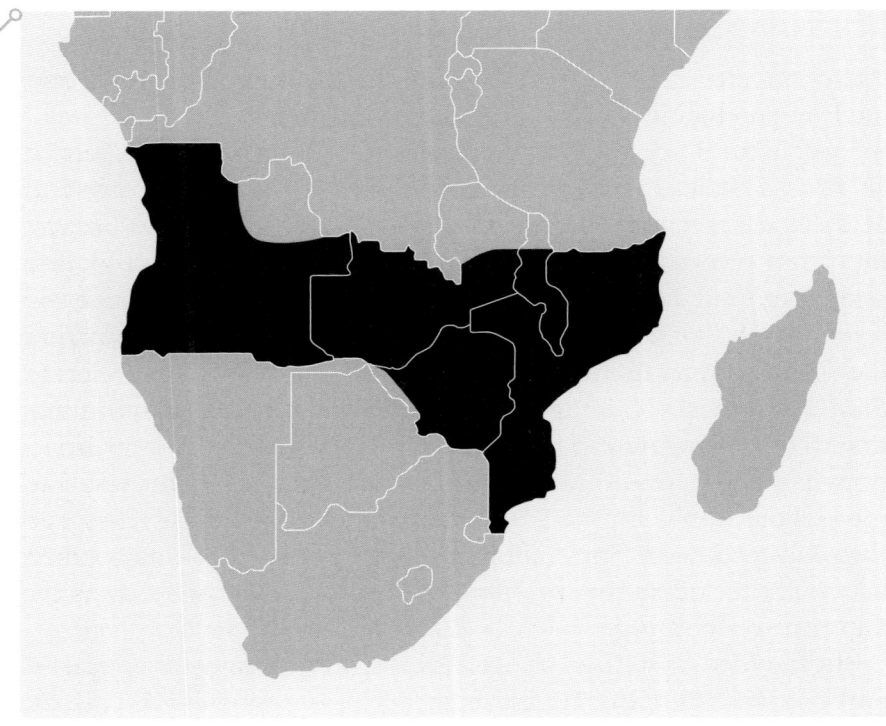

Fig. 4.5 Crumbling Dreams: *By putting forward the Pink Map, the Portuguese government tried to recover its status as a major colonial power, lost after the Berlin Conference (1884–85). Although with different geometries, the general concept of the Pink Map—a continuous strip of Portuguese sovereignty between the western (Angola) and eastern (Mozambique) coasts of Africa—was a project that dated back to the eighteenth century. The Portuguese Pink Map menaced Cecil Rhodes' plan for connecting the Cairo to Cape Town by an "all-red" railroad line. The Portuguese west–east axis and the British north–south axis could not co-exist.*

The first 178 km of railroads from Mossamedes to Alto da Chela was the jumping off point for the effective occupation of Angola. It was widely believed that the prosperity of the English territories at the Cape and Natal and of the states of Transvaal and Orange, inhabited by Dutch descendants, a prosperity that contrasted starkly with the poverty of Angola and Mozambique, reflected the absence of means to reach the fruitful inner plains, especially those of the district of Mossamedes where the climate was similar to that of southern Europe. In Orange, Transvaal, Natal, and Cape the first settlers "found themselves in a healthy and suitable environment, which would encourage them to spread themselves, if not so rapidly as rabbits in Australia, at least in a rather meaningful proportion."[20]

By 1906, Angola and Mozambique boasted 783 km of railroads with another 400 km under construction. Although in the end the dream of the Portuguese Pink Map was never realized, a ragbag of European tracks did finally link the Atlantic and the Indian Oceans, running from the Portuguese territory of Angola, through the Belgian Congo and British Rhodesia to Mozambique.

The Failure of the Trans-Saharan Railway

The Fashoda affair was embedded in the Egyptian Question, and in particular, in the digging of the Suez Canal, a major infrastructural project that unleashed latent British–French rivalry in Africa. This rivalry culminated in the armed standoff at Fashoda. French interest in a Suez Canal dated back to the Egyptian campaign of Napoleon I. The army of archaeologists, scientists, and cartographers he brought along discovered the remnants of an ancient waterway extending northward from the Red Sea and then westward toward the Nile. This inspired French engineers to imagine a modern north–south canal to join the Mediterranean with the Red Sea. This project was later abandoned when a preliminary survey indicated (as it turned out, erroneously) that the Red Sea was 10 meters (33 ft) higher than the Mediterranean. The elaborate infrastructure that was deemed necessary to tackle this problem made the project too expensive and time consuming.

But the idea continued to exercise the imaginations of engineers and entrepreneurs during the first half of the nineteenth century.

One of the first to come up with a realistic plan was the French explorer and cartographer Louis Maurice Adolphe Linant de Bellefonds. Commonly known as Linant Bey or Linant Pasha (indicating high rank or office in Arab society), Linant de Bellefonds had been appointed chief engineer of Egypt's Public Works in 1837.[21] Prior to that, between 1831 and 1837, he had been chief engineer of the public works of Upper Egypt, in charge of the major works associated with the network of irrigation canals. Decades later he worked closely with Ferdinand de Lesseps in the construction of the Suez Canal.

French Saint-Simonians, and particularly the famous Barthélemy Prosper Enfantin, were excited by the prospect of the canal. Through his *Société d'Etudes du Canal de Suez*, Enfantin encouraged a group of engineers, in close collaboration with Linant de Bellefonds, to study the feasibility of a Suez Canal. Among them were Robert Stephenson, the English civil engineer and locomotive builder, Alois Negrelli, the Austrian railroad pioneer, and Paul-Adrien Bourdaloue, the renowned French civil engineer, topographer, and expert surveyor. Bourdaloue's painstaking survey of the isthmus and in particular the accurate determination of its topographic relief convincingly demonstrated that there was no practical difference in altitude between the two seas.

In 1854, Sa'id Pasha, the Khedive of Egypt and Sudan, with whom Enfantin was on excellent terms, invited the aristocratic French diplomat Ferdinand de Lesseps to discuss the possibility of a French-built canal. About a month later, de Lesseps obtained a concession to construct a canal open to ships of all nations. De Lesseps had to maneuver his way through a diplomatic snakepit: the Ottoman Empire, France, and Great Britain were allies during the Crimean War against the Russian Empire. Although none of the allies wished to provoke an unnecessary incident, the French proposal did challenge British power. De Lesseps himself knew that the Suez Canal threatened the British monopoly of commerce around Cape Horn and its long-lasting domination of the seas. Nonetheless, he immediately set to work with the two famous French engineers in Egyptian service, Linant de Bellefonds and the hydraulics engineer Mougel (also known as Mougel Bey), an expert in irrigation and dams. Not surprisingly, they were able to present a project for a direct link between the Mediterranean and the Red Sea in rather short order.[22]

De Lessep's concession stipulated international participation in the design and construction of the project. To satisfy these terms, he mustered the support of the *Académie des Sciences* (French Academy of Sciences) to convene a group of thirteen experts from seven countries – the *Commission Internationale pour le Percement de l'Isthme des Suez* (International Commission for the Piercing of the Isthmus of Suez). The commission examined de Bellefonds' and Mougel's project and in 1856 approved the plan with only minor amendments. The *Compagnie Universelle du Canal Maritime de Suez* (Suez Canal Company) was founded on December 15, 1858. Only four months later, in April 1859, it started dredging the canal at the site of what would become Port Said. The canal was opened to international shipping on November 17, 1869, in a ceremony attended by the Khedive Ismail of Egypt and Sudan and the Empress-Consort of France, Eugénie de Montijo, wife of Napoleon III.[23]

Although the first ship to transit the canal, following the French Imperial yacht *Aigle*, was the British P&O liner *Delta*, Great Britain stuck to its guns in opposing the project. The United Kingdom feared that the canal might interfere with its India trade and preferred a connection by train from Alexandria via Cairo to Suez under clear British control, a railroad line which Stephenson eventually built.

The financial problems that from the first had plagued the canal-building enterprise did not disappear after its opening. In 1875, the Khedive's external debts forced him to secretly sell his country's share in the canal to the United Kingdom for £4 million. Despite this, French shareholders still retained their majority interest. For that matter, France and Britain exercised de facto control over Egyptian finances because their governments underwrote the banks holding the remainder of the Khedive's debt. This of course had an extremely destabilizing effect on Egyptian politics and society. In 1881, Egyptian peasants and an educated class of officers and civil servants led by Colonel Ahmed Urabi (known among his people as *El Wahid*, the One) began demanding an elected government. Tewfik Pasha, the new Khedive, relented and a chamber of deputies was established. However, these reforms jeopardized European interests, and in early 1882 France and Great Britain issued a joint statement asserting the primacy of the Khedive's authority. Urabi's followers immediately proclaimed a new government with Urabi as minister of war. When violence exploded in the streets of

Alexandria, the British fleet attacked the Egyptian city's garrison and, despite heavy resistance, succeeded in forcing the Egyptians to withdraw. The French fleet, also at Alexandria, refused to engage. These two different European strategies were the origin of a trajectory of conflicts that ended at Fashoda.[24]

Although British Prime Minister William E. Gladstone predicted that occupying Egypt without French collaboration would mean the end of a peaceful relationship between the two nations, the 1888 Convention of Constantinople firmly declared the Suez Canal a neutral zone under British protection. After the Alexandria crisis, Britain and France followed completely different approaches to Egyptian domestic politics. Gladstone was firmly supported by his parliament when he advocated further military interventions to silence Urabi's followers. French Minister of War Charles Freycinet, by contrast, was embroiled in a complex political chess game, built upon revenge and opposing lobbies, which made it impossible to secure political support for an intervention in Egypt. The loss of Egypt placed France in a difficult geopolitical situation regarding its ambitions in Africa. Its failure to secure an eastern path to the sea jeopardized the long-cherished dream of linking the Atlantic and the Indian Ocean coasts of Northern Africa by rail.[25]

But new actors were soon to enter the game and shift the balance of power. In Sudan, the British-led Turkish/Egyptian rule was about to meet its demise at the hands of an army led by the so-called Mahdi. Muhammad Ahmad ibn Abd Allah, the Mahdi (the Redeemer), set his revolt in motion in 1881, soon consolidating his status as a powerful spiritual leader. The Mahdi and his men, the Ansars (Helpers), known in Europe as the Dervishes, delivered a shattering blow to an ill-organized Egyptian force at El Obeid, near Kurdufan.

The debacle at El Obeid and the consequent loss of most of the Southern Sudanese cities made it clear that the Egyptian treasury could no longer bear the burden of maintaining control of Sudan. Under pressure from Britain, the government of Egypt opted to withdraw from Sudan, leaving the Mahdists in power. But withdrawing the scattered Egyptian garrisons in an orderly and peaceful fashion was a challenge in itself. To accomplish this, the British dispatched the engineer Charles Gordon, a well-known officer with a distinguished service record in Crimea, China, and Sudan itself. Gordon arrived in Cairo on January 24 and reached

Khartoum by February 18, 1884. Although Gordon's orders were to evacuate Sudan, he soon came to believe that it was possible to crush the Mahdi and thereby to ensure the stability of the whole region, Egypt included. Gordon's feeling of security was partly based on the fact that the main infrastructures linking Khartoum to Cairo, namely the navigation of the Nile and the telegraph, continued to be under European control, thus reinforcing his belief that it would be possible to control Sudan with only modest reinforcements.

However, in March 1884, the Sudanese tribes to the north of Khartoum, who had previously been sympathetic to or at least neutral towards the Egyptian authorities, rose in support of the Mahdi and blocked the Nile. At the same time, the telegraph lines between Khartoum and Cairo were cut, severing communication with the outside world. The siege of Khartoum had begun. While the Mahdi's army pressed the city, the British government reluctantly organized a relief expedition, which arrived within sight of Khartoum two days after the fall of the city, on January 25, 1885, after a siege of 313 days.

With Sudan lost, Great Britain once again focused on Egypt as the linchpin of its Africa policy, rooted in safeguarding the Canal and ruling the River Nile. Both these objectives were challenged. France had not forgotten the Suez Canal and continued to harass, more or less explicitly, British troops in Egypt. Meanwhile, Italy had aggressively positioned itself on the colonial chessboard by challenging British claims in the Upper Nile. In fact, in 1890, the Italian government, soon followed by King Leopold II of Belgium, proclaimed Sudan a *res nullius*, suitable for conquest by any European country. Now the battle for Sudan had begun in earnest.

The British were aiming at a solid block of influence from southern Africa through East Africa to Egypt. Meanwhile, the French were attempting to expand from West Africa along the southern border of the Sahara Desert in order to control all trade through the Sahel and to establish a thoroughfare to the Red Sea. The intersection of these colonial rights of way, which were to be materialized in railroad lines, passed through the town of Fashoda, where a standoff between armed expeditionary forces led the two countries to the brink of war.[26]

The French road to Fashoda was built on a strong domestic pro-colonization movement. The so-called *Parti Colonial* gained many

Fig. 4.6   From Dakar to the Red Sea: *In a clearly nationalist and pro-expansionist setting, Jean-Baptiste Marchand, an Africanist and experienced military officer, proposed that the French government launch an expedition to regain power in the Congo region and open the path for a continuous strip of French domination from Dakar to the Red Sea. After an epic 14-month expedition across the heart of Africa, Marchand and his small force of* tirailleurs *hoisted the French tricolour flag at Fashoda on July 10, 1898. The image shows a triumphant and heroic Marchand and was the cover of the first issue of a set of fascicles describing the expedition.*

seats in the 1889 elections and its deputies enjoyed the backing of the wealthy *Union Colonial*. This lobby, composed of financiers, entrepreneurs, journalists, and politicians with an interest in the expansion of French colonial trade, was founded in 1893. At the same time, Jean-Baptiste Marchand, an experienced military officer with a solid career on African soil and well known for a number of heroic actions, had for some years been nourishing a plan to bring Fashoda under French control. He was convinced that this would consolidate French power in the Congo region, thus clearing a right of way for a continuous strip of French domination from Dakar to the Red Sea. After an epic 14-month expedition through the heart of Africa, Marchand and his small force of *tirailleurs* hoisted the French tricolor at Fashoda on July 10, 1898.

The other side of the Channel was also under the sway of pro-imperialist ideology. The British Prime Minister Robert Salisbury and his Secretary of State for the Colonies, Joseph Chamberlain, actively supported British domination of Africa, pursuing the famous policy of "Splendid Isolation" and attacking all those who opposed their rule. Unlike Marchand's flamboyant expedition through uncharted bush, the English headed off to Khartoum from Cairo in a disciplined if somewhat plodding fashion, moving their army at the same pace as they were able to construct a 400 km railroad line. The officer in charge of the British offensive was the aforementioned Horatio Herbert Lord Kitchener, a commander with extensive knowledge of and experience in the British Empire, mainly in Palestine, Anatolia, and Egypt. Kitchener was a member of the Royal Engineers and understood the importance not only of modern guns but also of modern infrastructures in the appropriation and securing of African territory.[27]

A young Winston Churchill, posted to the 21st Lancers under Kitchener's command as war correspondent for the *Morning Post*, considered the railroad the keystone of a successful strategy, inasmuch as it ensured transport, supplies, reinforcements, and communication for the advancing army and eliminated dependence on the dangerous and temperamental Nile. Churchill praised the construction of the "Desert Railway" as the main weapon in securing British victory over the Mahdist Islamic rebellion:

> In savage warfare in a flat country the power of modern machinery is such that flesh and blood can scarcely prevail, and the chances of battle are reduced to a minimum. Fighting the Dervishes was primarily a matter of transport. The Khalifa was conquered on

the railway....On the day that the first troop train steamed into the fortified camp at the confluence of the Nile and the Atbara rivers the doom of the Dervishes was sealed.[28]

It was indeed a confident Kitchener who met Marchand at Fashoda. Both men knew the importance of that specific small point on the map, "an obscure swamp on the Upper Nile" in Churchill's words, disputed by Belgium, Italy and, in the end, by France and Great Britain.[29] All considered Fashoda as the critical nexus of their west–east/north–south opposed axes.

Kitchener and his powerful flotilla of gunboats arrived at the Fashoda fort on September 18, 1898, after their long and successful campaign against the Mahdi army, reconquering Sudan in the name of the Egyptian Khedive, a British ally. After sending a message to the fort, Kitchener received Marchand's reply in which the latter congratulated the Sirdar (the title used by the British Commander-in-Chief of the Egyptian Army) on his victory over the Mahdists and welcomed him to Fashoda in the name of France. The British officers disembarked and were received by Major Marchand, with a guard of honor, and they shook hands warmly. Both sides were polite but each insisted on its right to Fashoda and acknowledged the possibility of going to war. Having handed Marchand a formal protest against the French occupation of Fashoda, Kitchener, politely ignoring the French flag, and without interfering with Marchand's Expedition, hoisted the British and Egyptian colors in two outposts nearby the Fashoda fort.

Clearly, Britain was not about to relinquish Fashoda. They claimed the place because, first of all, Britain regarded the French expedition as deceitful, a secretive stab in the back from deep within the heart of the Dark Continent, certainly by comparison with the open, planned, fair, and civilized English advance through Sudan; secondly, because they refused to countenance any obstacles to a British north–south axis of domination in Africa. As Churchill put it: "They should evacuate Fashoda, or else all the might, majesty, dominion, and power of everything that could by any stretch of the imagination be called 'British' should be employed to make them go."[30]

The crisis persisted through September and October, and both nations began to mobilize their fleets in preparation for war.

Fig. 4.7 Fighting for "an obscure swamp on the Upper Nile": *Backed by the "Desert Railway," Horatio Herbert Lord Kitchener, a commander from the Royal Engineers with extensive knowledge and experience in the British Empire, crushed the Madhist Islamic rebellion, and headed to Fashoda to regain it for British rule. The front page of L'Univers Illustré depicts the moment when Marchand and Kitchener shook hands warmly and politely, although both insisted on their right to Fashoda and acknowledged the possibility of going to war.*

Meanwhile, in London and Paris the two governments pursued negotiations. The French minister of Foreign Affairs, Théophile Delcassé, soon realized that Britain would not give up Fashoda and inasmuch as France was not prepared to go to war, a peaceful resolution was the only way out. Delcassé ordered his soldiers to withdraw on November 3, 1898. Marchand's last words when leaving Africa—"Who can say that the Sphinx is not about to smile?"—intimated an uncertainty about the future. Be that as it may, the Fashoda Incident was the last serious colonial dispute between Britain and France, and its classic diplomatic resolution is considered to be the precursor of the *Entente Cordiale*, a set of three agreements which ruled colonial affairs between the two nations and marked the end of almost a millennium of conflicts.

Greening the Sahara

From a European perspective, Africa was both far and near, remote and accessible. While "darkest Africa" remained a distant mirror of European ambitions, Europeans often saw the continent's northern desert regions as their backyard. During the 1950s, for example, the Sahara became a theme in European integration. The July 1949 issue of the German weekly *Die Zeit* reported that:

> The projects to open up Africa—whether in the state of planning or already in the making—are terrific, not to say fantastic. Having lost European spheres of influence in South East Asia and India and due to the need to gain new contracts and marketplaces, colonial powers have multiplied their efforts. The goal of upgrading Africa to American standards has enthralled Europe, and it piles huge amounts of data and graphs of all kinds onto long-suffering sheets of paper...[31]

Among the many schemes postwar planners sketched out, flooding the Sahara was one of the oddest—but also the one with the longest tradition. Beginning with the French occupation of Algeria in the 1830s, the territories across the Mediterranean Sea were continuously targeted for improvement by European developers. Plans for greening the Sahara—partly intended to exhibit European ingenuity—appeared regularly in various publications. As early as 1832, Saint-Simonists like Michel Chevalier proposed a *Systéme de la Meditérranée*, a centrally-planned Mediterranean railroad network to unify the entire Mediterranean basin. After France had lost the Franco-Prussian War in 1870/71, French topographer François Élie Roudaire proposed to compensate the lost territories in Alsace-Lorraine by opening up the vast arid areas between Algeria and Tunisia. He planned to use canal systems to flood three large basins lying below the level of the Mediterranean Sea, thereby creating a 13,000 square kilometer lake such as had reportedly existed during the Roman Empire. Roudaire promised to "let the desert bloom again with the help of modern technology" and to revitalize a supposedly sterile and unproductive territory which, though it might have been a cradle of civilization in ancient times, had since gone to seed and been ignored by successive generations. Although French celebrities like Ferdinand de Lesseps and Victor Hugo openly supported Roudaire's project, it eventually fell prey to the indifference of French politicians.[32]

European engineers and visionaries nevertheless continued to develop similar large-scale plans and projects for both African and African-European development, like Cecil Rhodes' Cape to Cairo railroad plan or the French plan to build a Trans-Saharan railroad line described above. One of these plans was almost realized: the American inventor Frank Shuman, described as combining "Teddy Roosevelt-like passion and Thomas Edison-like ingenuity," got the backing of the hero of Omdurman, Horatio Herbert Lord Kitchener, and a number of British businessmen, for an idea to eliminate the European need for coal by building solar energy plants in the African desert.[33] In June 1913, the cream of British colonial society in Egypt was actually able to attend the opening of a pilot plant in the village of Maadi south of Cairo. The solar plant produced 55 horsepower, enough to power pumps for supplying Nile water to distant irrigation sites. Shuman managed to become the first engineer to address the German *Reichstag* (parliament) and was promised 200,000 *Reichsmark* to build an even larger plant in German Southwest Africa: "Flush with success, fame, and funds, Shuman envisioned solar power plants on vast scales, even projecting a 20,000 square mile solar power plant in the Sahara desert to generate 270 million HP—an amount, he noted, equal to all the fuel burned in the entire world in 1909."[34] The plans were shelved after the First World War broke out, and with Shuman's death in 1918 the entire scheme passed into oblivion. The same fate befell solar energy in general, the promise of which paled in comparison with the availability of cheap oil after the war.

The Munich architect Herman Soergel advanced the most ambitious of these plans, in a sense a synthesis of infrastructures already proposed by many of his predecessors. He proposed to create a Euro-African super-continent joined by hydraulic and transportation infrastructures, a project slated to begin in 1927. Soergel proposed to sever the Mediterranean Sea from the Atlantic Ocean and the Black Sea by means of giant dams at Gibraltar and Gallipoli. The level of this now truly inland sea would eventually drop by as much as 100 meters as more and more water evaporated. This would provide new agricultural land as well as hydroelectric power from plants located at the new mouths of ever more rapidly inflowing rivers. This would secure Europe's future food and energy requirements. Africa would be developed and bound to Europe by means of new transportation links over dams and bridges. According to Soergel's calculations, elaborate irrigation systems could make a large part of the Sahara fertile again and temper central Africa's tropical climate to the extent that Europeans could settle it. Soergel called his infrastructurally-forged Eurafrican continent "Atlantropa," touting it as a new rival for the equally extensive continents of America and Asia. Like Shuman's solar power plant, the scheme did not outlive its protagonist who died as a result of injuries sustained in a Munich bicycle accident in 1952.[35]

The only somewhat more down to earth French physicist Bernard Dubos succeeded the visionary Soergel. In the 1930s Dubos proposed to set up huge wind-power stations in the Atlas and Hoggar mountains of Algeria that would transmit energy to Europe by day and to Africa by night. The scheme also called for a number of tall chimneys for the purpose of improving the weather by stimulating the formation of rain clouds—an idea that by the 1950s was coupled to the formidable forces that had been unleashed in the atom. The irrigation of the Sahara, an idea already espoused in earlier decades, would now be arranged by forcing passing monsoon clouds to shed their rain due to atomic bomb blasts set off over mountain ranges. Such schemes sprung from the same fertile techno-utopian soil that moved the philosopher Ernst Bloch to rhapsodize about "a blue atmosphere of peace; out of the desert comes fertility, out of ice comes spring. A few hundred pounds of uranium and thorium would be enough to make the Sahara and

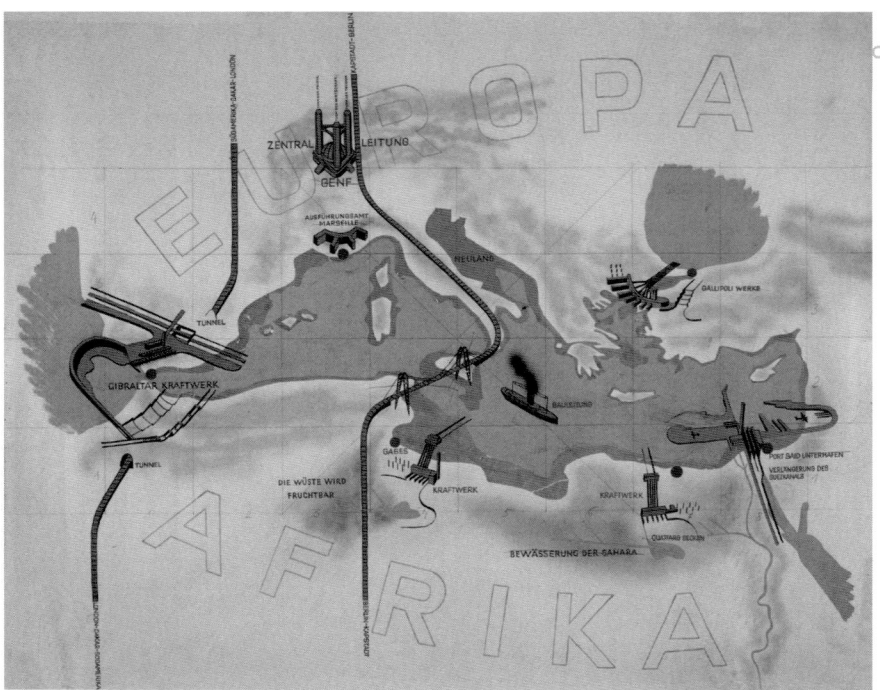

Fig. 4.8   Atlantropa:
Until his death in 1952, German architect Herman Soergel campaigned for his idea of a unified European-African continent, tied together by infrastructures and trade. In order to place it on an equal status with the Americas and Asia, he suggested calling it "Atlantropa." A whole generation of high-modernist planners and architects was fascinated by large building projects, which would balance out the "contingencies" of nature.

Gobi deserts disappear and turn Siberia, North America, Greenland and the Antarctic into the Riviera."[36]

Such effusive visions were often intertwined with new technological sublimes. These were the harbingers of European cleverness, ostensibly capable of opening up "savage" territories and rendering them habitable for "civilized" people, in the first place by applying technology to improving the extreme climate. If the greening of the Sahara can be seen as the ultimate test of European ingenuity, Africa in general served as a perennial drawing board for megalomaniacal engineering projects. Though very few of these ambitious plans were ever realized, to this day Africa has remained an arena for large-scale planning like the German "Desertec Solar energy Scheme" that was proposed in 2009 (and dropped a few years later). A contributing factor may be the *idée fixe* of Europeans that hot tropical zones tend to foster "idleness" because their inhabitants have not had to use productive labor to control nature as people of more moderate climates are forced to do. A recurring argument was that the native people were not interested enough in developing the land and its potentials, so that European intervention in Africa—and especially in "empty spaces" like the Sahara desert—was more than justified: to the point, even, of becoming a moral duty. Today, many Europeans once again believe that developing Africa is essential—this time to stem the tide of Africans crossing the Mediterranean Sea to ask for *their* share in Eurafrica.

5

From the Raj to the Yellow Peril

Introduction

The immense variety in the reception of Western science and technology in the different quarters of the Asian continent, the opposing and contradictory patterns of what is glibly and erroneously-called "technological diffusion," has been the topic of extensive debate. The general tenor of Western encounters with "natives" in Asia is very different from what we found in Africa. Western nations saw Africa as their backyard, in which it was both imperative and legitimate to institute an intensive colonial regime that ultimately covered the whole continent. This could be done because the fragmented African tribal system proved to be exceedingly fragile when confronted with European power; in addition to Europe's almost irresistible military resources, the small scale of the tribes or kingdoms turned African people into easy prey for Europe's strong centralized empires.

In Asia, Europeans and later Americans confronted ancient and highly-structured civilizations, some of them with a tradition of ruling their own empires and with strong—if by European standards archaic—cultural, scientific, and technical traditions.

In spite of these differences, however, in Africa as in Asia, the yardstick along which mutual evaluation of Europeans and "natives" took place was basically a technological one. Westerners advertised their superiority by exhibiting big guns and ships and by implanting civilizing infrastructures such as railroads, harbors, and telegraphs. Asians recognized the superiority of the invaders only in terms of a superficial technological prowess and aspired to meet those standards only in order to be able to assert what they themselves considered their innate—even racial—political and cultural superiority.

The ways in which India, China, and Japan reacted to Western technology were highly distinctive. First it is evident that the very different political status of each of the three regions was the main factor in shaping their response. India was a British colony with circumscribed autonomy. China was an empire that believed in its superiority, but that proved to be hopelessly vulnerable to Western military technology. In the end, it became a quasi-colony. Japan, despite being a small country, was able to maintain its sovereignty. Always on the lookout and suspicious of its powerful Chinese neighbor, it was confident and kept a sharp eye on the outside world. Japan teaches us that a clear vision of the *what, why,* and *what for* is vital to the success of any attempt to imitate or appropriate foreign technology. For that matter, Japan was the only one of the three that had any choice, as it was the only one formally in control of its own modernizing strategy.

Other factors help explain the contrast between Japan and all other non-Western territories in the nineteenth century: the recognition that innovation was a gain, not a menace; the decentralized feudal system of the Edo era that fostered different hubs for innovation and that encouraged competition; a high literacy rate that allowed easy incorporation of innovative processes in everyday life; the open door to European innovations through the *Rangaku;* military-driven industrial growth promoted by the Meiji restoration; the flowering of an entrepreneurial mentality during the Meiji period; and the all-inclusive Meiji modernizing agenda.

A final salient difference links Japan's modernization process to the outlook of independent India's first Prime Minister, Jawaharlal Nehru. Both Nehru and Japan embraced Western technology as the foundation of a modern society. Unlike Gandhi and the Chinese authorities, Japan's Meiji rulers agreed with the later

Nehru administration that did not consider railroads, telegraphs, and industrial manufactures as mere handy add-ons to the existing social order, a sort of cosmetic layer of modernity behind which traditional culture and ways of living could go on unscathed. Rather, these technologically-based infrastructures and factories were to be seen as the core of a new society, as well as the driving force behind more extensive changes leading to modernity. In this respect, Nehru's government and the Meiji shared the deepest Western industrial attitudes to technology and this probably explains much of their respective success in appropriating Western technology without forfeiting their specific identities.

India: From the Imperialist to the Nationalist Railroad

In a speech in 1917, Mohandas Karamchand Gandhi used railroads as a metaphor for the profound asymmetries characterizing colonial India. In practice, Gandhi rode the colonial railroad system to travel the length and breadth of India in order, as he put it, to "get a grasp of the life" of his colonized land. He traveled exclusively third-class because he wanted to share the experience of the Indian people. The same railroads that had been "tools of empire" and "tentacles of progress" were now enrolled into Gandhi's nonviolent independence agenda. Gandhi experienced the degrading conditions third-class passengers had to endure not only as humiliating and as an instance of savage imperial exploitation, but also as a strategy to secure European domination by undermining the health and morality of native Indians.

Gandhi ended his speech by suggesting:

> Let the people in high places, the Viceroy, the Commander-in-Chief, the Rajas, Maharajas, the Imperial Councillors and others, who generally travel in superior classes, without previous warning go through the experience now and then of third class travelling. We would then soon see a remarkable change in the conditions of third class travelling and the uncomplaining millions will get some return for the fares they pay under the expectation of being carried from place to place with ordinary creature comforts.

Fig. 5.1   A Train Called India: *Gandhi used the British imperial railway system as a metaphor for the inequalities of Indian colonial society. Presented by the Raj authorities as a symbol of progress, the experience of traveling in first-class or in third-class mirrored the luxury of the colonizers and the poverty of the colonized. For Gandhi the degrading and devilish conditions third-class passengers had to endure were not only humiliating as an instance of savage imperial exploitation, but also as a strategy to secure European domination by undermining the health and morality of native Indians.*

Thirty years later, in 1947, only two months after the declaration of independence, Gandhi returned to the issue of the British-built railroads, but now as the transport backbone of an independent nation:

> The passengers should consider the railways as their own property. They should keep the trains clean. They should not spit and smoke in the trains and should not pull the chain without real need. And not a single passenger should travel without a ticket. Then I would be able to say that we have attained true independence.

In 2004 the Eastern Indian Railroad commemorated its 150th birthday. The celebrations included a re-enactment of the journey of the first train from the Howrah station (West Bengal) to Hooghly, on August 15, 1854, a run of about 37 km taking 91 minutes. Just prior to this historic "Heritage Run," the Indian Minister for Railroads introduced a commemorative volume entitled *Symphony of Progress: The Saga of Eastern Railway. 1854–2003.*[1] Interviewed on the occasion, the railroad's General Manager considered the celebration a "momentous occasion" citing especially how "proud of their railway" the people in attendance were. The same commemorative and enthusiastic attitude toward infrastructural technology also pervades Krishnalal Shridharani's book, *Story of the Indian Telegraphs: A Century of Progress*, published in 1953 by the

Fig. 5.2 Iron Elephants: *This plate from the* Illustrated London News *depicts the arrival of a railroad locomotive in India in 1875. The first railroad on the Indian sub-continent was constructed in 1853, running from Mumbai to Thane (around 34 km); by 1880 the Indian railroad system had a route mileage of approximately 15,000 km.*

Mail and Telegraph Department on the occasion of the centenary of the Indian telegraph system.[2]

Both Gandhi's political use of railroads and the commemorative volumes on railroads and telegraphs echo a utilitarian idea of progress that is embodied both in technological innovations and in transformations of the existing social and political order. Nehru expressed the same vision in the Preface he wrote to Shridharani' book, describing railroads and telegraphs as the "great change...which has progressively altered the very texture of human life."[3]

These brief stories illustrate how railroads have formed contemporary India, not only materially but also ideologically, that is, as the material infrastructure of a modern national consciousness. However, when the Scottish historian, economist, political theorist, and philosopher James Mill equated civilization with improved transportation infrastructure in his *The History of British India* (1817) he hardly envisioned such an ironic outcome. In line with Jeremy Bentham's utilitarian principles, Mill had argued that the key to progress in India, which he portrayed as an almost savage society, lay in the introduction of Western science, knowledge, and material improvements. He dismissed Indian knowledge as "rude," basing his judgment in part on the opinion of the renowned British cartographer Major James Rennel:[4]

> If we apply the reflection, which has been much admired, that if a
> man were to travel over the whole world, he might take the state of
> the roads, that is the means of internal communication in general,
> as a measure of the civilization; a very low estimate will be formed
> of the progress of the Hindus: In India [says Rennel], the roads are
> little better than paths, and the rivers without bridges.

On the same tack, James Andrew Broun-Ramsay, First Marquis of
Dalhousie, a prominent Tory who became Governor-General of
India in 1847, envisioned the material improvement of India as
rooted in railroads, the electric telegraph, and mail services, which
he considered the "three great engines of social improvement."[5] He
also understood their role in maintaining the colonial order, identifying the telegraph as the main reason for Great Britain's victory
in the First War of Independence of 1857, in which the British East
India Company confronted part of its own army, the *sepoys*, a
combination of Hindu and Muslim soldiers.

Dalhousie saw himself as a Pharaoh, an Oriental monarch, and
an eighteenth-century enlightened ruler whose authoritarian
government would bring modernity and civilization to India. He
accordingly considered it incumbent on his government to extend
electric telegraphy to every nook and cranny of the subcontinent
as well as to institute a uniform postal system. Railroads, however,
unlike telegraphs and mail, could be profitable investments
and these, ordained Dalhousie, should be built, maintained, and
exploited by private capital, although under state regulation. In this
way Dalhousie encouraged private enterprise while maintaining
government control over the key infrastructures.

Dalhousie was convinced that material improvements were not
only the royal road to reforming the social order but also the instruments of choice for preserving British imperialism. The Indian
railroads project fused these two visions: the persisting idea of the
civilizing mission, that is bringing to India "the greatest invention
of all times,"[6] as well as greater effectiveness in imposing British
cultural values, religious beliefs, and political and military rule.
Colonial railroads also pandered to powerful economic and financial interests in England, mainly the cotton lobby that had vested
interests in cheap access to India's hinterlands with their rich
raw materials (coal, cotton) and their potential markets. In 1845,

following a short pioneering period, two railroad companies were founded to build and operate lines near Bombay and Calcutta, respectively the Great Indian Peninsular Railway (GIPR) headed by John Chapman and the East Indian Railway (EIR) led by Rowland Stephenson.

Scholars of British colonial railroads concur in estimating that British capital accounted for about 99 percent of the total investment, with only 1 percent attributable to Indian investors. The British, moreover, were guaranteed a 5 percent rate of return on their investments. It is hardly surprising that the construction and operation of this huge technical system was also in British hands, with British engineers, skilled workers, and foremen lured to India by high salaries.

John Chapman's GIPR aimed to develop a network of railroad lines radiating from Bombay. Chapman was well educated, an active Baptist, a skilled mechanic, and a journalist: he started out as a successful manufacturer of textile machinery, but when his company folded, he found employment as a mathematical instrument maker. He wrote on political topics in various newspapers and collaborated for many years with the well-known technical journal *Mechanics Magazine* and later with the *Railway Times*.[7] This Janus-faced career eventually led Chapman to promote a railroad line that would profit the strong English Midlands cotton lobby as well as contribute to the material improvement of India.

The first segment of the megaproject to link Bombay and the east coast of India (a visionary coast-to-coast line) was a passenger line inaugurated on April 16, 1853 between Bori Bunder (a warehouse district on Bombay's eastern shore) and Thane. The line was 34 km long and serviced by three locomotives, *Sahib*, *Sindh*, and *Sultan*. The GIPR's first chief engineer, James John Berkley, had been one of Rowland Stephenson's trusted men. Stephenson, highly successful railroad engineer and entrepreneur, was consultant engineer on the project.

Stephenson also headed the East Indian Railway project. In planning the railroad, Stephenson concluded that the transport of coal from the Raniganj mines, nearly 200 kilometers upstream from Calcutta, by means of traditional country boats (sampangs and the like) was far too slow and expensive. He proposed to solve this problem by routing his railroad through the Raniganj coalfields. Despite the fact that his scheme for such a railroad line stretching from Calcutta to Delhi via Benares was rejected by the Calcutta Government, Stephenson persevered thanks to support from the East India Company, a number of Anglo-Indian trading firms, and the shipping company P&O, recently arrived in India.

As described above, the first train left the Howrah station on August 15, 1854, for Hooghly, a distance of about 37 km. More than 3,000 persons vied for a ticket, but only 300 were actually able to make the trip. The first train ran to full capacity, with three first-class and two second-class carriages, plus three trucks for third-class passengers, all built locally by two British firms (Steward & Company and Seton & Company). The locomotive had been shipped to Calcutta via Australia.

The section Howrah–Hooghly was the first installment of a 160 km line that was to connect Calcutta with Rajmahal. This was later extended to Delhi via Mirzapur. Chief Engineer George Turnbull was in charge of the detailed surveys and, together with his fellow British engineers, planned the route and designed the necessary infrastructure like bridges and stations. The Raniganj coalfields were reached on February 3, 1855, and the bridge over the river Son, the largest tributary of the Ganges, was completed on November 4, 1862. The entire 870 km section from Calcutta to Benares was in service by December 1862. On February 5, 1863, a special celebratory train left the Howrah station for Benares, carrying George Turnbull, the Viceroy, the Governor, and other high members of the colonial administration and local aristocracy.

The third movement of the Indian "Symphony of Progress" was the railroad connection between Bombay and Calcutta. Once again, this was a British project, headed by the engineer Robert Maitland Brereton. Brereton was the offspring of a family of notable Victorian engineers closely related to the famous British civil engineer Isambard Kingdom Brunel. After studying practical mechanics at King's College London, Brereton joined Brunel's design team, where he participated in several major projects, including the construction of the Cornish railroad, Paddington Station (London), and the mammoth steamship SS *Great Eastern*, which would later play a key role in laying the first successful transatlantic cable.

In 1857, Brereton was dispatched to India for the construction of the Bombay to Calcutta Railway, which would form the trunk line of the Indian Railways. Having managed to survive the troubled period of the Indian Mutiny, he was appointed chief

Fig. 5.3 The Railroad Builders: *As railroads penetrated the Indian hinterland interior, classical imperialist relationships flourished. The extensive grid of infrastructures (in this case in India) was at the heart of the European way of understanding and dealing with nature by making the territory manageable and by rationalizing and controlling time, space, speed, and productivity. Rudyard Kipling captured this idea in his short story* The Bridge-Builders *(1893), by describing the sacred river Ganges spanned by a bridge built by British engineers.*

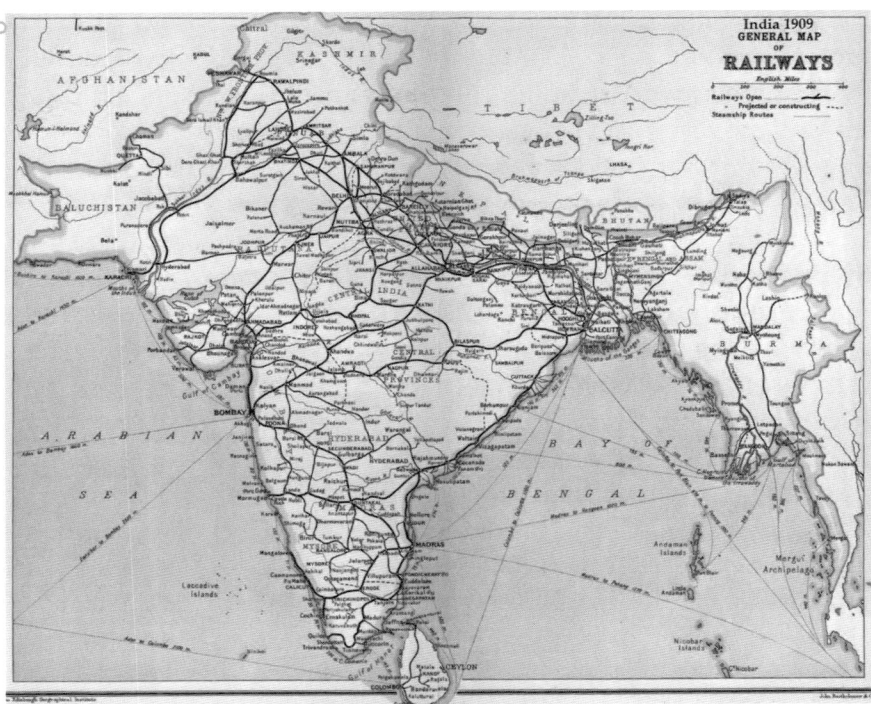

engineer for the Grand Indian Peninsular Railway. On March 7, 1870, ahead of schedule, the Bombay–Calcutta railroad was officially inaugurated, bringing the total length of the Indian railroad network to 6,400 km. By 1880, that network totaled about 14,500 km, mostly radiating inward from the three major port cities—Bombay, Madras, and Calcutta. By 1910, India had over 50,000 km of track, which put it just behind the United States, Germany, and Russia, and the density of its network was rivaled only by the American and European networks, being higher than comparable systems in South America, Africa, Asia, Canada, Russia, and Australia.[8]

The extensive railroad grid completely changed traditional Indian society. As railroads penetrated the interior, classical imperialist relationships flourished. All of India now exported raw materials to Great Britain and imported manufactured goods. Many local craftsmen lost their livelihoods and flocked to the industrial centers and, though Indian technicians did assist in building the railroads, British engineers and skilled laborers monopolized the remunerative "high-tech" positions.

Railroad construction in India was neither just a huge technical project, nor simply a "tool of empire." It was the very essence of the European way of understanding and dealing with nature by making the territory manageable and by rationalizing and controlling time, space, speed, and productivity. In his short story *The Bridge-Builders*, Rudyard Kipling projected the image of "Mother Gunga—in iron" (the sacred river Ganges spanned by a bridge built by British engineers) a powerful metaphor for the British subjugation of India, and a symbol of the triumph of modern rationalism over religious tradition.

Ironically, this colonial project was taken over lock, stock, and barrel by the founders of the new independent nation state of India. Both Gandhi and Nehru shared this concept of mastering the land through infrastructure, though now, instead of a productive colony, it was about creating a nation state on the basis of the uniform and domesticated territory produced by modern technological infrastructures (railroads, telegraphs, harbors, postage). That said, it was also obvious that these two leading actors of the Indian struggle for independence had different attitudes to modern/European technology and its significance for India's identity, differences that recalled the intense debates between traditionalists and westernizers. It was nonetheless inescapable that the nation whose independence they sought would be built on a territory created by British engineers.

Conventional wisdom presents Nehru as eager to integrate India into the world economy on the basis of aggressive modernization and Gandhi as an advocate of an idealistic traditionalism. Be that as it may, both always embedded their criticism of European technology in a critique of colonialism. Nehru considered modern technology a creative force, even if it was abused and distorted by imperialists. He argued that India could only make progress as an independent nation by pursuing a technology-driven process of modernization, a conviction he carried out with great enthusiasm during his 17 years as India's Prime Minister. Nehru initiated a large and diversified program of material improvements (dams, irrigation works, and hydroelectric power) and founded a network of engineering and technology-oriented institutes for higher education, specifically the Indian Institutes of Technology and the National Institutes of Technology. For Nehru, India's measure as a nation had to be taken on the basis of its command of modern technology.

Gandhi tended to the view that Western technology was an innately bad practice, inevitably leading to mass exploitation and dehumanization. Gandhi's position was close to that advocated earlier by the so-called revivalists during the Bhadralok Debate (1890–1915): the main political and economic unit should be not the nation state, but the village and the family; industries should be founded on a family handicrafts basis. Large-scale projects, though admittedly essential to the small communities, should be restricted to infrastructural networks, such as railroads and telegraphs.

Gandhi summarized this alternative to the Western vision of modern societies in his book *Hind Swaraj* (1909). The concept *swaraj*, self-governance or "self-rule," and "home-rule," suggested a new non-hierarchical, decentralized government, which rejected British political, economic, bureaucratic, legal, military, scientific, technical, and educational values and institutions. As far as technology was concerned, the *swaraj* ideals substituted traditional practices for Western technology.

Gandhi's struggle to institute *swaraj* was a failure. Nehru's vision of a powerful centralized interventionist state, based on a critical appropriation of a technologically-driven colonial modernity emerged as the winning strategy for an independent India. Fascinated by the USSR's socialism and planned economic reconstruction, Nehru came to believe that India should pursue a similar mix of socialist ideology, centralized governance, heavy industrialization, and techno-scientific development.

The Opening of China & the Rise of Japan as a Westernized Society

In 1912, the British writer Sax Rohmer published a series of short stories featuring the fictional character Dr. Fu Manchu. The next year he published the first novel featuring the perennial and highly successful villain: *The Mystery of Dr. Fu-Manchu* (published in the U.S. as *The Insidious Dr. Fu-Manchu*). Fu Manchu's countless murderous plots employed an arsenal of poisons derived from natural venoms (fungi, spiders, snakes) or brewed up in his own laboratory. This powerful master-villain, intimate with the opium underworld and mixing "western science and oriental magic," as the cover of an American edition proclaimed, epitomized the well-known Yellow Peril myth—the specifically Asian threat to Western civilization and values.

Although Fu Manchu was the most famous and dramatic popular incarnation of this myth, the seeds had been planted years earlier. The term *Gelbe Gefahr* was allegedly coined by emperor Wilhelm II

Fig. 5.4　The Yellow Menace: *Although the term* Gelbe Gefahr *(Yellow Peril) was coined in 1895, it only became part of popular culture in the early years of the twentieth century. Sax Rohmer's character Fu Manchu, a master criminal who mixed in his own laboratory Western science and oriental magic, epitomized the well-known Yellow Peril myth, that is, the Asian threat to Western civilization and values.*

in September 1895, in reference to Hermann Knackfuss's painting *Völker Europas, wahrt eure heiligsten Güter* (Peoples of Europe, guard your dearest goods). The painting shows the Archangel Michael leading European nations in making war on a golden Buddha, an obvious metaphor for an impending attack upon Europe by Asiatic powers, namely China and Japan. The concept of the Yellow Peril became common in the early 1920s both in Europe and in the United States, and it was ubiquitous in newspapers, novels, cartoons, and postcards. About 1905, a series of postcards entitled *L'Arc-en-Ciel* depicted the yellow—Chinese and Japanese—menace to traditional Western, Christian, Caucasian, and industrial hegemony.

It has been argued that Europeans have feared Asia since the Middle Ages, in particular following the Mongol invasions led

Fig. 5.5 "Make Way for the Yellows": *This expression used by the French artiste Mille in one of his picture postcards of the Russo-Japanese War (1904–1905) not only illustrates the Yellow Peril syndrome in general, but also brings it to the realm of politics and warfare. The Yellow Peril imagery summoned ancient fears and often used medieval-like characters (chimeras, dragons, ogres) to emphasize the vicious and extremely dangerous nature of Asian expansionism that threatened European world hegemony. This 1935 cartoon of E. Schilling shows an Orientalized octopus dominating the world with its tentacles.*

by Genghis Khan. However, something clearly changed in the nineteenth century, transforming a diffuse ancestral fear into a specific menace. This change is closely related to imperial disputes, mainly to the so-called Self-Strengthening Movement in China and the rise of Japan as an industrial westernized empire. Both cases exemplified the same urge for modernization through Western technology that we have witnessed in the case of post-colonial India, though the specific situations and approaches were vastly different.

China: Staging Modernity

The Chinese Self-Strengthening Movement followed on a series of military defeats and concessions to foreign powers by the Qing, or Manchu, Dynasty (1644–1911/12). Although the decline of the Qing stemmed from its own crisis of governance and legitimacy, it was certainly hastened by the Opium Wars that started out as a dispute over trade with the British Empire and later involved France and Russia.[9] The Opium Wars eventually sealed China's loss of sovereignty.[10]

The Empire of the Great Qing had rarely been stable, plagued as it was by local rebellions contesting the frequent new taxes, numerous religious and ethnic conflicts, and general political volatility. By the end of the eighteenth century rampant official corruption had thoroughly destabilized the empire. The White Lotus Rebellion (1794–1804), a tax protest by impoverished settlers led by a secret religious society, marked the beginning of the end, ushering in an era of uprisings that lasted throughout the nineteenth century.[11]

The Tianli Sect or Heavenly Principle Sect, a branch of the White Lotus, was behind the politically-inspired Eight Trigrams Uprising of 1813 and, from 1795 to 1806, an ethnic revolt—the Miao Rebellion—also challenged government authority. In the mid-nineteenth century, Hong Xiuquan, a convert to Christianity, led a major upheaval—the Taiping rebellion (1850–64)—against the Manchu Qing dynasty. Hong soon controlled large parts of southern China, united under the Taiping Heavenly Kingdom, and tried to implement social reforms based on shared property and on Christian values. This rebellion cost some 20 to 30 million people their lives, rendering it the bloodiest civil war in history and the second bloodiest war of any kind, surpassed substantially only by the Second World War. The rebellion heralded a period of pervasive social disorder. Numerous uprisings erupted all over China: the Dungan revolt (1862–77) and the Panthay Rebellion (1856–73), both ethnic surges led by the Muslim Hui people, the Miao Rebellion of 1854–73, also an ethnic revolt by the Miao in Guizhou province, and the devastating Nien Rebellion in northern China from 1851 to 1868.

At the same time, China was confronted by the growing power of European empires. Great Britain dominated nearby India, the East-Indies archipelago was the jewel in the crown of the Dutch Empire, Macau had become a Portuguese colony, and the Russian Empire was advancing into areas north of China. An early Qing response was the Canton System (1756) that enabled the Chinese government to exercise tight control over all commercial transactions between China and foreign nations. European and American traders had very limited access to Chinese ports. All commerce moved through an official trade guild with direct dealings between European and Chinese merchants being forbidden. China demanded hard currency or specie (gold or silver coinage) as the medium of exchange in international trade. Hence European countries had to pay in silver for commodities such as tea, silk, and porcelain, a practice that soon created a major trade imbalance, as China purchased few Western goods.

This situation was particularly onerous for Britain, China's main trading partner. To redress the imbalance, the British East India Company began to stimulate opium production in India in the expectation of being able to sell it profitably on the Chinese market. An intricate trading structure emerged that relied partially on legal markets—opium was allowed in China for medicinal purposes—but also leveraged illicit ones, based on smuggling by ambitious merchants. The opium routes between India and Southeast Asia had already been explored by the French, the Portuguese, and the Dutch in the seventeenth century, but the active monopoly by the East India Company turned the opium trade into a major business and a key factor in maintaining a balance of trade with the East.

The illicit opium trade grew steadily and continuously during the second half of the eighteenth century. By the early 1820s, China was importing perhaps 255 tons of Bengali opium annually; a decade later, in 1830, this had increased more than fourfold to roughly 1,120 tons. Two years after the revocation of the East India Company's 1834 monopoly, that is, with the restoration of free trade, China was importing an estimated 1,820 tons of opium.[12] The Qing authorities responded vigorously by repressing Chinese drug traffickers and confiscating British opium supplies. This eventuated in a British declaration of war and the landing of a large British-Indian force at Canton in June 1840 by the Royal Navy.

The rousing defeat of the numerically-superior Chinese forces once again demonstrated British military superiority, based on new technologies like iron, steam-powered warships able to move against the winds and tides and fitted with heavy guns, such as the famous *Nemesis*, as well as the Army's modern muskets and cannons. *Nemesis* was one of six iron gunboats commissioned by the East India Company, using cutting-edge technology: a 120 horsepower steam engine, auxiliary sails, a flat bottom for sailing in shallow waters, a retractable rudder, a hull divided into seven water-tight compartments—all wedded to significant firepower in the form of Congreve rockets. The *Nemesis* and her companions were clearly superior to the Chinese war junks in terms of maneuverability, navigability, and firepower. Moreover, the weapons used by British soldiers were quicker, more efficient, and more accurate than traditional Chinese cannons. Despite that fact that the Chinese had tried to copy or buy European military technologies, namely cannons and a ship, the British took Canton easily and sailed up the Yangtze River, quickly assuming control of China's main commercial artery. In August 1842, the Qing authorities were compelled to sign the Treaty of Nanking, the first of the so-called "unequal treaties." Its exceedingly harsh terms undermined the Dynasty's prestige and seriously compromised Chinese sovereignty: China had to pay Britain an indemnity, cede Hong Kong, and open five ports to foreign commerce. In 1844, the U.S. and France signed similar treaties.

In subsequent years Chinese authorities tried to circumvent the Treaty of Nanking by boycotting foreign traders and bullying native merchants who dealt with the British. Countless small skirmishes congealed into a major incident in 1856: the seizure of the *Arrow*, a schooner allegedly under the British flag. The Arrow Incident served as a general-purpose pretext for European imperialists

Fig. 5.6 Nemesis:
Referred by the Chinese as the "devil ship," Nemesis was the first British ocean-going iron warship. Launched in 1839 it was used in the First Opium War. The picture depicts Nemesis, Calliope, *and* Starling *destroying the Chinese war junks in Anson's Bay during the Second Battle of Chuenpee, on January 7, 1841.* Nemesis, *a long, narrow flat-bottomed steamboat admirably suited to navigating the river mouths of China, embodied the efficiency of British expansionism in the era of what later become known as gunboat diplomacy.*

further to open China to the world economy, in the first place by demanding stringent adherence to the "unequal treaties."

The Second Opium War (1856–60) was a re-enactment of the first one, but with an extended cast of characters, following successful British efforts to enlist France, the U.S., and Russia in an alliance against China.[13] The French Empire joined Britain in response to the execution of a French missionary, Father August Chapdelaine, and the U.S. used an attack on a U.S. Navy officer to legitimate its participation. Finally Russia, China's main continental neighbor, offered help to the British and French, although in the end nothing materialized from this quarter.

The technological superiority flaunted by Britain during the First Opium War had lost little of its luster. Hence the war ended badly for China in 1858, which was now forced to abide by the mercenary terms of the four Treaties of Tientsin, to which Britain, France, Russia, and the U.S. were parties. These treaties opened eleven more ports to Western trade, extended the European diplomatic network, ensured free commercial navigation on the Yangtze River, secured the right of foreigners to travel in China's hinterland, and stipulated an indemnity to be paid by China to Britain and France.

Although the Qing authorities, seriously weakened by internal rebellions and previous wars with foreign countries, had little choice but to sign the treaties, radical political movements within China resisted these new encroachments upon Chinese national sovereignty. Resistance peaked during 1859 and 1860 on the crest of several minor diplomatic and military incidents. In the summer of 1860 a large Anglo-French force sailed from Hong Kong and captured the port cities of Yantai and Dalian to gain the Bohai Gulf, then marched inland toward Beijing. Peace talks were inconclusive. On September 18, the allied army annihilated Chinese forces near Zhangjiawan and proceeded to Beijing, arriving on October 6. Although the invaders remained outside the city, they looted both the Summer Palace and the Old Summer Palace and reduced them to ashes.

On October 18, 1860, hostilities came to an end with the signing of the Treaties of Tianjin. Once again, Western military and technological superiority had been decisive in enforcing European interests in China. The defeat of the Imperial army by a relatively-small but technologically-advanced Anglo-French military force, together with the death of the Emperor and the burning of the Summer Palaces shook the once powerful Qing Dynasty to its foundations and left a thoroughly-humiliated China, now a quasi-protectorate of Western powers, to wonder how things could have come to such a pass.

The response was a major modernization effort known as the Self-Strengthening Movement (1861–95). It can be seen as an attempt by Chinese intellectuals and politicians to cope with Western technological superiority. However, the movement contained a number of intriguing internal contradictions that in the end made it impossible to implement the technical and political reforms that might have transformed China into a modern state.

Chinese intellectuals found themselves on the horns of a dilemma in their efforts to adopt the Western path to modernization while keeping faith with Chinese cultural and political institutions. Lin Zexu, a scholar and official of the Dynasty, commissioned Wei Yuan, a prominent member of the Qing intellectual elite and secretary to several statesmen, to compile information on foreign countries that he had gathered during his lifetime. In the preface to his 1844 report, *Illustrated Treatise on the Maritime Kingdoms* (*HaiguoTuzhì*),

Wei expressed the gist of the movement: "learn the superior barbarian technique with which one repels the barbarians."[14]

The goal of the Self-Strengthening Movement was thus to adopt Western technology in what amounted to a superficial, instrumental, and merely operational way, primarily in order to defend China from future humiliations. The "self-strengtheners" considered that China had very little to learn from Western civilization, apart from weaponry and, perhaps in the future, a limited number of industrial technologies. This point of departure, together with China's vast territory and poor communication systems, made appropriation and diffusion of foreign knowledge and skills extremely difficult.

Nonetheless, the Self-Strengthening Movement stimulated some efforts at modernization along the lines of the Western technological model. In the first phase (1861–72) attention was restricted to military technology, with the adoption of Western ships and guns, the building of dockyards and arsenals to secure maritime defenses, and the training of technical staff. Regional leaders, supported by the central government, sponsored new military industries;

Fig. 5.7 Learn the Western Technique to Defeat the West: *Cannon in the Nanking arsenal, in Jiangsu province. The adoption of Western ships and guns, together with the building of dockyards and arsenals in ports and the training of technical staff, was part of the Self-Strengthening Movement agenda for modernizing China. Although this modernizing momentum opened up China to the West, the Self-Strengthening Movement perceived the adoption of Western technology only as a means to avoid future humiliations and not as part of a new worldview.*

these included Zeng Guofan (Shanghai Arsenal), Li Hongzhang (Nanking and Tientsin Arsenals), and Zuo Zongtang (Fuzhou Dockyard). Foreign experts were treated like precious commodities and imported into China together with their machines and other materials.[15] Although these programs were pivotal for acquiring technical knowledge and skills, they proved to be quite expensive. Foreign materials and experts were extremely costly, despite which the artifacts produced in China were inferior in quality and efficiency compared to the Western originals.

In the second phase (1872–85), under the leadership of Li Hongzhang, commerce, industry, and agriculture received increasing attention. Students were sent abroad to France, Britain, and Germany for technical training. The new industries were "government-supervised merchant undertakings," that is, operated by merchants but controlled and directed by government officials. Examples include the China Merchants' Steam Navigation Company, the Kaiping Coal Mines in Tientsin, and the Shanghai Cotton Mill. Infrastructural investments began in this period, centered on the new industries: the Imperial Telegraph Administration opened two lines, Taku–Tientsin and Shanghai–Tientsin, and the first railroad line, 10 km long, was built north of Tientsin. Despite its inefficient management style, this merchant–bureaucrat combination was the main driving-force behind new industrial enterprises.

Finally, the third phase (1885–95) brought both consolidation and a decline in reformist élan, as conservative factions overwhelmed the movement's original politically-inspired protagonists. Nevertheless, military industries, infrastructures, and new industrial sectors, in the first place cotton-weaving, continued to grow. Although "government-supervised merchant undertakings" continued to dominate, like the Kweichow Ironworks (1891) and the Hupeh Textile Company (1894), attempts were made to implement new organizational practices. Joint government and merchant enterprises and "private enterprises" blossomed in this period.

The Self-Strengthening Movement did not really thrust China into a serious process of modernization. The measures that were actually implemented across three decades just barely scratched the surface: Western technology was imported and imitated on the basis of purely utilitarian motives. Chinese modernizers never viewed them as a civilizational structure capable of building a new

China. Western technology remained a collection of exotic objects and practices artificially attached to the traditional Confucian worldview, which the Chinese continued to consider as a superior—indeed universally valid—*corpus* of values, ethics, and general knowledge.

Japan: Full Steam Ahead for Modernization

Although Europeans finally stopped worrying about possible threats from the Chinese empire, a new powerful menace did emerge in Asia: Japan.[16] With its small territory, Japan had for centuries lived in the shadow of a China-centered Asian world. Unlike its continental neighbors, however, Japan managed to preserve its autonomy despite ambitious Chinese imperialism. This successful resistance to the Celestial Empire was very much indebted to the long period of stability and prosperity Japan enjoyed during the Edo period (1603–1867), under the rule of the Tokugawa family.[17] The Tokugawa dynasty reinforced the Shogunate (*bakufu*) by increasing the Shogun's military and political power, at the expense of the Emperor who nonetheless retained his symbolic and religious authority. In Western terms, and the analogy was made by Europeans who traveled in Edo Japan, the Shogun held authority over the material world, like European kings, and the Emperor over the spiritual one, like the Pope.

The success of the Tokugawa family was anchored in a very formal, hierarchical, and strict order, which governed both the internal and the external policies of the Shogunate. In this context, two major codes were promulgated: the *bukeshohatto* (1615), a set of edicts to control the *daimyō* (feudal lords) and the *samurai* warrior aristocracy; and the *sakoku* (1639), an isolationist policy that transformed Japan into a "closed country."

The seclusion policy derived from a more general social vision, according to which foreigners, inevitably seen as barbarians, should be kept at the margins of civilized Japanese society. Moreover, Japanese rulers were wary of European expansionism and wanted to prevent Japan from becoming someone's colony. Vigorous Spanish and Portuguese proselytization that had spread Catholicism in a systematic, persuasive, and efficient way had already resulted in a sizable Catholic population among both peasants and aristocrats. The 1637–38 *Shimabara* Rebellion, an uprising of 40,000 mostly Christian peasants, allegedly instigated by Catholic missionaries, did little to lay the suspicions of Japanese authorities to rest.

Religious and cultural purity rather than economic concerns were the driving force behind the *sakoku* policy. Thus, the Dutch East India Company was allowed to trade as long as it agreed to desist from missionary activities. Dutch merchants—ironically so, given the artificial origins of much of their own country—were established on a newly-built artificial island in the bay of Nagasaki called Dejima. Through the Dutch, Japan maintained an opening to Western science and technology. The so-called *Rangaku*, that is, Dutch studies and by extension Western studies, was a body of scientific and technical knowledge developed by Japan on the basis of Dutch texts imported at Dejima. These texts allowed the Japanese to keep a close eye on developments in Western technology and medicine despite the isolationism enforced during the Tokugawa Shogunate. *Rangaku* was in fact based on a thin but, in the long run, consequential channel of communication between Japan and the outer world that kept Japanese intellectual and ruling elites up to date on the deep transformations induced in Western societies by the Scientific Revolution and the Industrial Revolution.[18]

By the 1720s, a significant number of translated foreign books circulated in Japan.[19] One of the finest examples is the *Kōmō Zatsuwa*, literally translated as "Red Hair Chitchat," but better known as Sayings of the Dutch, authored in 1787 by Morishima Chūryō, a Japanese author of popular fiction and *Rangaku* essayist. The book covers a vast array of scientific, medical, and technical topics. It describes instruments such as microscopes and hot air balloons, and discusses the fundamentals of static electricity, the building of large ships, Western hospitals, and the state of the art in combating illness and disease. It also boasted a section on contemporary geographical knowledge.

Other books on specific topics like medicine, geography, biology, astronomy, chemistry, physics, and mechanics complemented Morishima Chūryō's "Encyclopedia." They included scientific and technical descriptions, illustrations of instruments and machines, and information on experiments made in Western countries. For example, Japan's first handbook on electricity, *Oranda Shisei Erekiteru Kyūri-Gen* (Fundamentals of the Elekiter Mastered by the Dutch), by Hashimoto Muneyoshi (1811), described experiments with electric generators, conductivity through the human body, and Benjamin Franklin's 1750 experiments with lightning. In 1840, Udagawa Yōan's *Seimi Kaisō* (Opening Principles of Chemistry) presented a cornucopia of chemical knowledge mostly based on William Henry's 1799 *Elements of Experimental Chemistry*, including a description of Volta's electric battery and an explanation of Lavoisier's theories. Kawamoto Kōmin's 1854 book *Ensei Kiki-Jutsu* (Odd Devices of the Far West), focused extensively on detailed descriptions of steam engines and steamships, as did Okata Kôan's translation of G.J. Verdam's Dutch treatise on steam machines, *Volledige Verhandeling over de Stoomwerktuigen*.

Many Japanese writers on *Rangaku* topics performed their own experiments and built the devices shown in their books. In 1776 Hiraga Gennai demonstrated his own electrostatic generator, the *Erekiteru*, based on the Dutch *elekiter* (electrostatic generator and Leyden jar) and a thermometer (*Kandankei*); Hosokawa Yorinao assembled and demonstrated his *Karakuri*, mechanized puppets or automata inspired on Western models. Some of these instruments and machines, like the *elekiter*, and its Japanese adaptations, could be purchased as conversation pieces in the so-called curiosity shops.

Fig. 5.8 Shopping for an *elekiteru*: *Around 1770 Japan had its first electrostatic generators obtained from the Dutch. Following the same principle as the Leiden jar (1745), the Japanese created the elekiteru. The picture shows an elekiteru being demonstrated and sold at a curiosity shop in Osaka. The sign at the entrance reads "Newest curiosities from foreign countries." The elekiteru is a fine example of the close link between Rangaku (Western Learning) and Japan's successful modernization.*

But it was European weaponry that above all mesmerized the ruling classes. The increasingly threatening presence of European merchant ships and gunboats in Japanese waters led to a renewed interest in Western armaments, especially cannons. Notable military incidents involved Russia (attacks on the Hokkaido, Sakhalin, and Etoforu islands) and Britain—in particular the 1808 assault of the British warship *Phaeton* on Dutch ships (an enemy nation in the Napoleonic Wars) moored in Nagasaki harbor. The traditional Japanese method of iron smelting was adequate only for casting small pieces; in order to manufacture the larger Western-style cannons it was necessary to employ so-called reverberatory iron furnaces—a smelting technology unknown in Japan. Several *daimyo* (feudal lords) took up the challenge to build and operate such a furnace, the Saga feudal clan being the first to succeed. The Saga furnace was constructed and operated by a small group of scholars, technicians, and iron smelting artisans under the direction of Yosuke Sugitani, the Saga chief-engineer and a *Rangaku* scholar. The group based itself on the 1826 Dutch handbook, *Het Gietwezen in 's-Rijks IJzer-Geschutgieterij te Luik* (Foundry Practice in the Government Armaments Foundry at Liege) by Huguenin. Initial experiments were undertaken in December 1850 and by April 10, 1851, after some failures, they succeed in casting a large cannon. A month later, when making another successful cannon, they were able to cast almost 100 percent of the smelted pig iron. It did not take long before other *daimyo*, using the same Dutch source, succeeded in erecting and operating their own reverberatory blast furnaces.

The technological achievements inspired by the *Rangaku* provided fertile ground for the radical and rapid modernization that followed on the opening of Japan to foreign trade in 1854.[20] The arrival of the Black Ships, in Japanese *kurofune*, the name given to Western vessels under the command of the American Commodore Matthew Perry, was the key event in intensifying, accelerating, and diffusing the technological transformations that were already underway. Perry's expedition, the way for which had been paved by earlier American ventures, reached the Japanese coast in 1852. In July 1853, four steamships of his flotilla entered Edo bay. The flotilla counted seven ships in all: four first-class steamships, two sloops of war, and one ship-of-the-line; two of the ships were equipped with the new and very destructive Paixhans shell guns.

Following this display of superior military technology, Perry handed Japanese authorities a letter from American President Millard Fillmore, demanding the opening of Japan to foreign trade. Perry then left for the Chinese coast, promising he would be back for an answer. He returned early in 1854, this time with the whole of his flotilla, and in March was able to sign the Treaty of Kanagawa that opened three ports to American trade. In 1858, the door was thrown wide open by the United States–Japan Treaty of Amity and Commerce, which secured an effective opening of Japan to the world. Similar treaties, all considered "unequal," were signed with the United Kingdom (Anglo-Japanese Friendship Treaty, October 1854), Russia (Treaty of Shimoda, February 7, 1855), and France (Treaty of Amity and Commerce between France and Japan, October 9, 1858).

The Black Ships episode showed that although *Rangaku* had enabled the Japanese to keep up with significant Western scientific and technical advances, it still lagged far behind Western countries in terms of innovative potential. Japanese leaders recognized the imminent threat to the country's sovereignty were they to fail in implementing a thorough program of modernization. They had only to look to China, a once-formidable power reduced to semi-colonial status by the West, to appreciate the dangers to which they were exposed.

The Meiji Restoration brought the answer to this complex political tangle. The Tokugawa leadership proved powerless in the face of to the menace from the West and the domestic unrest and dissidence it incited and began to collapse under its own weight. The subsequent Meiji revolution, led by a group of young samurai and court nobles, sought to restore direct imperial rule, aiming to fortify Japan against external threats. On February 3, 1868, the rebels seized Kyoto's Imperial Palace and proclaimed the Meiji Restoration. A triumvirate of leading samurai formed a provisional government that was to institute "enlightened rule" (the word *Meiji* means precisely this) a combination of Western-style progress with traditional values.

Although the Samurai revolutionaries promulgated the slogan: "Revere the Emperor, Expel the Barbarians" (*sonnōjōi*) in order to foment anti-foreign and legitimist sentiments and to undermine the Shogunate, the Meiji leaders were well aware that they needed foreign technology to industrialize Japan. Their motto "Enrich the

Fig. 5.9 Japanese Trains at Station: *Although there were previous attempts, under the influence of Dutch traders and other foreigners, to propose Japanese railroads, the first line did not come to reality before the Meiji Restoration. Japanese railroads were built using British financing; and around 300 British and European technical advisors were hired, including civil engineers, general managers, locomotive builders, and drivers. In 1872, the first railroad between Shimbashi and Yokohama was opened, with nine round trips daily. A one-way trip took 53 minutes in comparison to 40 minutes for an electric train in the present.*

country, strengthen the military" (*fukokukyōhei*) showed clearly that they understood that Japan's sovereignty and rank in the new capitalist and imperialist world depended on an efficient and modern economy and military sector. The Meiji agenda drew its inspiration from Euro-American technological success and economic vitality, making the "barbarians" in the end the pivotal factor in building up the new Japan. The Meiji program aimed to reform every aspect of Japanese society.[21]

The decentralized, feudal, *daimyō* mode of governance gave way to a unified, centralized national state. A westernized bureaucracy was instituted serving a constitutional monarchy inspired by Prussia, in which a politically-active Emperor shared power with an elected Diet (Constitution of the Empire of Japan, also known as the Meiji Constitution, 1890–1947). Although not immediately evident, this constitution also cleared the ground for the eventual emergence of party politics.

The Meiji rulers also eliminated the rigid status system, based largely on samurai privileges, and implemented a merit-based social order, much more fluid and appropriate to the new economic order. Four years of elementary schooling was made compulsory, for both boys and girls. Although literacy had already been widespread in Japan during the Edo era, the new mass schooling raised standards to a new level. At the same time, middle-level education and national universities were reformed to emulate American and French models. The government hired Western teachers and also sent students abroad to learn new skills, policies based on the fifth and last article of the 1868 Charter Oath (*Gokajō no Goseimon*, also known as Oath in Five Articles), which stated: "Knowledge shall be sought throughout the world so as to strengthen the foundations of Imperial rule."

In accord with their motto, the Meiji government also sought to overhaul the military establishment. The samurai-based system of recruitment gave way to universal conscription. Equipping the Japanese army and navy with up-to-date military technology was given priority. The state itself funded projects to expand and build new arsenals and shipyards and recruited foreign experts and expertise. In all this it is hard to miss the close ties between military modernization and the more general Meiji program of industrial renewal, particularly the way military upgrading served as a conduit to disseminate and appropriate foreign technology.

The new powerful centralized state required a financial system modeled along European lines to fuel its modernization plans and budget needs. Both the banking and monetary systems were reformed. The Yen was adopted as the unit of currency, valued at roughly half an American dollar, and the tax system was reorganized following a national land survey.

The Meiji leaders, in particular those who had traveled abroad, knew that efficient transportation infrastructures were at the heart of an industrial economy. The government itself took on the responsibility of funding and building the first railroad line. This 29 km long line connected Tokyo (Shimbashi) to its port at Yokohama. The Emperor attended the opening ceremony on October 14, 1872 and was among the first to make the round trip between the two stations. The Japanese government financed construction with a £1 million bond issue through its London-based Oriental Bank.

British advisors, entrepreneurs, and craftsmen, the so-called *yatoi* (technical staff), were still highly visible in this pristine stage. The first steam locomotive (called the *Iron Duke*) was brought to Japan by the Scottish trader Thomas Glover in 1868. He demonstrated the machine on an 13 km track in the Ōura district of Nagasaki. Later, Edmund Morel became the first foreign Chief Engineer appointed by the Japanese government, charged with guiding and super-vising railroad construction in Japan, and advising on railroad management and technology. Most of the materials and machines were British as well. British engineers were paid astronomical salaries; for example, the foreign general manager in the railroad office earned ¥2,000 per month, whereas the highest-ranking minister in the Japanese government earned only ¥800 per month.

But by the 1880s, Japanese engineers and technicians had achieved considerable autonomy, and foreign railroad engineers began to disappear from the scene. The Japanese had learned the art of railroad construction and management by dint of careful observation and the actual management of railroads under British tutelage. In addition, Japanese students who had studied railroad technology in Britain on government grants were returning home in order to apply their own foreign-learned expertise to domestic railroad construction.

The first Japanese citizen to be appointed to the post of Director of the Railway Board (1871) was Inoue Masaru, a former student of engineering at University College, London. He had returned to Japan in 1863 on board an English vessel, after surreptitiously leaving the country with four other men in violation of the late Tokugawa Shogunate's isolation policy. The "smuggled" friends, known as the Chōshū Five, returned home after the Meiji Restoration to participate in the modernization movement. All of them became members of the Meiji government, two of them— Inoue Masaru and Yamao Yōzō—in technology-related functions. Inoue, having participated in the construction of the first Japanese railroad line and in the planning of the national railroad network, founded *Kisha Seizo Kaisha*, the first locomotive factory in Japan, and became its first president in 1896.[22]

Yamao Yōzō, who studied engineering at University College London and later at the Andersonian Institute in Glasgow, established the Imperial College of Engineering, the Imperial College of Art, and the *Kobu Daigakkō*, which later became the Department of Technology at Tokyo Imperial University. Yamao, who viewed technical education as the *sine qua non* of Japanese industrialization, served as rector of the Imperial College of Engineering and as President of the Japan Engineering Society. He exploited his English and Scottish networks to hire British engineers as college instructors and to send Japanese youngsters to study engineering abroad.

In addition to modernizing national infrastructures, the Meiji government also founded a series of "model factories" to promote industrial growth in a variety of fields. However, the inordinate investments that were needed soon drove the government to the edge of bankruptcy. Then as now, privatization was the solution and the government turned to private entrepreneurs for the necessary investments. The policy shift proved successful and the new industries were soon able to pay their own way. The first successes in both corporate performance and technology transfer were in the cotton-spinning sector. Japanese production became competitive with the West by the early 1890s, and Japanese mills soon became contenders in the yarn export market. Other successes followed, in the railroad domain with the Nippon Railroad (financed mainly by members of the *daimyo*), in telegraphy with the Daihoku Telegraph Company, and in numerous other domains with firms manufacturing machines and other technical devices. Aside from promoting a successful class of small- and medium-sized enterprises, the Meiji system also enhanced the power of the *zaibatsu* firms (large family-controlled vertical monopolies), such as Mitsubishi.

The assimilation of Western industrial values was celebrated at the first National Industrial Exhibition, held at Kanei-ji temple main hall in 1877. The Exhibition aimed at spreading knowledge of Western technologies and encouraging competition among local industries. Displaying Western machines was intended to foster both the imitation of foreign artifacts as well as the invention of local versions, thus incorporating Western technologies into Japanese industries. Though the first exhibitions were technologically immature, they did fulfill the ideological purpose of bringing the Meiji modernization agenda to the people. In the fifth National Industrial Exhibition (1903) products were exhibited by the founders of renowned companies that still exist today, certainly an indication of the strategy's success.

In its first two decades, the Meiji government achieved significant industrial growth and rapid economic expansion. In 1898, Japan was able to revoke the last of the "unequal" treaties that it had signed with Western powers in the 1850s, heralding its new status as an emerging world power. The war with China (1895) redefined Japan as Asia's first modern imperial power, and the war with Russia (1905) proved that a Western power could be defeated by an Eastern state. Japan had moved decisively from a peripheral to a central position among the nations of the world.

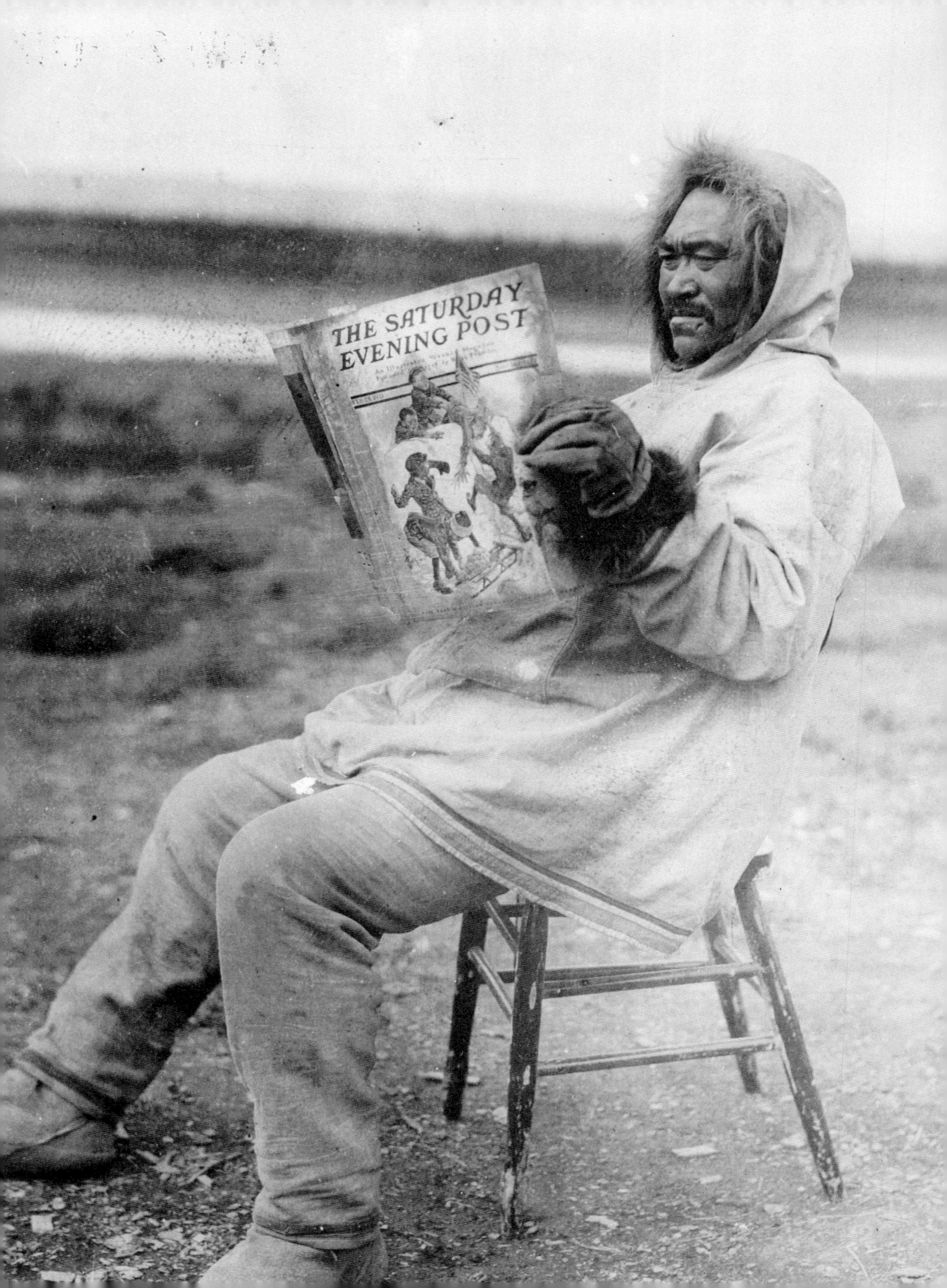
THE SATURDAY
EVENING POST

6
A New World Order & the Collapse of Colonialism

Introduction

The preceding three chapters focused on what might be called the Age of European Imperialism, a period lasting from the seventeenth century to around the Second World War. This surely overlaps with what we have called the Ages of Discovery, an overlap that underscores the persistence of "mapping" as a necessary adjunct of colonialism. Imperialism, however, only wanted to know the world in order to exploit it better. Scientific curiosity was tolerated in the wings and indeed led to some revolutionary advances, but practical technologies of domination, extraction, and transport had at least equal prestige in this world of colonial profits and imperial glory.

What finally shook things up were the two World Wars of the twentieth century—partly motivated by imperial competition itself. These technological horror stories not only debilitated Europe as a global power, a weakness the United States was quick to exploit, but also provided revolutionary movements and nationalist tendencies in the colonies themselves with windows of opportunity to rise up against their colonial and capitalist oppressors. The post-First World War Russian Revolution of 1917 set the tone and adumbrated the final downfall of European imperialism—though perhaps downfall is too strong a word. It would be better to say that blatant (and self-satisfied) imperialism morphed into various forms of structural dependency and assumed different guises, for example that of the Soviet Union itself, which we have described as a kind of Eurocentric "internal colonialism." And while the Russian Revolution may have sounded the death-knell of high imperialism it certainly did not—as Lenin and his comrades had fervently hoped—portend the end of global capitalism as an economic system. "The West"—thanks particularly to the U.S.—proved both too resilient and too alluring.

Though the shock waves generated by the two wars reverberated throughout the world, they hardly resonated the same way everywhere or with quite the same effect. It is simply impossible to do justice to these diverse histories within the scope of a single chapter. We shall therefore adopt a schematic expedient that will allow us to illuminate some of the variety of the emergent new global world without having to bushwhack through thickets of local details. Imagining Europe at the center of the world, we can proceed to describe in an exemplary way—sometimes delving into greater detail—what sorts of effects (or non-effects) the cataclysmic European wars of the twentieth century had on specific "others" located to the north, south, east, and west. In a way this hearkens back to chapter 2 where we discussed Europe's relations with a number of "significant others." After the First World War the cast of characters expanded, though to the East the Soviet Union (and now China) and to the West the United States continued to have the greatest impact on Europe's position in the world.

The First World War, also known as the "Great War," was in some sense a dress rehearsal for the really big one, the Second World War. Though it can lay claim to its own unique horrors, the First World War cannot hold a candle to the Second World War in terms of levels of technological violence, number of casualties and displaced people, civilian deaths, and the geographical scope of the devastation that was wrought. The aftermath of the Second World War was equally dramatic and superlative. Under the insane strategic standoff that was the Cold War, two nuclear superpowers, the U.S. and the USSR, faced off across Europe, dividing the old continent down the middle, each "taking" its share. "The West," but also "the East," had achieved its apotheosis.

This new bi-polar world provided new opportunities for former colonies in the South (Africa) and Far East (Asia) to emancipate themselves—playing off one side or the other in their often long-cherished bids for national independence. So, the anti-colonial agendas of the late 1940s onwards were much more confrontational than those that emerged in the wake of the First World War. The "revolt against the West" in fact fed on the weak responses to the challenges of the 1919 self-determination movements and was an answer to colonial development policies deployed by European empires in the periods immediately following the two World Wars.[1]

The chapter begins with a discussion of Europe's changing position in the world, from the uncontested center of world technology and standards to one of the pillars of a new "West" in which the United States increasingly played a dominant role. From there we move to Europe's new strategic presence in the northern polar regions and its particularly exploitative presence in South Africa, and, by extension, to the post-colonial global South. A subsequent section on the techno-social revival of post-colonial China and India is followed by a reflection on what remained behind, once Europeans had abandoned or been forced to relinquish their former colonies.

European Internationalism & the Making of the West

This section explores the shifting position of Europe in the international order between 1850 and 1940. During the phase of high imperialism up to the First World War, Europe was the uncontested metropolis of world trade, information, communication, science, and technology. It commanded this position partly by virtue of its ability to impose standards in a wide variety of technological domains—even though there was often bitter rivalry between especially France and Britain about units of measure and matters like the prime meridian. The First World War, won by France and England partly

Fig. 6.1   Fighting for the Europeans: *Soldiers from the 4th Australian Infantry Division at Garter Point, near Zonnebeke, Ypres sector, wearing small box respirators on September 27, 1917. Later on, Australians and New Zealanders remembered the traumatizing experiences of World War battlefields like Ypres or Gallipoli as catalysts of a separate national consciousness.*

thanks to the intervention of U.S. troops, marked the definitive entry of the U.S. into the international arena as a "great power." President Woodrow Wilson's *Fourteen Points Speech*, about which more below, confirmed this status and laid the basis for a new twentieth-century world community led by the West. At the same time Wilson's speech was interpreted by many native elites in colonized societies as a call to exercise their rights of self-determination and to rebel against their colonial masters. Though this did not immediately deprive European nations of their empires, it certainly did widen the cracks in the colonial edifice and prepared the ground for the wholesale collapse of overt imperialism after the Second World War.

Europe Sets the Standards

The decades between 1850 and 1880 "were a period when a growing number of transnational networks, both formal and informal, were created. Many of them, such as the Red Cross and the International Statistical Congress, had a strong European focus and can also be seen as motors of Europeanization—even if they were not referred to as such when they were created."[2] Internationalism and international organizations, foundational for a transnational civil society, were two consequences of increased trans-border communication.[3]

This created a need for language translation, convertible currencies and dependable customs duties, common standards of measurement, and shared terminologies and rules of the game—in time followed by copyright and patent laws and universally comprehensible pharmaceutical descriptors as necessary adjuncts. Fresh forms of visualization and the development of statistics as a cosmopolitan science converted this bewildering multiplicity into a coherent system of measuring and counting, reflecting the faith in numbers that undergirded Western science.[4]

Harbingers of this internationalization were the accords reached by the riparian states forming the Rhine and Danube Commissions, dating from the early nineteenth century.[5] Additional impulses came from telegraphy and cable and news agencies like Reuters, Wolff's Telegraph Bureau, and Agence France Presse that came into being after 1840. International cooperation agreements followed in ever more rapid succession: By 1870 there were 17; in 1880 there were 20; rising to 31 (1890–1900) and 108 (1900–1910).[6] Thereby

> together with state and political nationalism there came into being a technological, administrative and juridical internationalism that created a network of international institutions and treaties with the goal of extending as far as possible the juridical security and efficiency of the technological administration traditionally provided by individual states to the entire body of nations by means of international administrative legislation.[7]

This was neither a straightforward process nor one in which the most rational technological solutions necessarily prevailed. On the contrary, national political interests continued to shape the history of standardization and normalization. Still, the process was enacted by committees whose advisers were technical experts and scientists, and who had little truck with political taboos.[8] This is not to say that they were unaware that decisions about technological standards had long-term and far-reaching repercussions. They certainly knew that they created deep path dependencies, as exemplified in railroad gauges. Individual (sometimes national) railroad companies first built an infrastructure restricted to their respective territories. With a few exceptions like the French Saint-Simonists, no one initially contemplated trans-regional or transnational expansion. Simply put, many of the technical challenges faced by the new transnational normalizers derived from the nineteenth-century process of nation state development. Numerous countries long continued to use two or more different gauges. Indeed, there is a very "short list of places where the original gauge was not supplanted by one introduced later."[9]

An especially successful example of international cooperation was the *Allgemeiner Postverein* (General Postal Union) initiated in 1874 by the German Postal Minister Heinrich von Stephan and known as the *Welt-Postverein* (World Postal Union) after 1878. By 1897 every country that considered itself "civilized" had joined this exemplary institution—one that had succeeded in linking very disparate infrastructures into a single system that has moreover continued to be a reliable and trustworthy institution.[10] Since 1865 the World Telegraph Accord and since 1909 the World Radio Union have been similarly influential. Maritime law conventions and rail freight accords followed.[11]

In 1875 the International Bureau of Weights and Measures was founded in France, where the so-called "Prototype Meter" had been gathering dust in a Parisian repository since 1799. The successful "metrification" of France was a shining example for other countries to follow. But internationalizing norms could also be an affront to national pride. Whereas distant Japan signed the Metric Convention in 1885 the nations of the British Empire refused to adopt the metric system for nearly a century, and then only in part. Nowadays approximately 95 percent of the world's population officially uses the metric system.[12] By contrast, the British dominated the standardization of international time regimes— understandably so in view of their interests in marine navigation and their proprietorship of extensive domestic and colonial railroad systems. At the Prime Meridian Conference in 1884 the site of the Royal British Observatory at Greenwich was established as zero meridian for the purpose of synchronizing various systems of timekeeping. The Canadian railroad engineer Sandford Fleming had initiated a movement to standardize time during the 1870s. It is said he was annoyed that he had missed a train during a trip to Ireland due to contradictory timekeeping.[13]

The standardization of the 24 global time zones also facilitated the establishment of a weather forecast system and improved military surveying.[14] At the beginning of the twentieth century the Eiffel Tower was converted into a monumental antenna for broadcasting time signals and aligning clock hands across Europe. Henri Poincaré, then head of the Bureau of Longitudes, justified French investments in wireless transmission by pointing to the British monopoly over the global telegraph cable network, which put English companies in a position to cut all communications between Paris and its colonies.[15] This was just one significant example of how a technology developed for colonial rule contributed to the independence of one empire, the French, vis-à-vis another European empire, the British, but it also suggests how a technology such as the telegraph, first conceived for separating empires, could contribute to a hidden European integration through time synchronization.[16]

It is hardly surprising that smaller states were the most fervent advocates of internationalism, inasmuch as the machinery of international cooperation provided a means of asserting themselves against the great powers and their military superiority.[17] It was no accident that places like Geneva, Brussels, The Hague, and Stockholm became centers of international organizations. In the same way, the pioneers of internationalist aspirations were often more marginal members of their societies: immigrants with a multi-ethnic background, experts in international law, pacifists, idealist businessmen, scientists, and journalists who wrote for a global public. In addition, strong impulses came from movements promoting women's rights and human rights as well as from initiatives to combat diseases and epidemics.

Not every type of internationalization was global. The Marxist Second International was originally geared toward international solidarity of the exploited classes but did not immediately include the workers of the "non-civilized" world. On the other hand, the *Institut Colonial International* was founded in Brussels in 1894 for the exchange of—primarily technological and administrative—information relevant to colonial policies among the various colonial powers. Like the already existing *Institut de Droit International* and the *Institut International de Statistique*, The *Institut Colonial* provided private citizens from Belgium, France, the Netherlands, and Great Britain with a forum in which to debate myriad issues: legal questions, but also those related to the construction of railroads in the colonies, for example. Representatives from Latin America, the U.S., Germany, Denmark, Japan, Austria-Hungary, Portugal, Russia, Spain, and Italy later joined them. In 1948 it was renamed the *Institut International des Sciences Politiques et Sociales Appliquées aux Pays de Civilisations Différéntes* (after 1951 the *Institut International des Civilisations Différéntes*) until it was disbanded in 1982.[18]

In the era of European-dominated technological and organizational expansion everything appeared to have a global dimension; everything had to be interpreted according to international or intercontinental criteria. This despite the fact that some nations inevitably saw themselves as the progressive avant-garde, a perception that made them inclined to stick to their own standard weights and measures, and to presume them binding on others as well. But there was also a surge of adaptation and conformity to international norms. Strict adherence to national norms and standards also carried the threat of technological marginalization and exposed its adherents to the danger of losing touch with international advances because their local standards hindered trade. The technical committees of the League of Nations relied heavily on this spur to globalization.

The prime mover of standardization, however, was and is world trade.[19] Standardization and normalization were essential for the smooth functioning of free markets. At the same time they could also cause a great deal of mischief if companies and governments abused them to gain an advantage over their competitors. In transnational commerce, standards function not only to ensure comparable quality but, ideally, to achieve uniform regulation through price alone.[20] One of the most influential institutions has proved to be the International Standards Organization (ISO). Not coincidentally, it has its seat in Geneva and today manages over 18,000 standards. "The collective work of those standard setters may be as essential to today's global political order as anything done by the UN system. ISO's work is certainly as essential, perhaps more essential, to the governance of the global industrial economy."[21] However much the narrow interests of nation states and their standard-setting bodies frustrated the process throughout the twentieth century, the Geneva-based secretariat "was focused on creating standards that would transform the global transportation and communication infrastructure, thus removing a major impediment to the creation of a global market."[22]

Intensified face-to-face encounters among Europeans and non-Europeans fueled dreams of a common language. Some internationalists waged total war on the Babylonian confusion of the nations by designing "universal languages." Born of the same spirit that generated the metric conventions and the global postal system, they were intended to "synchronize" the "citizens of the world." Ludvik

Fig. 6.2 Negotiating Global Standards: *In May 1955, members of the International Standards Organization gather in Stockholm for the 3rd General Assembly. At the beginning of 1955, ISO had 35 members and 68 standards (called recommendations). From 1947 to 1965, American Henry St. Leger was the first Secretary General.*

Lejzer Zamenhof embodied this ideal in a very concrete way. A Jew born in Bialystok, then part of Russia, he spoke a good dozen languages. Starting in 1887 he distilled from them an easy-to-learn universal language called Esperanto. Together with Volapuek and IDO, Esperanto developed into one of the most successful artificial languages. Nevertheless, it did not manage to break the primacy of English, and this, even more than internationalism, must be seen as one of the most lasting consequences of European imperialism.

It should not be imagined, moreover, that the kind of internationalism pursued at Geneva or embodied in Esperanto met with universal approval. By the 1930s if not earlier it was becoming clear that such efforts at cultural and technological homogenization were arousing suspicions among those who were more interested in drawing boundaries between nations and highlighting differences between peoples and cultures. Where the League of Nations had embodied an open Europe interested in playing a leading role in promoting world trade and global unification, there was now a movement afoot that cherished a vision of a different, self-sufficient, and self-contained continental Europe under the leadership of Berlin and Rome.[23] To chauvinists and Fascists Geneva had become the epitome of everything they rejected: the unfettered movement

Fig. 6.3   World Peace through Esperanto: *In August 1936, four Dutch socialists cycled to Manchester for the Esperanto Conference, demonstrating international, cross-cultural communication. Almost 50 years before, Ludvig L. Zamenhof had first published a newly-designed global language under the pseudonym "Dr Esperanto."*

of persons and ideas as well as their intermingling and assimilation into a global civilization shaped by "the West."

Wilsonian Moments: A Technological Affair?

By the end of the First World War, Woodrow Wilson, then President of the United States, had achieved almost messianic status as a herald of the self-determination of peoples laboring under colonial rule. On January 8, 1918, Wilson gave a speech before the U.S. Congress stressing America's duty to promote moral justice and peace, which he singled out as the main reason for American involvement in the war. In the speech he proposed a "program of the world's peace," based on 14 points. This famous *Fourteen Points Speech* was a plea for political and commercial freedom, for a free world ruled by a balance of power.[24] It provided the ideological foundations for the League of Nations and the Versailles Treaty, which were respectively founded and signed in the following year. The U.S. itself remained outside the League because, while Wilson signed the treaty, the United States Senate never ratified it.

Wilson's speech had tremendous repercussions worldwide, a media success possible only in a world interlaced by new

Fig. 6.4   The Power of Radio: *In this picture, President Woodrow Wilson is talking to an airplane pilot over wireless telephone in front of the White House, November 21, 1918. Wilson's Fourteen Points speech, delivered on January 8, 1918, anchored a new world policy based on free trade, open agreements, democracy, and self-determination. Broadcast all over the world, it fed anti-colonial movements that also used new communication technologies—radio and telegraph—to coordinate their agendas. It was through radio that the newly-formed Soviet State offered an alternative to the same anti-colonial movements when the slow pace of self-determination disenchanted them.*

communications technologies. Following the declaration of war in 1917, President Wilson, fully aware of the strategic importance of both cable and radio as means of controlling the global flux of information, gave the U.S. Navy full control of all American radio operations.[25] Wilson's rhetoric, transmitted through telegraph cables, news agencies like Havas, Reuters, or Wolff, and by means of radio broadcasts, traveled quickly to Africa and Asia where it fomented all sorts of nationalist movements. The American Committee on Public Information recorded that Wilson's speeches, mainly the *Fourteen Points*, but also the *February 8th, 1918 Four Points*, and the *July 4th, 1918*, circulated in "England, France, Italy, Spain, Switzerland, Holland, Scandinavia, Russia, Australasia, Japan, China, Siberia, South America, Central America, Mexico, India, South Africa, Greece, Egypt, and Canada" and "were also broadcast, a new technology at the time, from the Navy's wireless stations."[26]

Despite their roots in domestic U.S. and European politics, the various "Wilsonian moments"[27] were embedded in a thoroughly international context, in which communication played a crucial role. Throughout 1919, Wilson's speeches galvanized a series of almost simultaneous anti-colonial revolutions, such as the Egyptian Revolution, the March First movement in Korea, the May Fourth movement in China, and Gandhi's *satyagraha*,[28] revolutions which ultimately forged a new world. These rebellions, which at first fed off the rhetoric of the *Fourteen Points* and afterwards on the disillusionment caused by a Versailles Treaty that had virtually ignored the emancipation of colonial peoples, initiated a process that led to the partial collapse of the imperial world, and to a new geography of independent nations.

References to Wilson's *Fourteen Points*, which molded this movement, surfaced in nationalistic speeches replete with rhetorics of modernization that echoed nineteenth-century declamations. In Egypt, Mustafa Kamil and later Saad Zaghloul pursued independence hand-in-hand with modernization. They were both well-educated men. Kamil had attended the Law Faculty at the University of Toulouse in France, and Zaghloul had studied at the Al-Azhar University in Cairo, the oldest degree-granting university in Egypt and one of the main centers of Arabic literature and Islamic culture. In India, Lala Lajpat Raj, Bal Gangadhar Tilak, and Bipin Chandra Pal (the Lal–Bal–Pal triumvirate), all well-educated and much-traveled men, led the nationalist Swadeshi movement, the most successful of the pre-Gandhian movements. Lajpat Raj had studied at the Government College in Lahore, affiliated with the University of Calcutta, and had travelled to England and the United States. Tilak had attended Deccan College, one of the oldest institutions of modern learning in India, and was himself part of the first generation of Indians to receive a college education. Chandra Pal, who came from a wealthy Hindu family, was a teacher, journalist, orator, writer, and librarian, and lived in London. The Lal–Bal–Pal triumvirate pursued *swaraj* (self-rule) within a reformed and modernized India.

In China, the May Fourth movement was anchored in a Beijing student rebellion against the Chinese government's weak response to a Versailles decision to grant the Shandong Province to Japan, despite promises of returning it to China. Finally, in Korea, it was also students who led the March 1st (or Samil) Movement. Some of them had studied in Tokyo, but now agitated against Japanese colonial rule and for development and modernization by Koreans for Koreans. Other nationalist leaders were highly influenced by the American way of life and Wilson's defense of the Principle of Self Determination. Ho Chí Minh, for example, after having lived in the United States and England, traveled to France for the Versailles Conference and petitioned President Wilson to recognize Vietnamese civil rights and a new nationalist government.

Some historians argue that Wilson's plea for self-determination was intended to apply only to Europeans like Poles, Serbs, and Czechoslovaks, perhaps even Albanians, Croatians, Estonians, and Ukrainians. Others judge that the concept of self-determination was a revised version of the nineteenth-century *mission civilizatrice*, in which the civilized colonizers would now bring the colonized to the higher evolutionary state of self-governance. Although Wilson's position on colonialism remains unclear, non-Europeans nonetheless appropriated Wilson's rhetoric and adapted it to their own agendas. Whereas nationalist leaders may be accused of a certain naiveté in taking the words of the U.S. President at face value, they ultimately succeeded in their purposes.[29] After the 1914 to 1945 cycle of war—depression—war (The First and Second World War, and the Great Depression), the majority of Asian and African independence movements exploited the window of opportunity provided by the emergent Cold War to take the final steps to independence.

Researchers on decolonization list seven major legacies of colonial rule: (1) the template of the nation state and constitutions; (2) a formal system of education; (3) legal system and bureaucracy; (4) military organization; (5) public health policies; (6) urbanization; (7) lifestyles and sports.[30] Although the powerful role that guns, railroads, ships, and telegraphs played in the making of European empires is generally acknowledged, technology remains paradoxically outside mainstream versions of the colonial legacy like the preceding list. The same goes for the role played by communication technologies such as telegraph and radio, first in the maintenance of colonial order and later in the dissemination of a broad anti-colonial agenda. Wilson's plea for self-determination could only become a global event with a massive effect on the international order because it was telegraphed and broadcast worldwide. The simultaneity of the 1919 Korean, Chinese, Egyptian, and Indian

Wilsonian moments derives not only from similar local and international political contexts, but also from the fact that anti-colonial nationalists were simultaneously apprised of the same information, thus creating a common language for their claims, based on the door opened by Wilson's speeches. On the other hand, the success of the post-First World War anti-colonial upheavals also depended on a tactical use of the new communication technologies.

Anti-colonial activists were able to influence European and American public opinion through the new media. Self-determination movements themselves were scattered across the globe and were able to interact and communicate in tactically relevant time-frames only thanks to radio and telegraph. These media also made it possible for the Bolshevik revolution, and in its wake the Third Communist International, to disseminate propaganda and coordinate their activities beyond Soviet borders. By 1922 the fledgling Soviet State was operating the world's most powerful transmitter and by 1925 Moscow was transmitting on short waves, a development that it would take the West another four or five years to bring to fruition.[31] This was a royal road to aiding and abetting alternative anti-colonial movements propagated by those who had become disenchanted by the slow pace of progress under the terms of Wilson's liberal anti-colonialist rhetoric.

Technology was certainly at the heart of European imperialism, but it was also a cornerstone of anti-colonial movements, and remained undisputed as a valuable asset after European colonizers eventually returned to Europe. We argue that technological infrastructures and artifacts were as crucial an aspect of the relationship between former colonizers and freed subjects, as they were of the relationship between earlier colonizers and the colonized. This space of exchanges is highly diversified, ranging from "technologies of poverty"[32] to Indian techno-nationalism, from the *bidonvilles* to the atomic bomb, but it still reveals technologies as a powerful binding force in contemporary world history.

The Geneva Global Community

In 1926 the Swiss architect Le Corbusier entered the lists in a design competition for the League of Nations Palace in Geneva. The 39-year-old architect was inspired by the idea that the city on the lake would need to be developed into a world capital and a global hub of commerce. Capital cities, especially planned capitals erected on greenfield sites, were in fact to occupy a great number of prominent architects throughout the twentieth century. Whether it was New Delhi, Chandigarh, Canberra, or Brasilia; whether it was Berlin—re-formed into "Germania"—or the new Soviet capital of Moscow (for which Le Corbusier had also submitted designs): in all these places revolutionary architecture was mobilized to create living spaces for the future. Planners expected that the new purportedly more-rational environment would influence the new inhabitants accordingly. Le Corbusier's thoughts on the Palace were characteristic:

> The most important issue of the Palace: ... An unimpeded view and [clear] acoustics in a location where—as at the Tower of Babel—people of all nations ... assemble to hold debates in which world peace is at stake. The path traversed by ear and tongue is the only one that can also be taken by heart and reason. ... One must [be able to] see clearly in order to determine world affairs—to utilize the optimism of the sun's rays.[33]

At this time it was still considered appropriate for architects and members of other technological professions—perhaps even their calling—to express views on ideological questions. A kind of "technocratic internationalism" (see *Making Europe* series volume 3) evolved that explicitly distanced itself from the realm of politics and the economy. Still, its protagonists claimed to have solutions for the immense problems facing the postwar global community—above all the preservation of world peace. These were invariably based on the pursuit of efficiency and instrumental reason.[34] This addressed a certain pathos slumbering in the concept of world government—and by extension in the League of Nations.[35] In this connection Le Corbusier's allusion to the Tower of Babel was very revealing. His project's overblown aspirations came to grief precisely because of the "Babylonian confusion of tongues"; the plethora of incommensurable norms that made it impossible to complete the edifice. The catastrophe of the First World War had laid bare European disunity for the entire world to see. The League of Nations was intended to create, if not a common language, at any rate common norms.

Hence, between 1919 and 1946 something akin to a European globalism existed; the entire world met in Europe. The choice of Geneva, a neutral location in the heart of the Continent, symbolized

Fig. 6.5 Governing the Globe: *Opening session of the League of Nation at Geneva, Switzerland, on November 15, 1920. Often assessed as an outright failure after its demise in 1946, the League later on was credited for attempting collaborative approaches to global governance and for maintaining expert networks.*

efforts to secure world peace and to facilitate international attempts at crafting an intertwined and entangled history. The League of Nations refocused all the efforts made since the mid-nineteenth century to reach agreements by international consensus and to secure internationalism as an autonomous political level distinct from the flux of national rivalries.[36] Technical and economic cooperation and standardization that transcended national boundaries were designed to prevent future wars between nations. The experience of two World Wars would confirm that this strategy first had to be applied to Europe itself.[37]

The imagined "Global Community" in Geneva was in fact already dominated by European powers and their concerns.[38] Europe provided the lion's share of funding as well as the leading actors: the Secretaries General of the League of Nations were, in succession, the British Eric Drummond, the Frenchman Joseph Avenol, and the Irishman Sean Lester. A primary reason for promoting technological cooperation was to keep international collaboration separate from the sphere of "professional" politics. Using a European venue and European languages and styles of interaction also had the effect of familiarizing non-Europeans with the conventions of modern diplomacy, in effect promulgating new standards for international relations.[39]

Even if the League of Nations as measured by its own political aspirations is generally judged a failure, it was quite successful in promulgating technological conventions and standardization. It inaugurated forms of global decision-making and cultivation of expert knowledge that became standard operating procedure for the rest of the twentieth century.[40] The League developed its own global communication networks to pursue pragmatic projects like defining standards, completely bypassing the usual diplomatic channels. Working in the shadow of the First World War, a *Götterdämmerung* that had put an end to optimistic musings that moral and technical progress would bring the world together of its own accord, experts now considered themselves the last best hope for world peace—seeing themselves as agents of the elaboration of non-political transnational networks in science, health, and entertainment.

Non-members of the League were expressly invited to participate in the agenda-setting for international cooperation, although as it turned out the long-term consequences of these agreements would affect primarily Europe.[41] Concrete bargaining points dealt with issues of medicine and health organization, transportation and communication, economic, legal, social, and financial questions, the administration of mandated territories and, finally, the humanitarian treatment of minorities, refugees, and stateless people. They also extended to disarmament, intellectual cooperation, child welfare, and containment of the opium trade. Finally, the International Labor Organisation (ILO) was created as an adjunct to the League, an organization that continued to play a major role in international labor law and the consolidation of human rights after 1945.[42]

One striking example of the far-reaching effects of a widely-overlooked accord was the activity of the International Container Bureau established in February 1933. Its Executive Committee under the chairmanship of Silvio Crespi was assigned nine expert delegates to represent the various modes of transport. The Bureau functioned "as an organ of investigation and liaison between railroads, road and water transport firms, and clients in the several countries interested in developing the container service."[43]

It was based on an initiative debated during the World Motor Transport Congress at Rome in September 1928. The "container" was there introduced as a means of ensuring closer cooperation between road and rail transport, or "combined transport," as it was called. "In January, 1930, a competition was held to determine the most suitable container system for international traffic.[44] That September, competitive experimental tests with containers were held at Venice."[45] Practical demonstrations followed at annual meetings.

Later, when the container had become the great success story of modern logistics and global trade, many legends would circulate about its origin.[46] Individuals like the American Malcolm McLean, who launched the first container ship in 1956, certainly deserve their prominent position.[47] But as is true in most cases, this innovation was a collaborative undertaking that depended on international cooperation among numerous experts.[48] The furious pace and solid work of the "container committee" was typical of most of the other committees and earned the League of Nations a reputation for running like clockwork during the years between the Wars; contemporaries even described it as a "machine." Although challenged during the 1930s by upstart Fascist states that strove to establish competing bureaus of standards in Berlin and Rome, the "machine" continued to function even during the Second World War. It was not dissolved until 1946, when the founding of the United Nations at San Francisco and New York signaled the power shift from Europe to the U.S.[49] As late as 1929, only 24 of 444 international organizations registered at the League of Nations were located outside of Europe.

But for that matter the League had never been a purely European affair anyway. India, for example, was one of the founding members of the ILO in 1919. Its independence movement had flexed its muscles during this "Wilsonian moment": Nationalist advocates saw India's membership as a forerunner of national sovereignty. When the divided border province of Punjab needed a new capital after Indian independence in 1947, Le Corbusier was approached. Chandrigarh was the only one of his revolutionary plans for urban design and development that was actually realized.[50] Jawaharlal Nehru had requested him to "let this be an innovative city, a symbol for the freedom of India. Undeterred by the traditions of the past, ... an expression of the nation's faith in the future."[51]

North & South

The "Wilsonian moments" of the previous section occurred only in China, India, South Korea, and Egypt. The rest of the imperialized world, in particular most of Africa, was clearly not yet up to demonstrations of concerted resistance to colonialism. At the other end of the spectrum—and indeed of the habitable globe—the North Pole remained to all intents and purposes—at least in the perception of Europeans—a *res nullius* which was up for grabs and whose sparsely-scattered inhabitants were more to be pitied than feared or exploited.

   This section takes up where the discussion of Africa and the North Pole left off in chapter 1. There the emphasis was on exploration in the service of mapping, on discovery and claiming; here the story is about the twentieth-century exploitation by Europeans of the resources these territories had to offer, be it diamonds and gold, or living space and strategic positioning. In South Africa, the discovery of gold and diamonds had awakened the most avaricious and repressive tendencies of European imperialism. This mineral wealth partly accounted for its early formal independence from Europe but also for its downward spiral into the specifically racist form of "internal colonialism" that became known to the world as apartheid. In the North Polar regions, quite by contrast, the avaricious search for natural resources and strategic location did not particularly enslave native populations so much as marginalize them. In the end, the imminent demise of the fragile polar ecology due to global warming has aroused more sympathy than the human tragedy of the Inuit.

North to the Future

In the course of the twentieth century the Arctic regions drew ever more international attention. This was prompted by greater accessibility by airplane, the emergence of global meteorology, and the lure of natural resources. Eight northern countries claimed Arctic territory: Russia, Canada, the United States, Denmark (with Greenland), Norway, Sweden, Finland, and Iceland—all of them technologically-advanced states. Many disagreements about the northern borders erupted around mining and research, but also in relation to fishing, whaling and, finally, travel and tourism. When the Cold War broke out, the Arctic Sea became the *Mare Nostrum* of the major rivals who sought to monitor what was effectively a no-man's land by means of a string of northerly radar stations deemed capable of providing early warning of submarines operating below the ice or ballistic missiles crossing it above.

Following the Cold War and under the influence of more literal climatic changes, the Arctic region once more became an object of rival interests vying for reserves of raw materials and energy resources like coal, oil, and gas. This also now extended to the seabed. The expanding claims of nation states to territorial waters have created additional problems such as the conflict between Norway and Russia over the Barents Sea continental shelf, an area presumed to contain rich deposits of hydrocarbons. In August 2007 the Russians drew accusations of polar imperialism and strong denunciations in Europe when they laid claim to an area of 1.2 million square kilometers beneath the Arctic ice by placing a titanium flag 4,261 meters below the surface. Indeed, if there is any region in the world where geopolitical competition for territory and resources is still very much alive, the Northern circumpolar region is it.[52] The new endeavors of the rich nations point "North to the Future"—as the state motto of Alaska would have it.[53]

Propelled out of their isolation by these techno-economic invasions, the indigenous people of the Sami, Inuit, and Northern Siberia have at long last rediscovered their ethnic coherence. In 1977 a Pan-Eskimo movement, the Inuit Circumpolar Conference, was founded and promoted the idea of a "Fourth World." This enabled northern territories to acquire money, advanced technology, and also jobs, the latter unfortunately not primarily for the native people, but in the first place for skilled immigrant workers. Today the Inuit participate in social welfare programs but not necessarily in the workforce. This is precisely what Colin Ross (see chapter 1) had predicted in 1934: "When the white man has covered the Arctic Circle with his instruments and machinery he will not need the Eskimo any longer, just as in the tropics where the primitive races were dismissed—except as workhorses."[54]

Ross considered the Inuit to be "outrageously healthy" and content, but in danger of being proletarianized. "All the polar explorers who met these Eskimos untouched by civilization ended their narrations by uttering the fervent wish that civilization should

Fig. 6.6 Saturday Evening's Rest: *This picture from the arctic region, showing an "Eskimo" reading the American magazine* The Saturday Evening Post *on November 26, 1913, has become a favorite collectible on the Internet and elsewhere. Although it is most likely to have been carefully arranged, for Western eyes it was one of the popular pictures which contrasted the exotic with the accustomed. Its message was: even the remotest regions are open to Western divertissements.*

never reach them."[55] For Ross the impact of European civilization was calamitous, because

> it befell primitive people at a time when we were convinced that it was a blessing and a superior way of living. Today we know much better that it can be a blessing as well as a curse and that it is not at all a superior way of living but a different one, and that an unrestricted invasion of primitive people cannot be but disastrous.[56]

In fact, a recent overview states:

> The development of industrial fishing, forestry, mining, oil and
> natural gas exploration, and military installations, along with the
> necessary transport, communication, and administrative infrastruc-
> ture, has required the introduction of a large immigrant popula-
> tion. ... Many of these immigrants ... or their descendants have since
> remained. These non-native groups form a predominantly urban
> population, inhabiting the new towns around industrial centers.
> The indigenous populations, by contrast, continue to live primarily
> in the rural areas. ... Native people have become socially and
> economically marginalized in their own homelands.[57]

From Boas to Ross, encounters with the nomadic people of the
North almost always evoked great admiration for the way their
living conditions were adapted to the environment, as well as
sadness about the changes that modern civilization and technology
would entail for Inuit culture. From the onset of European expansion, but especially after the turn of the twentieth century, many
European intellectuals took an increasingly critical view of material
and cultural diffusion. Paradoxically, modern technology not only
enabled colonizers to marginalize or even exterminate native populations, but also made it possible for detractors of arctic imperialism
to conserve "authentic" living conditions (framed as "unspoiled"
by modern civilization) by recording them in photographs, with
phonographs, cameras, and all the other media technologies used
in ethnographic research. Nonetheless, while today's North Pole
has come to symbolize global warming, it is not the plight of the
Inuit that figures in the publicity, but rather a desperate Polar Bear
clinging to a melting ice floe.

Blood Diamonds

This is how one author describes the aftermath of the second Boer
War:

> The British army won the war, but the British government lost
> the peace. When the Union of South Africa was formed in 1910,
> the Dutch settlers—or Afrikaners, as they called themselves—
> were a majority of the white population, and because only whites
> could vote, Afrikaners took control of the colonial government.
> Afrikaner governments ensured a high standard of living for their
> white constituents. A "colour bar" reserved high-wage skilled jobs

for white workers, and the Native Lands Act of 1913 set aside 93 percent of the land in the country for white ownership. New industries were established to produce steel, military arms, and other goods, to create more jobs for whites, and to reduce South Africa's dependency on foreign trade. Apartheid, or "apartness," was introduced in 1948 to enforce the economic domination of whites over blacks, Indians, and "coloureds."[58]

The discovery and mining of diamonds (see chapter 4) had a huge impact on European–South African relations all across the board. For one thing, the mines needed masses of workers, and their rapid influx into nearby cities and new settlements brought together people from various regions and backgrounds. Even with the "colour bar" in place and an exploding population, different non-European cultural imprints still mattered; however, these internecine differences paled in comparison with the overarching distinction dividing black from white people, distinctions reinforced by comprehensive rules of segregation. The relationship between the "Europeans"—which was what white South Africans generally called themselves during apartheid—and non-white people in South Africa remained strained, to put it mildly, throughout the twentieth century. Racial segregation has rarely been practiced as rigorously as it was there and then, including the establishment of completely separate infrastructures: signs reading "Europeans only" denied "blacks" and "coloreds" access to facilities such as public toilets, railroad carriages, hospitals, sports fields, beaches, schools, prisons, offices, and churches.[59]

Because of this harsh policy of segregation, South Africa became internationally isolated, at first within the Commonwealth, and later within virtually every other organization with which the nation was associated. In spite of this, Europeans did not refrain from making profitable business deals with the country. In the eyes of South African business associates, on the other hand, such deals increasingly assumed a moral *haut goût*, a bitter taste of wheeling and dealing, until European enterprises such as Volkswagen, Mercedes Benz, and BMW were subjected to boycott threats. With an estimated one-third of global mineral resources, however, South Africa remained one of the richest countries in the world and a respected trading partner.

But South Africa was also in the international vanguard when it came to organized resistance by indigenous people. In 1912 the

South African Native National Congress was formed. It served as a forum for challenging the Native Land Act of 1913, which had prevented Africans from buying, renting, or using land, except in the so-called reserves. Land ownership and property issues remained pivotal in the following decades, contributing to the attraction of socialist ideology after independence.[60] The matter continued to fester elsewhere in Southern Africa as well: countries like Rhodesia, later renamed Zimbabwe, have been struggling with the postcolonial re-expropriation of land until just recently.

As had been the case in Europe during the time of Manchester liberalism, businessmen in European settler colonies had to be subjected to inordinate pressure before they would concede an inch to organized labor. South African trade unions were among the first in Africa to demand better working conditions from their "European" employers. Peaceful and militant strikes, boycotts, and other acts of defiance gave teeth to these demands. Actions were strategically aimed at essential services such as the sewer system and garbage collection, the targets of a strike in Johannesburg as early as in 1917.[61]

Trade unions were often harbingers of independence movements.[62] In 1943 the All-African Convention, established in 1935, together with other colored organizations, initiated a "Non-European Unity Movement." Adherents refused to cooperate with the government, championed full democratic rights for all South African men and women, and strove for full equality between white and colored people.[63] Its ideological leader was an Indian doctor from Cape Town named Goolam H. Gool, who advocated an unceasing revolution on the part of the proletariat.[64] An Indian-born lawyer from South Africa, Mohandas Karamchand Gandhi, drew different conclusions: his *Satyagraha* tactic, which aimed at a peaceful boycott of European commodities and services, went on to become a "traveling concept" on a global scale (see chapter 5). However, the emancipation movements still measured progress by European yardsticks of "civilization,"[65] at least until the Black Consciousness movement of the 1970s proclaimed a global mission for Africanism. In Steve Biko's words:

> We reject the power-based society of the Westerner that seems to be ever concerned with perfecting their technological know-how while losing out on their spiritual dimension....The great powers of the world may have done wonders in giving the world an industrial

and military look, but the great gift still has to come from Africa—giving the world a more human face.[66]

Its rapidly-accumulated wealth allowed South Africa to emancipate itself from British colonial rule as early as 1910. Independent rule was financed by the same extraordinary profits that also enabled the extension of transport and communications infrastructures into the country's interior. South Africa was also one of the first former colonies to be granted a mandate by the League of Nations; in 1919 they were charged with the governance of South West Africa—a colony that had been taken from Germany as a prize of the First World War. South West Africa had been annexed in 1915 and was subsequently incorporated into the new Union of South Africa. It is not without significance that German miners had discovered diamonds in South West Africa shortly before the First World War. The resistance movement in the former German colony, now named Namibia, did not succeed in gaining independence from South Africa until 1990. And with South Africa claiming the entire Oranje River on the basis of a British–German treaty signed in 1890, the dispute over the border function of the river corridor remained unresolved.[67]

Both the First and Second World Wars stimulated the South African economy by creating a huge and profitable demand for industrial goods and munitions. The country became a major manufacturing power and the size of its industrial labor force and its industrial output increased by leaps and bounds. Within less than a century, the South African economy had undergone a second major transformation.[68]

Though their industrial value remained insignificant until the second half of the twentieth century, diamonds and gold exemplified the far-flung ties that characterized Europe's relationship with distant countries rich in resources. Metropolitan control and disposition of peripheral resources has remained pivotal for international relations between Europe and the rest of the world until the present.[69] The lure of resources drove Europeans to extend economic or political control over remote territories; the subsequent methods of extraction and exploitation generally led to the immiseration of native populations within these countries as well as to generally unsavory trade relations. This applied to such diverse commodities as ivory and rubber (for example, in the Congo), petroleum (for example, in Venezuela or the Near East), or bananas and other tropical fruits (for example, in Central America).

The process of decolonization was intrinsically shaped by what Western industries deemed indispensable, like Uranium (or Coltan today) in the Congo's Katanga region. Here, U.S. and Belgian intelligence services collaborated in liquidating the independent-minded new president, Patrice Lumumba.[70] In the spirit of Lumumba, elites in young and independent nation states quickly seized control over their new nations' resources. Sometimes they even managed to build up monopolies and cartels themselves, the most successful of which was probably the Organization of the Petroleum Exporting Countries (OPEC), which was founded in Bagdhad in 1960. In most cases, however, former colonial powers managed to consolidate economic ties without having to exercise formalized political control. This led to the enduring impression that Third World countries had succumbed to some sort of chronic postcolonial *dependencia*. Uncomfortable resource dependencies were also instrumental in stimulating European attempts to find artificial substitutes for "tropical resources." For example, by the mid-1950s General Electric in the United States and *Allmänna Svenska Elektriska Aktiebolaget* in Sweden simultaneously succeeded in producing artificial diamonds, mainly for industrial purposes. But alas, no large-scale R&D complex has yet managed to realize the alchemist's dreams of producing gold.

At mid-twentieth century, though just for a brief moment, Southern Africa produced almost 99 percent of the world's diamonds. It lost this dominant position not only due to artifi-cially, produced diamonds, but also as a result of the opening up of new diamond fields in Liberia, Sierra Leone, Congo, Angola, Siberia, Australia, Canada, and Brazil. South African monopolies were challenged, and the diamond trade became much more diffi-cult to control.[71] Europeans and Americans faced a market with more players, resulting in fierce competition and harsher trading methods. At the beginning of the twentieth century the inhuman behavior of the Belgians in the Congo had inspired the term "blood rubber." Now it was applied to the "blood diamonds" that were excavated in exchange for military equipment or—as in the cases of Jean Bedel Bokassa, Charles Taylor, or Robert Mugabe—for bribery.[72]

The hunger for resources (often referred to as the "curse of resources") time and again stimulated international efforts at regulation. A conference in Kimberley in 2002 attended by diamond-producing countries and consumer nations, including the EU, worked out a system of controlling the international diamond trade to prevent abuses. But the international movement of small high-value resources like diamonds is extremely difficult to monitor. The classic European diamond processing centers like London, Amsterdam, Antwerp, or Idar-Oberstein are now facing competition from exchanges in New York, Dubai, Tel Aviv, or Mumbai. The exclusive access to raw materials, once a cornerstone of European wealth and prosperity, has successfully-been "globalized." In view of the history of the South African diamond trade, it is certainly ironic that the program launched in 1985 to reduce European dependencies on resource-intensive industries by means of advanced technological research and development was called "Eureka"—just like the first diamond that was found in South Africa in 1867.[73]

The Global South

In his book *The Dual Mandate* published in 1922, British governor-general of Nigeria Frederick Lord Lugard declared that the tropical regions should be seen as a "heritage of mankind," and that no one was entitled to prevent Europeans from exploiting their resources. This statement confirmed the accord, based in international law, which Europeans had signed during the 1884/85 Berlin Conference, making colonial possession contingent on an "effective" utilization of land and its resources. This was one major reason for European colonial powers to embark on development programs in their colonies. These became ever more ambitious, albeit not necessarily more "effective," after both the First and Second World Wars. Increasingly coordinated by central colonial administrations, modern infrastructures in Africa largely served point-to-point colonial exploitation and did not network into coherent systems very well. Yet for almost a century colonialists and geo-politicians perceived Africa as a European responsibility. For some, this responsibility took the form of a technocratic dream of "Eurafrica," a common infrastructural future for the two complementary continents.[74] Herman Soergel, the inspired and ingenious marketer of *Atlantropa*, printed a poster that depicted Europe riding not on a bull but on an African elephant.

With the notable exception of South Africa, where a settler colony imported intra-European tensions and came to internecine blows during the two Boer Wars, Africa as an imagined "supplementary space" for Europeans stayed remarkably free from armed imperial conflicts. Only the Fashoda, the Pink Map incident or the Moroccan crises between 1898 and 1911 disturbed European unity in their "civilizing mission" in the southern continent. Mentally, Europeans often distinguished a northern Africa that was geographically and culturally relatively close to Europe from a Sub-Saharan Africa that was perceived as being absolutely dependent on European cultivation in the form of infrastructures like railroads, harbors, schools, or hospitals. Here colonial development and developmental aid remained notoriously incoherent. Africa became synonymous with the "Global South" and chronic shortfalls in development indices. These qualifications were, of course, based on European benchmarks.

Europeans themselves had lessons to learn from transferring technology to hot climates and very different natural and societal settings. One of the outcomes was the idea of "tropical" technology adapted to the climate, which later became known as "appropriate technology." Another lesson was the realization that contrary to initial European suppositions, Africans proved to be keen and industrious in both the adoption as well as the creative adaptation of Western technologies, especially when viewed from the vantage point of smart usage.[75]

What has remained of European colonialism in Africa? A great deal of "hybrid" technology has already been developed in Africa, and human and material resources have long been intensively exploited. Urban agglomerations have arisen and have grown into megacities that are hardly comparable with any other cities in the world. However, with a few local exceptions, no "Industrious Revolution" resembling anything like the dramatic development that took place in Europe has taken off.[76] After decades of decolonization everyone agrees that Africa is still in a deplorable state, though its vast potential ought to have enabled different outcomes.[77] The fervent hopes that Africa would "Europeanize," initially cherished by colonialists, and then adopted by theorists of "modernization," proved futile. The famous statement by the Khedive of Egypt, Ismail Pasha, in the late 1870s: "Egypt is no longer in Africa; it is part of Europe," did not inspire other African countries to take the same course. Neither Francophone ties, French proposals to build up a Mediterranean Union, nor a few Spanish relicts in Northern Africa like Melilla and Ceuta can belie the fact that Africa shied away from serving as a laboratory for European technology transfer. As to the Sahara, none of the large-scale engineering projects was realized, and it remained a desert, and also a space for the projection of European fantasies. Even while decolonization was proceeding apace, it was abused as a test ground for yet another European technological sublime, atomic energy.[78]

The Far East

In the chaotic aftermath of the Second World War, most of Europe's former colonies became independent, but they still depended on foreign aid. The term "developing countries" indicated their need for active support in "catching up" with developed countries.[79] The terms of this support, however, unleashed the specter of neocolonialism. The "invisible hand of neocolonialism" defined national and global economic priorities, for example by denying or granting basic infrastructures.[80] In fact, the desire for Western technology in the former colonies only fostered neocolonial aspirations in the metropolis—through the involvement of capitalist firms in the economy of developing countries, through the exploitation of natural resources and populations, or through the training of scientific and technical elites with loyalties to the West. On the other hand, refusal to grant access to critical state-of-the-art technologies has sometimes triggered powerful nationalistic sentiments in pursuit of Western technology using exclusively local expertise and materials.

As noted, the aftermath of the Second World War also meant the end of Europe's unequivocal world leadership. For the first time since the Renaissance, Europe was no longer the leading player in the new globalized world: the Continent was devastated and some of its most prominent scientists and engineers had fled to America. Technologies developed by the new world leaders, the U.S. and the USSR, now began to outclass European technology—one of the pillars of its world dominion. After the Second World War European technology merged into the broader melting pot of Western technology, dominated by contributions from the U.S., and forming a package that was immensely attractive for nations seeking to become autonomous and "civilized."

The following two cases from the Far East show how euro-Western technology played a formative role in framing both India's and China's postcolonial national identities, despite their different political models. They also show the decisive influence of small Western-educated elites—reminiscent of the Soviet Union—on the policy choices confronting the new nations. These elites were more often than not educated at metropolitan universities (Ho Chi Minh and Leopold Senghor, both in Paris; Jawaharlal Nehru and Gamal Abdel Nasser in England; Amílcar Cabral and Agostinho Neto in Lisbon). Many were also educated in local Western colonial schools, for example the Nankai Middle School, at Tianjin, Northern China, or the Bandung Technical College on Java. Whatever the case, these westernized elites perceived progress and modernization very much along the same lines as the former colonizers, that is, attributing enormous developmental potential to euro-Western technology and adopting the latter as keystones in their developmental agendas. That said, local political, economic, and cultural circumstances were decisive in determining precisely how such technologies were assimilated: how selectively they were appropriated, whether they were purchased or copied, and how they were actually put to use.

The Empire Strikes Back

In a speech on November 17, 1955, Jawaharlal "Pandit" Nehru, Prime Minister of India since 1947, speaking in Hindi, described the Bhakra Dam as one of the "symbols of our progress." The Bhakra–Nangal network was one of the seminal post-independence projects, part of Nehru's modernizing agenda. Nehru considered hydroelectric power vital to his economic strategy and Bhakra, a site he visited ten times during its construction, represented "something tremendous, something stupendous, something which shakes you up when you see it. Bhakra, the new temple of resurgent India, is the symbol of India's progress."[81]

Nehru envisioned independent India as a highly centralized interventionist state, based on technologically-driven modernity. Like many of the leaders of colonial independence movements from all over the world, Jawaharlal Nehru studied at the "heart" of the Empire, at Trinity College, Cambridge, where he graduated in natural science in 1910. He then studied law in London at the Inns of Court School of Law. He was admitted to the English bar in 1912.

Fig. 6.7 The New Temple of Resurgent India: *The Bhakra dam was among the earliest river valley development schemes undertaken by India after independence and was central to Nehru's agenda for the development and modernization of the Indian economy, heavily based on infrastructures and key industries—steel, iron, coal, and power. In 1955, Nehru poured the first bucket of concrete into the foundations of Bhakra. The dam was completed by the end of 1963, standing as an icon of nationalism and modernization.*

During his English period, Nehru studied the work of numerous writers, economists, political scientists, and philosophers such as George Bernard Shaw, H.G. Wells, John Maynard Keynes, Bertrand Russell, Meredith Townsend (the controversial journalist, owner, and editor of *The Spectator*), and Beatrice Webb, wife of Sidney Webb, Baron Passfield, with whom she founded the London School of Economics and Political Science. Many of these were active in the Fabian Society, a British socialist organization that deeply influenced Nehru.

His economic agenda for India was based on state-owned, -operated, and -controlled means of production, namely key industrial sectors such as steel, telecommunications, transportation, electricity, and mining, and was framed along Fabian lines. The Fabians also influenced Annie Besant, the British socialist, theosophist, freemason, women's rights activist, and supporter of Indian self-rule—Nehru's mentor and later president of the Indian National Congress. Muhammad Ali Jinnah, the founder of Pakistan and Nehru's fellow activist in his struggles against the British Empire, as well as other independence leaders such as Obafemi Awolowo

(Nigeria), Lee Kuan Yew (Singapore), Michel Aflaq (Syria), and Salāmah Mūsā (Egypt) were all influenced by Fabian thinking, which promoted moderate socialist reforms.

On his return to India in 1912, Nehru joined up with the nationalist movement. By the time the First World War broke out, he had become an advocate of Besant's campaign for Home Rule. Under the mentorship of Gandhi, whom he met in 1916, Nehru rose to prominence, ultimately becoming independent India's first Prime Minister. Nehru's belief in state planning and control over the economy had an enormous impact on traditional leading sectors like mining, electricity, and heavy industry. But it also poised India to acquire and domesticate new post-Second World War technologies like atomic energy, space flight, and computers.

By 1948, Nehru's advocacy of atomic energy had resulted in the establishment of an Indian Atomic Energy Commission, chaired by a Cambridge-trained nuclear physicist, Homi Jehangir Bhabha. In 1945 Bhabha had founded the Tata Institute of Fundamental Research as a center for Indian nuclear research and development. This research was supported by Canada and the United States, both of which countries supplied India with reactors and technical and design information under the Atoms for Peace program. The deal was that India refrain from any military use. Despite this, Bhabha, supported by many fellow scientists, began to advocate a military nuclear weapons program, a project that by the 1960s had received Nehru's consent. Bhabha went public, giving a famous radio speech on October 24, 1964, in which he argued that "atomic weapons give a State possessing them in adequate numbers a deterrent power against attack from a much stronger State" and giving an account of the economic advantages of atomic bombs when compared with other weapons.[82] His cause eventually carried the day and on May 18, 1974, "Smiling Buddha," the code name assigned to India's first nuclear test device, was detonated under the watchful eyes of Bhabha's successor, Raja Ramanna. Like Bhabha, Ramanna was a Western-educated physicist, holding a PhD from London University's King's College.

Although India's nuclear weapons program has generally been treated as nothing more than a strategic response to China's nuclear menace (China had detonated its first nuclear device on October 16, 1964) it also turned out to have played an important role in forging India's post-colonial identity.[83] In the first place, India's success in carrying out a nuclear program and building a bomb proved that Indian scientists and engineers were just as knowledgeable and competent as their Western colleagues. Not only this, but their choice to serve an independent India, rather than "the West," showed that they were thoroughly emancipated from the former colonial power, despite having been educated in England. Overall, it has become clear that India's opposition to the U.S. and its Western allies, particularly its violation of the non-proliferation treaties, reinforced rather than weakened its economic potential and its status as a leading post-colonial nation.

India's disobedient nuclear strategy led the United States, based on an arms embargo directive, to deny India a Cray supercomputer that could be used for further development of nuclear weapons. This embargo and the consequent unavailability of a whole range of Western technologies deemed critical for national modernization, compelled India to develop its own technological alternatives based on the mobilization of its own national resources, although in close collaboration with the USSR (and later Russia).[84] The aim was now to develop a network of national research programs and infrastructures that would enable India to develop indigenous alternatives to international strategic technologies.[85] Like the nuclear program, India's supercomputers program, its civil aviation project, and its space research program all reveal how India's emulation of Western technologies was used to assert Indian nationalism. This was ideologically reinforced by framing the denial of Western technology as a neocolonialist ploy to reinforce Western hegemony by other means, and the capacity to resist it and find a workaround as a nationalist triumph.

The design and development of PARAM 8000 (*param* means "supreme" in Sanskrit), India's first supercomputer, was driven by the U.S.'s denial of supercomputer technology to a "renegade" India. When the export of Cray supercomputers to India was forbidden, the Indian government ordered the national Centre for Development of Advanced Computing to focus all its efforts on achieving national self-sufficiency in the field of computer science and to take steps to build a supercomputer *Made in India*. The first prototype took three years to design and develop. The project was headed by Vijay Bhatkar, an engineer trained at the Indian Institute of Technology at Delhi. The PARAM 8000 was operational by 1991 and boasted the very respectable speed of 1 GigaFLOPS, exceeded at the time only by the newest Cray computers. It was the first of a whole series of ever more powerful PARAM computers that formed the backbone of the Indian electronics industry and were purchased by a number of foreign states, including Russia and Germany.

The so-called SARAS project envisioned the construction of an Indian civilian aircraft by the National Aerospace Laboratories (NAL). The project suffered a delay of two years because India was once again the victim of an embargo affecting one of the main components, the starter generator. The embargo was imposed in response to India's Operation Shakti, the codename for a set of nuclear test detonations that took place in May 1998 (*shakti* is the Hindi term for cosmic energy). Since the starter generator is an essential part of the high-performance energy system needed in modern military vehicles, the SARAS project too was hit by the 1998 embargo, despite the fact that it was not itself a military project. In 2004, when the inaugural flight finally took place, Raghunath Anant Mashelkar, the Director General of the Council of Scientific & Industrial Research at the time the SARAS project was launched, said: "Today is the 22,498th day of my life, and I've not known a happier morning."[86] Mashelkar's quip expressed his relief at finally having a flying plane after the two-year delay imposed by the U.S.

India's first moon orbiter project *Chandrayan-1* (Sanskrit for Moon vehicle) was also indirectly the outcome of the U.S.'s refusal to sell India a cryogenic rocket engine in the early days of spaceflight. The orbiter, inserted into lunar orbit on November 8, 2008, was the culmination of a series of satellite launches using indigenous rockets at domestic launch facilities. These had been designed and built at the Indian Space Research Organization, established in 1969 to develop advanced space technology at the instigation of Vikram Sarabhai, another Cambridge physicist. India's first satellite, *Aryabhata*, was launched in 1975 by the Soviet Union; in 1980, *Rohini* was launched, the first satellite to be placed in orbit by an Indian-made launch vehicle. This was followed by two more advanced rocket types that delivered numerous communications satellites and earth observation satellites: the Polar Satellite Launch Vehicle for polar orbiters and the Geosynchronous Satellite Launch Vehicle for geostationary satellites.

The seduction of Western technology was not limited to high-tech sectors, but also influenced everyday consumer choices, as for instance the dominance of British cars in India or French cars in Tunisia, or in the brand-loyalty of the Royal Riders Club.[87] The latter Club was founded in 2008 to celebrate the spirit of riding a particular motorcycle, the Royal Enfield Bullet, known simply as the Bullet, but also nicknamed "the thumper" because of its distinctive one-cylinder sound. The Bullet was originally manufactured by Royal Enfield in Redditch, Worcestershire, United Kingdom. Like the iconic Harley Davidson, the Bullet is experienced not just as a motorcycle, but also as a philosophy and a way of life; and "bulletiers" accordingly join up in clubs to share motorcycle adventures on the open road. Although this has traditionally been a male prerogative, Indian women have also shown themselves to be passionate bulletiers and several female clubs have been established recently as, for example, the Regals (in Mumbai) and the Hop on Gurls (in Bangalore).[88] In a British cable TV program, when asked if riding a Bullet could not be seen as bowing to neocolonialism, the bulletiers answered that the Royal Enfield Bullet, even when it was still being built only in England, stood above the colonizer.

The story goes back to 1949 when, shortly after independence, the Indian Army ordered 800 350cc Royal Enfield Bullets for border patrol use. Because it was such a large order, the Enfield Company decided to partner with Madras Motors and open a factory in Tiruvottiyur, Chennai, in the state of Tamil Nadu. In 1955, the 350cc Bullets were sent from the Redditch factory in unassembled "knockdown" form for final assembly at Tiruvottiyur. By 1957, machine tools and dies were also being sold to Enfield India. The growing popularity of the Enfield Bullet in India, mainly due to its simplicity, eventually led to the manufacture of parts in India, and by 1962 the manufacture of complete Royal Enfields.

The Bullets were made in the Madras factory in very much the same way as they had been made in England in 1955, with part of the work still done by hand. The Enfields soon acquired the status of cult bikes, especially in the 1990s with a growing market for retro-vintage motorcycles. In 1986, Raja Narayan, an Indian civil servant living in England, returned to his homeland and initiated a "prodigal son strategy" to bring the Bullet back to England. In that first year he succeeded in exporting a number of 350cc Bullets to Great Britain, the first step in a wider internationalization plan. Today, more than 20 countries import the 350cc Bullets, including the U.S, thus closing a circle that began with the travel of the Royal

Fig. 6.8   The Bullet: *When British Royal Enfield shut its doors in Great Britain in the 1970s, Enfield India continued to manufacture and sell the Bullet in its original, vintage 1955 form. As in the case of Harley-Davidson, the Enfield Bullet is not just a brand of motorcycle but a way of life; and Royal Enfield Bullet Riders can be found all over the world. First a European motorcycle in India, the Bullet is now an Indian cult motorcycle in Europe and the U.S.*

Enfield Bullet from England to India and ended with its return from Madras to the United Kingdom and to Europe.

The Chinese Variations

In 1920, Lenin launched the GOELRO plan for the electrification of Russia, to be completed in a 10–15 year period.[89] His claim to the Gubernia Conference of the Russian Communist Party that "Communism is Soviet power plus the electrification of the whole country" became a powerful propaganda slogan throughout the Stalin era (see chapter 2).[90] Lenin was well aware of the vital role played by railroads and the telegraph in the Revolution that overturned the tsarist regime in Russia, and firmly believed that electrification was a precondition for the new regime's modernization agenda. This explains why he established a state commission to design an overall plan to bring electricity to every village and house in revolutionary Russia. The Russia Electrification Commission was directed by Lenin's long-time friend, the engineer and economist Gleb Krzhizhanovsky, and employed about 200 scientists and engineers, including the renowned Genrikh Graftio, responsible for the first Russian hydroelectric plant at Volkhov, near St. Petersburg (1918).

Following in the footsteps of proponents of railroads in the mid-nineteenth century, Lenin presented electricity not simply as a technical milestone, but rather as a harbinger of civilization that would "provide a link between town and country, put an end to the division between town and country, ... raise the level of culture in the countryside and ... overcome, even in the most remote corners of the land, backwardness, ignorance, poverty, disease and barbarism."[91] There was an intimate relationship between the technological system that powered Ilyich's lamp, the Russian nickname for the electric lightbulb, and the emergence of modern socialist Russia, a relationship that was critical in making Russia—and a little later the Soviet Union—a world power.

Fifty-eight years later, Deng Xiaoping used the very same rationale—socialism/modernization/technology—to introduce his Four Modernizations program, part of the so-called "Socialism with Chinese features."[92] China and Western technology have historically been restless bedfellows, as we already saw above when describing the Self-Strengthening movement. China's "Wilsonian moment," the May Fourth movement (1919), was led by a set of Western-educated intellectuals, who had studied both in Europe and in the U.S. and who aimed to modernize China by appropriating Western science and technology.[93] In 1914, Chinese students at Cornell University had already founded the influential Science Society of China, which published *Kexue* (Science), a journal close in spirit to the journal of the American Association for the Advancement of Science. Under the Republic of China (1912–49) with Chiang Kai-shek as president, foreign-trained scientists and engineers did high-quality research at Chinese universities and research institutes, either funded by the government or foreign organizations such as the Rockefeller Foundation. In 1919 Mao Zedong, then employed as assistant librarian at Bejing University, became acquainted with leading members of the May Fourth movement such as Hu Shi, who had been a student at Cornell and Columbia and Chen Duxiu, the later Communist Party's first General Secretary, who was studying at Beijing University. Mao soon became a Marxist-Leninist and a fervent supporter of Western science and technology in service to the revolution.[94]

The establishment of the People's Republic of China marked the onset of a large-scale program of technology transfer from the Soviet Union to China.[95] China not only organized its scientific and technical institutions along Soviet lines, but also initiated large-scale transfers of personnel, expertise, and machinery to and from the Soviet Union. During the 1950s it is estimated that China sent about 38,000 people to the Soviet Union for training and study, mostly technicians from key industries (28,000), but also students (7,500) and college and university teachers and postgraduate scientists (2,500). In turn, the Soviet Union dispatched around 11,000 workers, scientists, and technical experts to China, mainly workers in heavy industry (around 9,000), and about 850 scientific researchers and 1,000 technicians in education and public health. The Soviet Union participated in some 100 major Chinese research projects, including several in nuclear science.[96]

Transfers of Soviet technology and technical expertise to China jump-started Chinese production in such fields as electric power, steel, basic chemicals, machine tools, and military equipment. Among other things, Soviet input was essential for teaching Chinese workers how to operate imported or copied Soviet technology and industrial plants. By the end of the 1950s, however, Sino-Soviet political cooperation began to sour, leading the Soviets to withdraw from China and to terminate their scientific and technical aid programs. At about the same time, and probably not coincidentally, Mao launched the Great Leap Forward, a national effort to encourage large-scale, distributed, low-tech, labor-intensive—though often poorly planned—programs, such as the famous backyard blast furnaces, built without input from trained engineers in the expectation of being able to combine westernized modernization with local domestic self-reliance.[97]

The Cultural Revolution that followed exacerbated smoldering tensions between the regime and Chinese scientists and engineers, who were increasingly marginalized as counterrevolutionaries. The combined effects of an obsession with short-term problems, the deprecation of theory, and the belief that real creative power emanated from the people, had a devastating impact on the Chinese economy—leading to the Great Famine—and on Chinese scientific and technical capacity.[98]

For more than a decade China trained no new scientists or engineers and isolated itself from foreign scientific developments. The only exceptions were nuclear weapons and space technology, under the banner of the *liangdan yixing* ("two bombs, one satellite") program.[99] As soon as Soviet engineers, scientists, and technicians

Fig. 6.9 Pots and Pans: *Employees of the Shin Chiao Hotel build small, rudimentary steel blast furnaces during the "Great Leap Forward" in October 1958, in Beijing, China. Mao encouraged the establishment of small backyard steel furnaces in every commune and in each urban neighborhood as a demonstration of technological nationalism. Although Mao knew that high-quality steel could only be produced in Western-like large-scale factories using reliable fuel and Western technology, he decided to continue the low-quality backyard steel project to feed the revolutionary enthusiasm of the masses.*

left China, Mao adopted a techno-nationalist stance—"We stand for self-reliance."[100] His first target was the nuclear program, which up to then had been developed under Russian guidance, and which was now to be placed under Chinese leadership and carried out by Chinese personnel. A large number of engineers and scientists were gathered from all over China to the Nuclear Research Centre in order to achieve Mao's goal. On October 16, 1964, China successfully exploded its first atomic bomb.

At the same time, China's space program culminated in the launching of *The East Is Red* satellite in 1970, equipped with a transistor circuit that broadcast a song by the same name, one of the anthems of the People's Republic of China during the Cultural Revolution. China's space program was led by Qian Xuesen, a scholar educated at Shanghai University and Massachusetts

Institute of Technology (MIT) and one of the founding members of Caltech's Jet Propulsion Laboratory. During the McCarthy era Qian was accused of communist sympathies and deported by the U.S. government.

Deng Xiaoping's Four Modernizations shifted the focus from Mao's techno-nationalist project to a more techno-globalist one, enrolling science and technology policy to buttress economic development.[101] In his famous speech at the March 1978 National Science Conference in Beijing, Deng presented Western science and technology as the core of modernity and as key players in China's New Long March, stating: "The key of the Four Modernizations is the mastery of modern science and technology. Without the high-speed development of science and technology, it is impossible to develop the national economy at a high speed."[102]

In fact, Zhou Enlai had already proposed the Four Modernizations as early as 1963, with the aim of bringing Chinese agriculture, industry, national defense, and science and technology up to Western standards. Both Zhou Enlai and Deng Xiaoping were educated in a Western ambience: the first attended the Nankai Middle School, at Tianjin (Northern China), a prestigious private school under the aegis of the elite Phillips Academy at Andover, Massachusetts, and the second studied in France under the *Mouvement Travail-Études* program. Like Zhou, Deng became acquainted with Western European culture quite early in life, first through France and later through Russia.

Although continuing to support indigenous R&D, Deng's Open-Door policy encouraged China to acquire vast amounts of know-how from foreign companies, universities, and governments, expanding international cooperation both directly through scientific and technological exchanges and through foreign economic investment. Various channels were used to support this flow of expertise, ranging from cooperative scientific and high-tech ventures with foreign corporations to enrolling Chinese expatriates returning from studies at major American and European universities—and even including government-sponsored corporate, cyber, and industrial espionage. Although science and technology policy was seen as subservient to economic and foreign trade policy in general and hence a government responsibility, it was not so rigidly planned as to exclude input from firms or from scientists and engineers.[103]

It is beyond doubt that the celebration of *zizhu chuangxin* (indigenous innovation) has been a perennial feature of Chinese science and technology policy and is rightly seen as a symbol of technological sovereignty. Still, it is clear that China's "leapfrogging" aspirations, that is, the desire to jump ahead to state-of-the-art technologies without traversing the intervening stages, continues to depend heavily on the existing European and American matrix: on technology transfers, the training and development of Chinese talent in foreign universities and research institutes, and the employment of substantial numbers of foreign consultants.

What is Left after the Europeans have Gone?

When taking leave of their colonies in Africa, Asia, or South America, voluntarily or otherwise, Western powers left behind infrastructures and technological styles that shaped the new nations through the force of path dependencies. Although post-colonial rulers exploited the technological legacies of colonial rule in different ways and under different economic and political conditions, these legacies inevitably became the cornerstones of emergent rhetorics of modernization and economic sovereignty.

The colonizers' Euro-Western concept of development, rooted in technology and productivity, was generally adopted lock, stock, and barrel by the ex-colonized and soon empowered a technology-driven vision of modernity. Whether by propagating self-reliance or by importing foreign technology, whether under a democratic or a dictatorial regime, whether under U.S. or European influence or in the Soviet orbit, former colonies were convinced that accessing and mastering Western technology was critical to maintaining their independence and to establishing themselves in the post-Second World War world.

However, the assertion that new nations embraced Western technology in its various ramifications—that is, machines, organization of research and production, expert training—is not to suggest that this was in any way a uniform process. On the contrary, widely-divergent strategies were deployed to appropriate Western technology and to apply it to specific objectives. Even in cases where ready-to-wear technology was imported from abroad, local strategies and actors played an important role in defining the terms of its use. India and China learned important lessons when the technological assistance they took for granted disappeared and they had to develop the necessary resources themselves. Self-reliance, sovereignty, and power were the watchwords of emerging postwar nations.

The worldwide network of technology transfer and/or appropriation is complex, involving international, national, and regional powers that are all but unmanageable. The contrast between so-called "dual societies" and "modernizing" societies is one of the most striking of the many unintended consequences of this unruly technological globalization.[104] Most countries in Africa and some in Asia and in Latin America belong to the category of "dual societies," that is, those that assimilated contemporary Western technology but have restricted its benefits only to a political and economic elite, the sole proprietors of power, knowledge, know-how, and *savoir faire*. The rest of the population lives in improvised housing made from scrap materials, such as plywood, corrugated metal, and sheets of plastic and they are excluded from public services, facilities, and basic infrastructures, such as water supply, electricity, and sanitation. The (waste) materials that sustain this context of exclusion and poverty are the new "technologies of poverty."[105]

Striking examples of sites sustained by technologies of poverty are the Dhaka slum, in Bangladesh; Dharavi, in the suburbs of Mumbai, India; Kibera, the largest slum in Nairobi, Kenya, associated with the flying toilet (polythene bags used to dispose of human waste); Cité Soleil, in Haiti; and the Brazilian *favelas*, such as *Cidade de Deus* (City of God), popularized in the well-known and Academy Award nominated (2004) film by Fernando Meirelles. Modernizing economies, on the contrary, benefit from laws, regulations, and government programs aimed at avoiding this skewed relationship between access to technologies and infrastructures on the one hand and wealth and political power on the other—thus preventing the social abyss that characterizes dual economies. Outstanding examples are South Korea, Taiwan, and Singapore.

Like China, Latin America followed a strategy of importing European and American technology, even though the historical and geographical conditions could not have been more dissimilar.

Fig. 6.11 **Hybrid Landscapes:** *Western technology often acts as the great equalizer. This picture shows a dense agglomerate of houses in the old part of the city of Fez, Morocco, Africa, most of them with satellite dishes. Consuming technology is a shared experience in most of today's societies, regardless of wealth, education, religion, gender, and age. Even radicalisms that condemn and appear to ban Western technology use updated weapons, media, and communication tools to impose their beliefs.*

South and Central Americans expelled their colonizers in the early nineteenth century, although they maintained a strong European identity and cherished close ties with Spain and Portugal. Nevertheless, indigenous technological competence was generally not up to international standards. Imported technologies were consumed as commodities, with limited knowledge of how they worked and with limited development of local expertise. Maintenance, too, was imported—"substitution of R&D by a telex machine"—or depended on low-tech skills, closer to *bricolage* than to expertise.[106]

The specter of neocolonialism hovers over the relationship between former colonizers and newly independent states, with "neutral" technologies hiding new forms of dominion and new ways of exploiting natural and human resources. Nevertheless, economic and technological globalization has changed the rules of the game by redefining the terms of access to world power. In recent years a "de-Westernization" of global economic power has gained momentum, with Eastern countries like China becoming the main actors in the current process of globalization, challenging traditional Western hegemony over the global economic system.[107] The BRICS countries are all throwing down the techno-economic gauntlet to Western economies and R&D policies: China, for example, has radically increased its percentage of GNP devoted to research and education, with venture investment increasing 50 percent; and this while rates of venture capital investment are dropping both in the U.S. and Europe.[108]

7

The Reconstruction Period

Introduction

This final chapter brings the story of Europe globalizing up to the present day. It paints a picture of the reconstruction of the global order after the Second World War: the establishment of the Cold War held in place by murderous arsenals of atomic weapons and characterized by the definitive breakthrough of the U.S. as the uncontested leader of the Western camp, the gradual reorganization of Europe along federal lines, and the emergence of new key technological domains like the mass automobile and the expanding infrastructures of leisure time, both of which contributed to European mass mobilization. At the margins of the new bipolar world, the so-called Third World nations, keen to pursue their own economic and social development, eagerly sought new Western technologies of modernization but were frustrated time and again by the persistence of *dependencia* and other neocolonial arrangements. Meanwhile, at the other end of the welfare spectrum the "startup nation" of Israel showed itself a model of rapid economic and military development despite highly disadvantageous climatological and geopolitical factors that would have predicted a different outcome. This exemplary case is described in a concluding section.

Ambivalences of Americanization

Though America's influence in the world had clearly been waxing since the end of the nineteenth century, the final victory of

Fig. 7.1   Europa and the Bull: *From the first canned tomatoes in the 1840s to the Care packages of the late 1940s, from American canned beer to Andy Warhol's "Campbell's Tomato Soup"—American culture seems to be symbolized by a tin can from the vantage point of Europe. Under the aegis of the Marshall Plan in 1949 the motif of Europa riding on the bull was reinterpreted, with the fear of becoming an outlet for American mass-produced goods thrown in for good measure* (Die Zeit, *February 3, 1949).*

Americanized production and its consumerist lifestyle came only after 1945. Then the U.S. *imperium* engulfed Europe as an *emporium* that subjected the Old Continent to slow conquest by a million U.S. consumer goods.[1] Once again, many Europeans were forced to conclude with Stefan Zweig that Americans were not "like us."[2]

Since 1850 the mental (self-)mapping of the West, and in particular the U.S., had fastened onto core features like youth, ambition, enterprise, innovation, and flexibility. Ironically, it was America's ex-Europeans who as often as not heralded the U.S.'s perceived mission and who took it upon themselves to instruct the rest of the world in civilized behavior and the fundamentals of political freedom. The concept of the "West" crystallized and acquired a new political valence as the Cold War fixed new ideological and strategic boundaries at the Iron Curtain. "Western Civilization" increasingly lost its neutrality and became an object of love or hate. This confrontation of ideologies revived ancient dichotomies: Orient versus Occident, "Western Roman Christianity" versus "Eastern Orthodoxy," Western "traditions of freedom" versus "oriental despotism." Now "the West" became an ideological rather than a geographical concept. Being "Western" meant cultural, political, and economic alignment with Western Europe and the U.S. and included mandatory alliances (like NATO) and treaties. Pressures to align with the West were particularly strong for strategically-located countries like Brazil, Argentina, Mexico, Canada, Australia, and Oceania.

The Cold War doubtless contributed more to the worldwide dissemination of the Western cultural package than any other global configuration since 1850. Those who professed allegiance to the West could expect to receive financial, economic, technical and, above all, military support. This had its equivalent in the Eastern Bloc, but the latter commanded far fewer resources and consequently had nothing to oppose to the mystique of welfare boosters like the Marshall Plan. Moreover, the arms race tied up huge amounts of money and personnel. Impressive numbers of American and Soviet engineers were employed in military research and development during the Cold War. This moved even the American President and former general Dwight D. Eisenhower, in his farewell speech of 1961, to warn of a military-industrial complex that in his view was leading a life of its own within the North Atlantic Alliance.

Another factor that made the American corporate model so irresistible, even for Europeans, was the fact that the U.S. government was allied with and energetically supported initiatives by private industry and other non-government organizations. This institutional constellation was so manifest that it repeatedly led to accusations of American craftiness and dominance.[3] In fact, if anything, the retreat to ideological trench warfare after 1945 only damaged the U.S.'s global image: "A fervent anti-Communism justified much illiberal conduct after The Second World War. There could, American liberal-expansionists believed, be no truly enlightened dissent against the ultimate acceptance of American ways, and this faith bred an intolerance, a narrowness, that was the very opposite of liberality."[4] These are certainly the authentic roots of the infamous "Ugly American" syndrome.[5]

The polarization of the Cold War often led the "West" and in particular the U.S. astray in judging political situations. An extreme example was the Congo. Under Leopold II the country had already been pillaged for its ivory and rubber resources. After 1908 the Belgians continued to rule the colony in a kind of half-hearted manner yet continued to exploit its raw materials.[6] As early as the Second World War, the U.S. imported most of its copper from the mineral-rich province of Katanga. After the war, Katanga also produced 90 percent of the cobalt and 60 percent of all the industrial diamonds processed in the U.S.; and starting in the 1950s Katanga also exported uranium ore for the production of atomic weapons. Eight out of every ten American warheads contained uranium from the Congo. This was of such critical importance to the U.S. that it took it upon itself to revamp Congo's transportation infrastructure to facilitate the export of these strategic materials. After Belgium precipitously left the colony in 1960 and Patrice Lumumba formed a post-colonial government sympathetic to the Soviet Union, he was "removed" in an action that has never been fully clarified, but that was most probably carried out jointly by the CIA and the Belgian secret service.[7] In his successors, Mobutu Sese Seko and Laurent and Joseph Kabila, the U.S. subsequently supported dictators who, though far from representing the values of the "free West," had the undeniable advantage that they were able and willing to guarantee the crucial supply of raw materials.[8] This cynical trade-off, re-enacted in a wide variety of settings, did much to undermine the moral standing of the West, especially among the nations of the emerging Global South.[9]

Yet at the cessation of the Cold War in 1989/90 it seemed that, despite everything, the "West" had triumphed over alternative ways of life and that a downright post-historical age was in the offing. Having eliminated feudalism, the class system, cannibalism, slavery, and racial conflict, after defeating Fascism and colonialism, the West now prided itself on having defeated communism as well. Moreover, following the Maastricht Treaty of 1992, "good old Europe" seemed to be moving towards something like a "United States (of Europe)" including the gathering of post-communist nations into the common fold. Capitalism seemed so inevitable that even China made some tentative moves in the direction of a market economy.

Just at the moment when the Internet and other technical networks seemed once more to hold out the dream of "a world with just a single heartbeat" (Stefan Zweig), a world reconstructed in the image of the West, the dreamer was rudely awakened. Al Qaeda terrorists, acting in the name of a "fundamentalist Islam," renewed the "anti-imperialistic" struggle by flying two hijacked passenger jets into New York's World Trade Center on September 11, 2001 ("9/11"). This mind-boggling feat of kidnapping Western transport systems to sow death and destruction in the "homeland" struck the purported "colonizer" in his most vulnerable spot: his unshakeable faith in the reliability of infrastructures. The attacks exploded the conception of the "West" as a clearly circumscribed geographical entity, and initiated intense reflection on its shared values: the separation of powers in a civil society, the idea of the rule of law that recognizes human rights as inalienable, the protection of religious freedom, and adherence to cultural tolerance, democracy, and the ideal of popular sovereignty.[10] Western culture was reframed as a de-localized set of literary, scientific, political, artistic, and philosophical principles that set it apart from other civilizations.

In consequence, at the outset of the twenty-first century the "West's" confidence has been profoundly shaken. This is especially true of the U.S., which perceived "America's gifts to the twentieth century ...: advanced technology and mass culture"[11] as an unshakeable foundation for power and prosperity. Anticipating a new golden age after the Cold War era, many U.S. citizens already imagined their nation in the role of an imperial world power justly ruling over a "Pax Americana." Since then, differences with "Old Europe" have increased, and there are even some Euro-chauvinists who contest the notion that the U.S. developed any kind of dominant culture even in "its" century. America, they argue, may have brought us the glamorous world of pop music and Hollywood, fast food, the Internet and multimillion-dollar media sports events, but Europe continues to dominate in the realm of high culture. And, they add, the U.S. must admit defeat even in the lowbrow sectors in which it seems so strong. The leading sport worldwide is European soccer, and as far as cuisine, style, or fashion goes, "local" cultures are the rage. Italy, for instance, is present around the globe with her pizza, pasta, and ice-cream parlors; Italian coffee culture and Tuscan wines are recognized globally as characteristically Italian no less than Milanese haute couture, haberdashery, or quality footwear. This *Italianità* flourishes without being embedded in power politics and, for this reason, some argue, one cannot find anti-Italianism anywhere in the world.[12]

Constructed as these "ethnicized" traditions may be, they perform on the global stage along with other "national" European contributions: British pop-rock icons, Spanish molecular cuisine, French fashion, or Scandinavian design and furniture. But even if the U.S.'s dominance can be challenged along these lines, there are now other contenders in the wings. Former "provinces" outside Europe have caught up. The Chinese or the Indians have come to add much more to global mass culture than mere flavorful spices, meals, or handicraft. Latin America contributes more than music, meat, or wine. Although still a technological leader, from a global political and cultural perspective Europe—and for that matter ultimately even the U.S.—is on the verge of becoming just another "provincial region" among others.

La Mutation Humaine

Europe—when we utter this word, we do not think of a continent, that small offshoot of Asia: neither do we think of this or that people, or of six or ten European states. Europe first and foremost stands for a particular mentality; it characterizes our relation to all things. Europeanness does not come with birth but is generated rather by education. Europe was the first "open system." ... Even today, the interchange of information, the "friction of brains" that raises the collective intellectual temperature, is practiced nowhere as intensively as it has been in Europe. Especially the fact that Europe consists of various language areas has led to an exchange of ideas that has proven more fruitful than any that has taken place in any other part of the world.[13]

The French scholar Pierre Bertaux wrote these self-congratulatory sentences in the year 1964, a point in time when in fact Europe had little reason to burst with pride. The Second World War had been even more devastating than the First, and it had durably weakened Europe's standing in the world. Fascist Germany and Italy had championed the idea of a different Europe organized on hierarchical principles, an idea that had met with sympathy all over the Continent. One of its many offshoots had been the industrialized form of genocide that had occurred in the hidden wings of the war. The Holocaust, too, had been a product of pan-European collaboration. Finally, especially in the case of France, there was the shamefully ignoble course of decoloniza-tion. Europe had lost much of its standing in the world: from 1945 onwards, the "liberators" of Europe transformed themselves into new superpowers. The United States and the Soviet Union simply asserted their intensely rivalrous political, military, and technological authority over the rest of the world. In the preface to his book *La Mutation Humaine*, Bertaux noted the fact that he began writing on the day the Soviets launched their first satellite, *Sputnik*: October 4, 1957.[14]

Pierre Bertaux was an exemplary twentieth-century European. He had received an international education and his outlook on life had been shaped by both World Wars. In addition, he was an expe-rienced colonial administrator and was convinced that the techno-logical and scientific dynamism originating from Europe and the United States should and would captivate the whole world.[15] In his view, education was at the core of what it meant to be European. It manifested itself in a "scientization" of all areas of life, not only in the form of Big Science (nuclear weapons, high energy physics, missiles, satellites, and computers), but also in mass motorization, washing machines, and workplace automation. Bertaux argued that the communications revolution in particular was causing a fundamental *mutation* of humanity and hence futurist research and perspicuous planning were needed to manage this process and keep it from going awry.

So, less than twenty years after the Second World War, Europe found itself in the throes of a kind of euphoria about its future. While this was doubtless in part a reaction to its onerous past, it was widely credited in the first place to its startling economic resurgence. The new prosperity was seized upon as the foundation for a new species of Western and Eastern European welfare states that were expected to emerge by the early to mid-1970s, marking the start of a new era.[16] By the mid-1960s, the fantasy of a prosperous and highly-mechanized future had reached its zenith.[17] Stimulated by Cold War competition and the successive loss of their colonies, all European countries had morphed into interventionist states that were anxious to secure and increase their prosperity.

In hindsight, *les trente glorieuses* seems like an exceptional period. Technology played a key role and was a defining topos of the European self-image. While Europe was mostly a pawn on the chessboard of the Cold War, it was also its beneficiary. Drawn into the maelstrom of ideological polarization, Western and Eastern Europe were respectively "Americanized" or "Sovietized." However, both were also beneficiaries of material and technological aid that enabled them to arm themselves with advanced weaponry and to become showcases of the scientific-technological potential of the respective systems. At the same time, the bitter antagonism between the superpowers also provided an opportunity for Europe to rediscover its identity or rather to reinvent itself once more. Despite the Continent's split loyalty to its two military and economic backers, "project Europe" also created a third road that partly bypassed the confrontation of the blocs.[18] Western Europe took the first steps along this road by forging transnational networks of scientific-technological collaboration. These embodied the renewed European ambition for superpower status, which was to be achieved by appropriating the necessary means (for example, nuclear power, missiles, prestigious aircraft such as the Concorde) and creating loci for innovation (for example, think tanks, large-scale research institutes, Big Science facilities). It also reflected the pride in European higher learning that Bertaux expressed in 1964. That year, in fact, saw the opening of the Guiana Space Center, a new tropical outpost of European science and high technology.[19]

Following wartime destruction with its millions of displaced people, Europe long remained in political and demographic turmoil. Flight, expulsion, emigration, exile, and asylum as well as migrant labor—migration in every sense of the word—became much more of an abiding feature of Europe than it had since 1850.[20] Although the Iron Curtain was certainly an obstacle to the untrammeled flow of ideas, material, and people between East and West, even this barrier proved more porous than it first seemed.

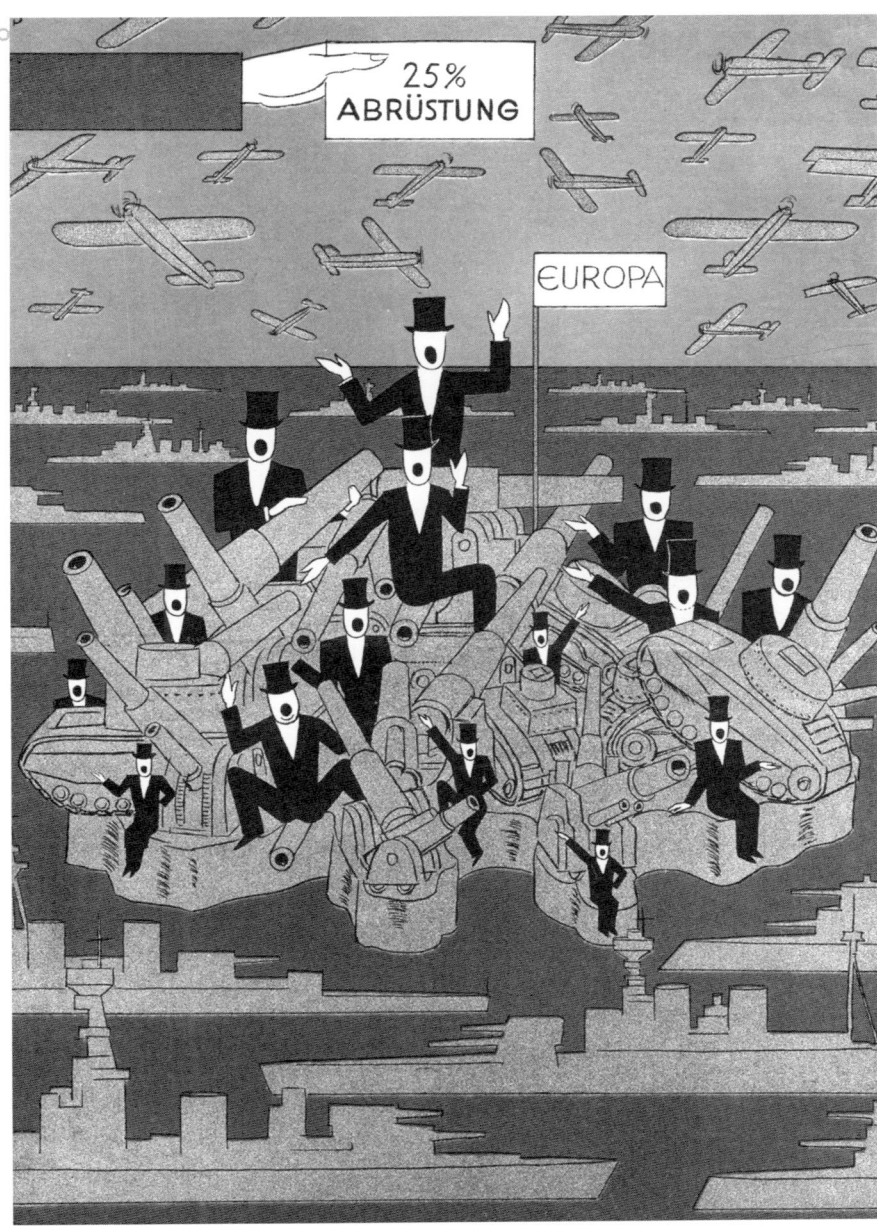

Fig. 7.2 Europe in Arms: *In 1932, graphic artist Erich Schilling commented on the inability of Europeans to make a 25 percent cutback in armaments, caused by economic depression, in the satirical "Simplicissimus." The motif was an early example of what in the 1980s was called Fortress Europe, which circumscribed Europe as a well-fortified island of wealth and prosperity. Later on, Schilling himself subscribed to the National Socialist version of a unified Europe.*

And while people moved from place to place in the East and West (in Eastern Europe, freedom of travel in particular remained the focus of endless debate), Europe also once again demonstrated its "absorbtive" capacity as well as recovering its position as a focus for the global circulation of people and commodities: in the best cases enacting Bertaux's notion of "intellectual friction."[21] At times, Europe seemed unable to cope with new waves of immigration and tried to profile itself as a "fortress." Nevertheless, on balance it possibly derived more benefits than losses from the increased exchange. In addition to immigration, the rise of global media and increased international travel within Europe itself also contributed to the broadening of European (educational) horizons.

European efforts after 1945 to pursue a "second colonization" in the shadow of two dominant "anti-imperialist" powers can be seen as attempts by former colonial nations to compensate for their loss of global authority during the war and its aftermath.[22] As noted above, however, the imperial era had left behind a legacy of Western science and technology. In view of the widespread post-colonial rhetoric of self-determination, it is astounding that the newly-independent nation states (as a political construct itself a successful European export) not only adopted the material "achievements" that European colonial rule had left behind, but also continued to develop them. In offering their help the Europeans, like the Americans and Soviets, acted not only from altruistic motives but tried to implement what was euphemistically called "development cooperation." Hence, the formal distinction between the colonial and neocolonial period was at times hardly noticeable, being limited to frequent changes in styles of interaction. At the very same moment that Bertaux began writing his book, the French philosopher Alexandre Kojève was advocating a shift in European colonialism from "taking" to "giving," arguing that expansion needed to be geared specifically toward Third World needs in order to accommodate and stimulate the growth of production in Europe. As Kojève put it, "poor consumers" are "bad consumers."[23]

Kojève had put his finger on the factor that in large measure accounted for the rapid economic resurgence of postwar Europe (or at least of Western Europe): the establishment of a "Fordist" production regime and an "entrepreneurial shift."[24] Fordism was predicated upon the doctrine of paying workers a wage that would enable them as consumers to afford the products they had manufactured. This could only be done within the constraints of capitalist logic by cutting production costs drastically and hence required companies to change over to serial and standardized mass production, classically symbolized by the assembly line. Factories enhanced productivity by, among other things, optimizing work-places and labor processes according to "Taylorist" dictates, hence becoming more efficient and ultimately making their products affordable enough to enable better-paid workers to buy their final products. Hence, Fordism was the antechamber of consumer culture, marked by democratized access to consumer goods and hyped-up fantasies of "everlasting welfare." More immediate returns also materialized as labor processes became increasingly automated, allowing for more leisure time thanks to reduced working hours, and in turn generating further consumer needs.

European Mobilities

Car Cultures

The automobile was perhaps the most culturally charged artifact of the twentieth century. It both symbolized and enacted the indi-vidualized mobility of commuter and tourist, economic prosperity, changes in forms of production, the utilization and wastage of fossil fuels, landscape consumption, and finally that which the Swiss environmental historian Ulrich Pfister once termed the "1950s syndrome," the assimilation of town and country and the fundamental transformation of life styles, leisure, and forms of consumption.[25] In the emergent "car culture" that spanned the entire twentieth century, there were "myriad ways in which use of the automobile shaped social relations, human geographies, national economies, and individual experience around the globe, making the car one of the most defining material artifacts, industrial products, and cultural symbols of the last one hundred years."[26]

Less well known is the role of the automobile in mediating relationships between Europe and the rest of the world. Towards the north, west, and south of course, it is impossible to leave European territory directly by car, while, during the Cold War era, any such undertaking towards the east promised little in the way of diversion, even if one managed to make it past the Iron Curtain. Notwithstanding, the automobile conquered the world on the basis of European inventions as well as American—and later Japanese—production and marketing methods. The U.S. automobile industry remained chiefly focused on the admittedly huge American domestic market and became adept at creating an endless succession of *derniers cris* to keep up demand for new models. By comparison with the European car industry, however, the Americans were laggard on the point of technological innovations. The exceptions were a number of safety technologies introduced in the 1950s and the 1970s (running gear, crumple zones, seat belts, and so on) that were required by new federal regulations. But across the board and in the long haul the highly-competitive European automotive industry proved more innovative.

European cars also dominated the luxury market—only Cadillac was able to compete, and then only for a few decades. American vehicles were no match for the international—and notably non-European—reputation of Rolls Royce and Bentley, Ferrari and Mercedes, BMW and Citroën. As early as 1900, Indian maharajas were the largest definable customer group for the British Rolls Royce factory.[27] High-end European cars symbolized luxury

Fig. 7.3 Learning to Change Gears: *At the end of 1943, the East African Army Service Corps in Nairobi trained young men from Kenya to become motor transport drivers. The stones on their heads stopped them looking down. Not without racist undertones, Africans were often described as skillful and keen in adopting automobile technology. After the Second World War, when Africa became a burial ground for European cars that needed sorting out, African engineers had to cultivate their skills at maintenance.*

worldwide. Such car ownership became an effective means of consolidating economic, political, and cultural hierarchies and boundaries between rulers and subordinates in colonial settings.[28] Even today, in all corners of the world, the possession of European luxury motorcars is a way to flaunt one's wealth and is an unmistakable fount of "symbolic capital" (Pierre Bourdieu) sending the message that the automobile owner has honed subtle distinctions of taste and become a participant in European "civilization."[29]

Europe and the United States also had a rather different take on mass consumption. In Europe, the concept of so-called "people's products" (*Volksprodukte*) that had emerged in the interwar period was taken up again in the postwar era. Its two most prominent representatives were the German *Volksempfänger* (people's radio receiver) and the *Volkswagen*. Thanks to the standardized and simple though technologically-advanced VW, large segments of the population that until then had been limited to bicycles and motorbikes, were now able to master automobiles. Numerous car manufacturers in other countries followed suit. England's Morris Minor and Mini, France's Citroën 2 CV/Dyane and Renault 4 CV/R 4, as well as Italy's Fiat Topolino/Cinquecento all copied the root idea behind the Volkswagen. These popular cars were also exported to other European countries. Even manufacturing plants were transplanted across borders. While the Italian company FIAT, for instance, established licensees and manufacturing plants in Spain (SEAT), Poland (Polski-Fiat), and the Soviet Union (Lada), motorbikes and automobiles that had gone out of fashion in Great Britain continued to be produced in India (Royal Enfield), where in 2009 the Tata Nano exemplified a reincarnation of the cheap people's car.

Volkswagen, however, was also a pioneer in the construction of production plants for its standard model in non-European regions, in capturing local markets, and in regard to its promotion of mass motorization, especially in newly-industrialized countries (or in corporate parlance: "future markets"). In each case a number of technical details were modified to meet local requirements, particularly local price levels. In South Africa the "Beetle" was produced between 1951 and 1979 after which it was replaced by other car models. *VW do Brasil* had manufactured the Fusca since 1953, even setting up a long-term development department so that they could open up new markets in Latin America and Africa on the basis of specialized models. In 1981, the production share at *VW do Brasil* had risen to 100 per cent of the national market and the Fusca took to the streets in Nigeria and Peru, as well as Venezuela, the Philippines, Chile, Uruguay, Saudi Arabia, and Zaire. Everywhere, the vehicle contributed to the basic motorization of transport.[30] Eventually, the Brazilian plant even exported certain models, such as the VW Fox, to Europe. Factories in the United States, Argentina, and China followed. Yet, Volkswagen achieved its longest-lasting influence in Mexico, where the Vochito, built at Puebla between 1967 and 2003, long dominated the Mexican streetscape. Supplying over 100 countries, 90 percent of the vehicles manufactured in Puebla today are exported.[31]

All the countries just mentioned developed a Fordist mode of production and established corresponding "car cultures." The construction of streets, gas stations, garages, and other service stations "mobilized" all of life and allowed it to flow more smoothly. "Suburbanization" flourished everywhere, introducing new dimensions to the circulation of people, goods, and ideas. Whereas the railroad had propelled the economic upsurge of the

Fig. 7.4 Motorizing Mexico: *The Volkswagen Beetle decisively contributed to making non-European regions, such as Latin America, mobile. From the early 1950s it was exported to, and from 1959 produced in, Brazil. Volkswagen de México followed in 1964. Nicknamed Vocho and painted in yellow, the car often served as a metropolitan taxi. In 1971, the first 500 of them were presented on the* Plaza de la Constitución *in México City. Production of the Beetle in Puebla ended as late as 2003. The "New Beetle" succeeded it—one of the postmodern lifestyle cars activating nostalgia.*

nineteenth century, the automobile was the motive agent in the twentieth century.

Volkswagen also served as a model in other respects: since the late 1950s it was the largest employer of foreign "guest workers," encouraging migrant labor within as well as outside of Europe. Furthermore, since the early 1970s the company was compelled to face the challenges posed by Japanese suppliers as well as the efficient production methods of "Toyotism": its just-in-time principle, its diversified quality competition, computerization, its flexibility in reacting to changing market conditions through specialization, its rationalized concepts of production, and the influence of clan structures on work processes.[32] "Globalization" required or rather enabled the search for efficient production locations, leading to specific sites within and beyond Europe responsible for the development and manufacture of specific parts. These could eventually be assembled to create the finished product thanks to complex logistical protocols. Such "lean production" via "supply chain management" soon took effect outside of Europe, and "global sourcing" developed as a result of gathering components and modules from all over the world.[33] "Simultaneous engineering" is not only helpful in responding flexibly to customer demands and market shifts, but also serves to put competitive pressure on individual suppliers, increasing efficiency and productivity while at the same time lowering production costs.[34]

As a matter of fact, the enduring success of Volkswagen's Beetle nearly caused the company's ruin.[35] Management hung on to its basic 1930s technological concept even after it had become clear that the focus on a single basic car design was clearly outdated. The modern VW Golf appeared on the market just in time (1974) for Volkswagen to avert total catastrophe and to build on its previous success with the Beetle, this time with a new understanding of how to meet customers' ever-changing demands. But with the saturation of Western and Eastern European markets in the 1980s and 1990s respectively, exports and overseas production became more important. At present, Volkswagen operates manufacturing plants in the United States and Mexico, Brazil and Argentina, South Africa, Kenya and Angola, Russia, India, China, Indonesia, Malaysia, Kazakhstan, Vietnam, and Japan.

Volkswagen has developed markets worldwide and has further acquired a whole array of additional brands—these include rather exotic niche products such as Bentley and Lamborghini, primarily intended to add luster to the company's image. Today, it portrays itself as a global player and aspires to become the world's largest car manufacturer by 2018.[36] Once a public enterprise with strong political ties and charged with producing a *Volkswagen*, the company has grown over the course of several decades into a multinational corporation that meets the challenge of "global competition" with modern production technologies, a diversified product range, and militant management methods. It is all the more fascinating that it has been able to maintain the reputation for quality enjoyed by German vehicles, given the fact that nowadays so few Volkswagen parts are actually produced in Germany.

Hospitality Industry

Europe's postwar industry became ever more productive and while this meant more profits it also facilitated increased leisure time for workers—a trend that was fast becoming the focus of individual and social life in Europe. Leisure time used to refer to the time spent on explicitly-recreational activities, time required to recuperate from and to be able to sustain intensive work routines. Gradually, though, "work and play" came to be seen as separate worlds and leisure increasingly came to be understood as spare time that could be devoted to "what is really important": self-discovery, socializing with like-minded people, family, as well as avocations. New technologies both created the preconditions for the increasing amount of free time and shaped how it was spent.[37] Media like the cinema, radio, and television shaped people's perception and mapping of foreign spaces; these were the major channels for propagating European and American ways of life, including the concept of leisure time.

But "leisure time" was not only about spare time in the evenings and weekends, it was also about vacations and tourism, and here too technologies became decisive. When mass tourism was inaugurated by British travel agencies in the 1820s and 1830s it quickly acquired all the features of an "industry" and soon became "big business."[38] After 1945, the globalizing effects of improved communication and transport technology, which had initially impacted the economy, the circulation of goods, and labor migration, also began to affect the so-called "hedonizing technologies" on a large scale.[39]

A striking example was the success of the famous *Club Méditerranée*, founded in 1950 by a descendant of a Belgian diamond dynasty, sportsman and yogi Gérard Blitz. He had a fine nose for the latest contemporary trends in upscale mass tourism and his Club Med was all about "Freeing people from their restrictions, allowing them to get together, recharge their batteries and return to basic pleasures: happiness by Club Med was born." Its founder summarized the club members' particular "spirit" more succinctly: "Our purpose in life is to be happy. The place to be happy is here. And the time to be happy is now."[40]

The first village to provide this hedonistic escape from the daily grind was built in 1950 in Alcudia on the island of Mallorca. Its improvised tent-like huts proved hugely popular with the several hundred pioneering guests and the idea spread rapidly. Just five years on, the first Club Med bamboo huts were erected outside of Europe, namely on Tahiti. At the time, however, it took four months to get there, enjoy the sojourn, and journey back so that the number of guests—despite the fact that it was already possible to pay in instalments—was necessarily limited. The *Club Méditerranée* was already having to deal with the core paradox of modern tourism, that is, framing travel as an exclusive experience while being commercially forced to upscale in order to keep the "exclusivity" affordable. The demanding tourist's desire for something special and the industry's simultaneous effort to democratize travel created an odd but perhaps useful tension that energized a ceaseless search for affordable novelty.

Another contradiction resides in the concept of modern travel: escaping from what is subjectively perceived as the increasing "automation" of everyday life into an exotic refuge of illusory "authenticity." The irony, however, was that almost every "pristine" experience depended essentially on technological arrangements and infrastructures borne by a kind of "gentle imperialism." Incrementally, these "tentacles of leisure" crept into every nook and cranny of the world, opening it up and making it accessible and comfortable for tourists. This included new transportation infrastructure by ship, railroad, automobile, and especially by airplane; a dedicated hotel industry and restoration technology; entertainment, leisure, and sports facilities; as well as communication technologies that enabled travelers to keep in contact with home. Specialized charter companies offered holiday packages, which included transport, hotel accommodation, and meals.[41] Even cultivated idleness in the shape of "chilling," wellness, and entertainment was (and is) predicated on the silent functioning of carefully-disguised technologies that run in the background.

But leisure is not only about relaxation or laziness. For some, leisure is the ostentatious "experience" of modern or sometimes nostalgic technologies like airborne sports, motorboats, sailing, motor sports, mountain biking, and thrilling fairground rides. In the course of the last century, such technological tourist sublimes have become an abiding feature of leisure activity. They are also unusually amenable to being photographed and filmed, enabling travelers to relive past exciting experiences at a later time. This, too, reminds one of nineteenth-century expeditions: the more unfrequented and exotic the location, the more attention the preserved scene was likely to receive. "Tech-talk" among the like-minded and those knowledgeable of both the country and its associated technologies was inextricably linked to one's leisure.

Even the first European pioneer to offer vacation packages was caught on the horns of the dilemma between culturally-required exclusivity and commercially-inspired inclusiveness. After the Paris Exposition of 1855, Thomas Cook began to organize trips to European destinations such as France, Switzerland, and Italy. In the early 1860s Cook began selling his first overseas travel tickets. A so-called *Fremdenindustrie* (tourist industry)—a term that first emerged in Switzerland—came into being, that offered tourists almost as many creature comforts as they had at home.[42] In addition, and perhaps paradoxically, travelers generally expected to experience things they would not encounter in their usual environments: living a slower, less "modern," and more carefree life; being surrounded by what was perceived as pristine nature; escaping the strictures of middle-class life (as well as escaping one's compatriots); educating oneself in art, languages, or sports, and attending cultural events; gravitating towards unknown and unfamiliar exotic and erotic experiences; and, lastly, pursuing improved health. This bundle of motifs slowly set Europe in motion and, in the course of the twentieth century, increasingly at a global scale. While mass tourism piggy-backed on successive emergent technologies like steam ships, railroads, cars, and airplanes, it also demanded specifically-new infrastructures, for instance sleeping cars, restaurants, and grand hotels.[43]

Modern mass tourism brought Europeans together as it enabled them to go abroad, turning foreign travel into a routine experience for people of all social classes.[44] Another peculiar paradox of tourism has not changed since the early nineteenth century: the idea of being able to set foot in another place and experience something different while at the same time expecting to feel safe and at home almost anywhere in the world. A tourism industry emerged which managed to wed new technologies to venerable human desires and curiosity. Furthermore, tourism stimulated a new look at the safety and reliability of travel conditions: nowhere have debates on "standards" (serving as a reference point for the level of comfort offered by hotels, the diversity of entertainment, cleanliness and hygiene, and so on) been so ubiquitous as in the travel business. This has created a huge market for interpreters, cultural intermediaries, tour guides, guidebooks, and standardization. Nowadays, the standards developed by Europeans, Americans, and Japanese in the course of the nineteenth and twentieth centuries are labeled as "international" and effectively recognized all over the globe: air-conditioned rooms, TVs that broadcast CNN, flushing toilets, showers, exercise rooms, and restaurants that offer a globalized cuisine. From within the confines of a hotel room, there is no longer any way of telling whether it is situated in Europe or anywhere else on the globe.

Club Med has taken a totally different tack: while exploiting the most exclusive locations in the world, it also camouflages the downsides of tourism with remarkable cunning. The vacationer is not even to be reminded of the fact that tourism is a for-profit business: in the club villages, money is replaced by *bar beads* as currency. Moreover the company's award-winning ads—addressing the global awareness of its "informed" guests—emphasize precisely those elements that nowadays would be considered distressing if not kept hidden. Playing on the image of ugly tourist ghettos on southern beaches, exemplars of the disastrous aesthetics of mass tourism, the club's village resort concept promises facilities that blend with the location and harmonize with the surrounding landscape, without sacrificing visual sophistication. As regards the ecological ramifications of tourism, the club asseverates ostensibly scrupulous and environmentally sound practices, particularly in the form of "eco nature" resorts. Additionally, it counters the accusation that its members are passive consumers in an exotic

environment by insisting that they are athletes and, since 2003, also multiculturalists. This makes the self-perception of Club Med-ers of a piece with jet setters and top achievers, those who tend to live "transnational" if not "globalized" lives.[45]

The first permanent club village was established in Agadir, Morocco, in 1965. Pallid north-Europeans could now book this sun-soaked North African resort all year round. As the first generation of club members grew older and started raising families, special childcare facilities were introduced. And from the 1960s on, with the opening of Fort Royal Village on Guadeloupe, Club Med also started to recruit American customers. The village chiefs were thus finally forced to bow to the *lingua franca* of international tourism—English. In subsequent years, additional resorts were built in all parts of the world: on Bora Bora, in the Bahamas and Brazil, Mexico and the Maldives, New Caledonia and Saudi Arabia.[46] Competitors soon appeared of course: in 1966 *Club Soleil* was founded, followed in 1970 by the Robinson Club and in 1973 by Club-Aldiana, all of which promulgated specific, individual "philosophies of leisure."

Though different in character, adventure parks serve as an equivalent escape for a less affluent clientele. Modeled on Disneyland, they often present their visitors with exotic surroundings and offer brief but intense experiences in the shape of "fun," "action," and "events." Such theme parks advertise themselves as concentrated mini versions of the *Gegenwelten* (alternate worlds) cultivated so artfully by *Club Méditerranée* and other tour operators.[47]

Since 1945, mass tourism in Europe developed into a prominent economic sector, as it became a signal feature of modern industrial nations and a token of cosmopolitan sophistication.[48] Its common denominator is its temporal, spatial, and socio-cultural boundlessness.[49] Together with the global accessibility facilitated by the mass media (and nowadays Internet) travel has enabled Europeans to adopt the more congenial elements of exotic lifestyles, elements they can "identify" with: Eastern spirituality and the search for meaning, Japanese manga, henna from India, tattoos and piercings from the Pacific region, judo and taekwondo from East Asia, surfing from Hawaii, or kayaking as practiced by the Inuit.[50] For many Europeans, travel has become the hallmark of a good and rich life, particularly for older people and in remote and exotic regions. The pioneering spirit of European expeditions in the nineteenth and early twentieth centuries has been reincarnated in the guise of modern and especially "alternative tourism"; yet this is not a simple imitation but a new individualized mode of travel that exploits radically new technologies of transport, communication, and creature comfort. Jules Verne's fictional hero Phileas Fogg astounded his contemporaries in 1873 by traveling around the world in 80 days, making use of all infrastructures available at the time; nowadays it is possible for almost anyone to reach virtually any place on earth within 24 hours.

Conversely, it has become an increasingly attractive development option for poor countries outside of Europe to attract tourists from the wealthy industrialized countries by investing in tourist infrastructure. Given budget charter airfares, the main things that need doing are the provision of hotel capacity and some form of remunerative packaging of the local culture. Safely ensconced in a familiar technological environment, the "alternative" or "mass" tourist participates in local traditions and customs that are more often than not "aestheticized" so as to excite but not upset or bore the international audience. In their globalized form, they only partly reflect actual local practices.

In these ways, modern mass tourism has established new relations of exchange and new dependencies between Europe and the rest of the world. And at the same time it has also created new vulnerabilities: everyone now knows how easily a plane can be hijacked, or how little it takes to carry out a terrorist attack in the relaxed atmosphere of dream holiday locations such as Egypt, Tunisia, or Bali.[51] For the tourist infrastructures of Shangri-La demand a leap of faith—a psychological investment which makes the traveler vulnerable in proportion to his or her desire to push aside all unwelcome thoughts at these oases of relaxation.

Hence, in the Club Med resort Les Boucaniers on Martinique everything is attuned to the passionate goal of leaving everyday life behind. Under coconut palm trees and in crystalline underwater worlds, the place offers "Caribbean insouciance" and a "Creole love of life." Leisure activities are "personalized" according to the latest vogue. All-inclusive is the word: meals and even tips are included in the lump sum fee.[52] However, the guests are not the only relaxed beneficiaries of what Napoleon once scoffingly called France's "Confetti Empire": Martinique itself has been transformed into an integral part of Europe and its inhabitants have adopted the inclusive entitlement mentality appropriate to European welfare states. When in 2010 the French state held a referendum to assess popular opinion on transforming Martinique into an overseas territory (instead of a Department), 79 percent of the island population voted against it. Despite the fact that the island has bred some of the most passionate critics of European colonialism, such as René Maran, Aimé Cesaire, and Frantz Fanon, the majority of Caribbean inhabitants prefer to enjoy the benefits bestowed by close links to France, including subsidies from the European Union. One third of the 400,000 inhabitants depend on welfare payments, another third are employed by the state.[53] Only a small proportion of the remaining citizens work at the local *Club Méditerranée*.

Startup-Nation

In 1961, *Club Méditerranée* opened a new village in Arziv, a town in northern Israel. In view of the fact that the country had existed for just 13 years, that hostile neighbors surrounded it and that it commanded almost no natural resources, Club Med's initiative was surprising to say the least. In the flood-prone Netherlands, there is a proud proverb: "God created heaven and earth, except the Netherlands, which was created by the Dutch themselves." Israel can justifiably make the same claim, though it was wrested from the desert rather than the ocean. Modern technology transformed much of Israel's territory from a desolate countryside into an exemplary "Old New Land," as the Austrian writer and pre-eminent Zionist Theodor Herzl had titled his 1902 novel. He expected that clever government and the implementation of technology would enable the imagined "Jewish State" to resolve its economic as well as social challenges.

In greater measure even than in other "colonies," European settlers in Palestine depended on new technologies to master the country's uniformly hostile nature. Irrigation at numerous scales was absolutely essential for the survival of the new population, and failure was not an option. Nobody in Israel could afford "White Elephants." Unfortunately in the interwar period European Jews had largely been denied access to technical knowledge and expertise, and so this had to be acquired *in situ*, in the Land of Zion itself. Help came from Europe, though not without the intercession of European Jews. The "Aid Society of German Jews" (*Hilfsverein*

Fig. 7.6 Connecting an Emerging Nation: *Inauguration of Tel Aviv landing ground in September 1938 by Palestine Airways Ltd. The 5–6-passenger Short Scion monoplane shows inscriptions in Arabic, English, and Hebrew. Zionist Pinhas Rutenberg not only founded this airline, but also what later became the Israel Electric Corporation.*

deutscher Juden), for instance, was able to win over Emperor Wilhelm II to the idea of founding a technical college in Palestine, one that was actually built between 1912 and 1924 as the *Technion* on Mount Carmel near Haifa. In various countries in Europe *Technion* societies were founded to support this venture. As chairman of the German *Technion* Society in 1923, Albert Einstein personally planted two palm trees in front of the first buildings. The *Technion* developed quickly. By 1954 a department of aerospace engineering was established, and by the late 1960s a faculty of computer science was opened.

To date the *Technion* has produced three Nobel Prize laureates, and its surrounding technology parks are home to internationally-renowned IT companies. The state of Israel invests more money in military and civil engineering research than most other countries. The flash drive, the mobile mailbox, the chat program ICQ, and the Centrino computer chip were invented in Israel, in addition to leading medical technology and pharmaceutical products.[54] At the New York based Nasdaq stock exchange for technologies there are more companies from Israel than from all of Europe together:

> There's a fight taking place among some European capitals to be the *primes inter pares* of tech hubs. London, Berlin, Paris, Dublin are

all trying to claim the throne. But there is already an occupant, and it is unlikely that any of the pretenders are going to dislodge it: Tel Aviv. Israel is the start-up nation, and Tel Aviv is its hub.[55]

This assessment in the *Wall Street Journal* of January 27, 2012, is not just remarkable because it casually defines Israel as a European country. That point was already obvious from Israel's participation in the "Eurovision Song Contest" or in the EU's Eureka program. More startling is the point that Israel has in no time become one of Europe's leading technology innovators. Pundits on economic affairs have suddenly revealed the startup-nation Israel to be a model case of an innovation culture: in the analysis a permanent state of emergency and the implacable threat to national existence are coupled to the workings of a vigorous free market. Both endow Israelis with an entrepreneurial proclivity for risk-taking and improvisation, not to mention chutzpa, team spirit, and a certain obstinacy. Moreover, compulsory military service leads men and women from Israel to assume responsibility comparatively early in life.[56]

The history of Israel brings several strands from preceding chapters together. Leaving aside specific religious aspects and also the Holocaust as particular catalysts of Israeli nation building, we argue that the country exhibits several typical features of a European settler colony. In the course of time, it has developed into a multicultural immigration society, and in the process manifested an inordinate vitality. In many ways, Israel pioneers a globalized society with all its potentials and dangers. European influences are still dominant, but they have become diluted and increasingly amalgamated with other traditions. Some Israeli traffic lights still remind one of the fact that a major part of German reparations paid after the Holocaust consisted of technical installations *in natura*. The "White City" in Tel Aviv with its Bauhaus style buildings from the 1930s and 1940s, a Mecca for architecture aficionados from all over the world, is a purely European hangover from the early stages of Jewish settlement.[57]

Two additional preconditions for the Israeli takeoff were, firstly, the continuous influx of highly qualified "human capital"—since the collapse of the Soviet Union many mathematicians and engineers have emigrated from Eastern Europe to Israel—and, secondly, foreign capital, most of which came from Europe and the United States. Israel maintains close ties with "Western" countries, comparable to Commonwealth countries; both have interests in sustaining a lively exchange of technology and information. Scientists from Israel commonly spend parts of the year at European or U.S. universities. The constant threat from all directions encouraged the development of numerous companies specialized in IT security. The omnipresent Israeli army, devouring almost 20 percent of the gross national product since the 1960s, has created high-tech production facilities that produce impressive spinoffs for civilian use. It is an unverified certainty that Israel, in addition to its nonpareil antimissile defenses, also commands atomic weapons. As is the case in China, India, or South Africa, an arms industry that includes an atomic sector not only serves as an ostentatious display of national independence and invulnerability, but also is a salient contribution to an export industry that challenges the United States, Europe, and Russia on world markets.[58]

Despite its very limited natural resources, Israel has brought its agrarian and industrial sectors to improbable heights of productivity. With respect to food it is, with the exception of beef and grain, almost entirely autarchic. Israel has strong capacities in oil refining, in diamond cutting, and the fabrication of semiconductors. It exports jewels, high technology, military supply, software, drugs and medicine, fine chemicals, and agricultural products like fruits, vegetables, and flowers.

Israel has evolved in a way resembling other onetime new European settlements like Canada, Australia, or New Zealand. Colonizers encountered what seemed to them a sparsely populated country, which was seen as almost "natural" and "empty." This encouraged a rapid colonization in which the new settlers took little heed of people already living there, more often than not driving them from their lands by force. Having achieved agrarian autarchy, an efficient, highly industrialized and export-oriented agribusiness evolved. To ensure (military) independence, this was complemented by industry and high technology. For Israel, however, the initial conditions as well as the political environment are particularly challenging, and this has made the Israeli nation and its people extraordinarily self-assertive.

Despite enormous internal tensions that have arisen among the diverse cultural and religious groups comprising its immigrant citizens, Israel is still defined as a bridgehead of "the West" by its Arab neighbors. The Cold War contributed to this impression.

Initially Eastern European countries supported Israel's development—Czechoslovakia, for instance, delivered arms and trained Israeli pilots until 1948—but the Soviet Union terminated the programs and obliged its satellites to support the Arab countries instead.[59] The more energetically Israel manifests itself, the more occasions there are for conflicts. Israeli irrigation, for example, has especially raised the ire of its near neighbors and become a transnational problem for the whole region. Dams in the river Jordan not only generate electricity, but also impound precious water.[60] Novel methods of efficient irrigation or of desalinization have been developed to tackle the notorious water shortage, a chronic bone of contention that time and again leads to tensions with neighboring countries like Syria or Jordan. In the Near East, the war for water now seems to be replacing the older war for oil. It is one of the material substrates of the Mideast conflict, which has been smoldering for decades. One wonders at the accuracy of Theodor Herzls 1902 description of a highly-engineered nation in the future, in which it might eventually be possible to eliminate frictions like these by means of technology.

The current state of permanent tension, however, prevents Israel from capitalizing on one of its greatest potential assets: becoming a startup-nation for the hospitality industry. Conditions are highly favorable due to its many touristic highlights and its intriguing history. For Europeans, the country and its culture illuminate crucial features of their own identity: religious and cultural origins, formative experiences in the context of the crusades, the fruitful as well as problematic relationship, and finally the fatal and almost successful attempt at the "industrialized" annihilation of Jews during the Second World War. Since 1948, Israel has been a strategic ally of "the West," but it also suffers from unfinished European business in the Near East region since the demise of the Ottoman Empire, not the least of which is the unresolved question of where and how the Palestinians should live. This tragic bequest is definitely at odds with a flourishing tourist industry. In 2007, the Club Med in Arziv was finally forced to the same conclusion and closed its gates.

Epilogue: Europeans Globalizing

In the Introduction to this volume we posed a number of research questions. Now, seven chapters later, it is time to take stock and see what kinds of answers we have been able to provide. Of course, the answers themselves are embedded in the many twists and turns of the preceding text, so the best we can do here is to provide a concise resume of what we have accomplished there. The stories told on the preceding pages hardly exhaust the rich variety of interactions between Europe and the rest of the world, despite their complexity and distinctiveness. Nonetheless, we consider them representative and revealing, especially insofar as they emphasize the key role of European technological prowess since 1850.

However, in contrast with more heroic narratives, we have made every effort to place traditional assumptions about European dominance into perspective. We chose to focus on stories and episodes that are usually marginalized when writing about Europe or writing about technology. On this count it would have been advantageous had we been able to tell more stories from the vantage point of non-European objects of colonization. However, such testimony is sparse and not easy to come by. Despite this lack, we feel we have been able to argue convincingly that technology was a decisive field of interaction between Europeans and non-Europeans, rather than some kind of one-directional imposition or diffusion. That, if anything, is the leitmotif of this volume.

This is not to say that Europeans themselves were wont to see things this way. During the nineteenth and early twentieth centuries, the mastery of science and technology were vital to Europe's

Fig. 8.1 A Hymn to Technology: *The 1900 Exposition Universelle held in Paris was a celebration of progress and modernity based on European science and technology. The fair, visited by nearly 50 million people, displayed impressive new machinery such as the Grand Roue de Paris (a 100-metre tall Ferris wheel), diesel engines, talking films, escalators, and the telegraphone (the first magnetic audio recorder). The Palace of Electricity was the heart of the exhibition and was fitted with five thousand multi-colored incandescent lights, eight monumental lamps, and a set of steam driven dynamos. The World's Fair included various exhibitions of French colonies designed to show African primitiveness versus French culture and technology, thus legitimating the civilizing mission.*

self-conception. Flushed with their impressive achievements in the field of material culture, many Europeans conceived of themselves as privileged agents of historical progress. European technology gradually evolved from its own proper realm—*techne*—to become a new *episteme*, a means to understand and evaluate the world. From the sixteenth century onwards, European technology became not just a question of *doing*, it also became a new dominant mode of *thinking*. In this sense technology not only provided the building blocks for human social evolution but also embodied the concept of progress as conceived by Immanuel Kant and Claude Henri de Saint-Simon, that is, as a cumulative process of civilization that never ceases. By contrast, outside of Europe technology by and large remained confined to a *techne*.

Since early modern times, technologies like shipbuilding and skills like navigation, soaring to great heights thanks to the incessant and intensive circulation and accumulation of knowledge, enabled Europeans to explore, map, and open up remote territories relatively cheaply and with tolerable risks. In the Age of Imperialism between 1850 and 1940, a drive to make remote regions accessible and to exploit their material and human resources went hand in hand with the conviction that European technological mastery should be extended to every corner of the globe.

The technology of steam power, for one, took Europeans almost everywhere in the world. Their mastery of other technological domains allowed them to survive there: for example medicine, earth-moving equipment, and building and urban technologies. It was technology that enabled relatively-small numbers of Europeans to conquer and to control huge territories using rifles, machine guns, railroads, telegraphy, and so on. It was technology of extraction that enabled Europeans to exploit resources and to develop agricultural or industrial production in the colonies. "Modern" technology and the attendant idea that it should be continuously optimized and implemented became the very heart of European colonialism and European identity in colonial contexts. It was utilized to map non-European territories, and in an exploitative mode, to expand and control the flow of humans, resources, goods, and information.

At the same time, colonial territories were used and abused as experimental settings by metropolitan engineers and as field laboratories for European scientists. Enhanced professional or political status were added perks for European technical experts with experience in colonial technology and science. Moreover, colonial territories sustained a very dynamic job market that supported the circulation of technologies, experts, and expertise both between Europe and its colonies and among colonial powers. A number of scientific and technological institutions were created to support this worldwide dissemination of science and technology, the most impressive being Imperial College in London and the *Institut Colonial International* in Brussels. State administrations themselves were reorganized in order to accommodate burgeoning corps of colonial engineers and physicians. In some cases it was hard for these technicians—as it sometimes was for missionaries—to distinguish between colonizing Africa and Asia and colonizing their own European motherlands. In both cases, they must have felt like "mandarins" of a future characterized by a continuous and inevitable process of *modernization*.[1]

New technologies, new patterns of mobility, and diversified market mechanisms connected previously-isolated economies across continents and oceans, enabled the rapid circulation of information and inspired fantasies of a new world order. But utopian visions by nineteenth-century liberals of a peacefully globalized market economy sustained by untrammeled flows of people and goods never materialized. Globalization proceeded in anything but a linear fashion, it was contentious and exploitative in the extreme and the "global condition" brought as many uncertainties for Europeans as it did for non-Europeans.[2]

A key point was that technology and knowledge transfer were far from unidirectional. Many non-European cultures also commanded advanced technologies and these often proved much better suited to local conditions or even functionally superior to imported technologies. Where this was the case, Europeans often hybridized their own technologies with native elements. Conversely, "Western" technologies and skills, wherever implemented, were creatively appropriated by native populations and adjusted to local usages and needs. These "creole technologies" were often more successful than technologies that were simply transplanted in an unaltered state. In contacts between Europeans and non-Europeans something technologically unexpected was more often the rule than the exception. While some nations and individuals recognized the new opportunities and sought to pursue them as cultural brokers, intermediaries, or interpreters, others shied away or felt threatened by the unknown. This duality in response to new technologies has remained a stable feature of the dynamic of globalization from 1850 until now.

But it is beyond doubt that since the late nineteenth century new technologies in the fields of steam, railroads, and armaments delivered the tools for what then was called the unification of the globe and what later on became "globalization." And it is certainly true that in the majority of cases, the process was initiated and driven by Europeans. However, it developed in more complex ways than Europeans had intended and often with startlingly-unexpected outcomes. The process resembled mutual formatting much more than a mere rubber-stamping of European technologies across the globe.

In the "modern" era, technological innovation was often glorified as the high road to peaceful development, alignment, and unification, and this was supposed to apply to relations among European nations as well as to non-European territories. In practice, however, new technologies tended to reinforce rather than mitigate existing societal imbalances and asymmetric power relations—in

Fig. 8.2 Installation of First Overland Telegraph Pole in Darwin, Australia:
Although the photograph was taken on September 15, 1870, to immortalize a technical event—the installation of a telegraph pole—it goes far beyond it by portraying Western civilization itself: in the middle of the Australian wilderness, a group of white settlers, dressed as if they were in a small European village, embodies the nineteenth-century values of the mission civilisatrice *and clearly associates technology to civilization.*

colonial settings to an even greater extent than in Europe—with some people being connected to the developing global networks and others being left out. The resulting conflicts could generally be quashed by means of advanced technologies of control (for example, firearms or airplanes). But they could also be mitigated or exacerbated by the introduction of other technological twists. Thus, global tensions themselves created a new landscape for the creation of new technological forms and functions. Surfing on the waves of both cooperation and rivalry, these new global technologies acquired an irresistible vitality and pervasiveness. After the Second World War cosmopolitan science and technology inspired a new world culture, shaped around global infrastructures of traffic, communication, tourism, and trade. Optimized logistics accelerated flows of goods and people, imposed new time regimes, and produced a cornucopia of widely-standardized services.

By the turn of the twentieth century, the pre-eminence of Europe—or at least "the West"—in the technological transformations of the past 50 years quite naturally led Europeans to see themselves as having a cultural, if not a racial, predisposition for successful competitive science and technology. There was even talk of a European "technology race." Of course, imperialism and colonialism were predominantly national endeavors, but in many respects colonizers felt themselves to share a common Europeanness, possessing common knowledge, skills, and dispositions that set them off from their surroundings, the colonized "others." Consequently, Europe's highly-industrialized countries felt themselves able and even obliged to "civilize" the rest of the world, in the first place to globalize "superior" European technology. The religious mission had been replaced by the material mission of elaborating colonial infrastructure: building railroads, telephone lines, and electrical networks together with the detailed knowledge and expertise that enabled its use: writing and education, scientific knowledge, organizational and management skills.

Even prior to the First World War, Europe had already felt herself threatened by new arrivals on the global scene. American as well as Asian competitors challenged European hegemony on many different levels. This, however, was hardly a novel situation. It was in effect a restoration of the reciprocal interaction that had existed prior to industrialization, and that had merely been suppressed by the technological assaults of the colonial era. Subsequently, throughout the twentieth century the tremendous dynamism of science and technology in Europe continued unabated, despite extreme variations in the political landscape. The shocks of the First and Second World Wars unhinged Europe and weakened her global influence. In the aftermath her colonial empires were gradually amputated, leaving a few of the more vulnerable Europeans to suffer "phantom pains."

As they achieved independence, the formerly colonized peoples eagerly seized European technologies (such as telephone lines, traffic facilities, broadcast and television infrastructure). These technologies were widely seen as a valuable asset after the European colonizers had gone home. New elites embraced these technologies and developed them further: for example, by building power plants and transmission lines, modern cities, and better transport networks. And although the enjoyment of European material artifacts was initially the prerogative only of the post-colonial elites, bit by bit broader segments of post-colonial populations came to covet Western goods as well—even though this was often ideologically masked by creative adoption and labeling. In a broader sense, too, European technology and knowledge inspired the former colonies to manage and organize themselves along European lines, at least in theory: with well-defined borders, modern governmental and administrative institutions, and the rule of law. Of equal importance was the appropriation of "low tech," even obsolete, technology (by European standards). In the hands of Africans, Asians, or South Americans, such technologies experienced a second life in new, creative, and sometimes unexpected social and economic contexts.

This rebirth of "old" technologies challenges traditional conceptions of innovation and reveals global peripheries as active sources of innovative usage, thus introducing a new set of actors into the technological drama. Meanwhile, of course, the European post-colonial agenda, like the colonial one, also had a strong technological dynamic—though now under the aegis of "exchange" rather than outright "exploitation." The majority of the cooperative projects undertaken by European countries and their former colonies were and are technological projects. Even today there is a shared Western rhetoric of "underdeveloped countries" and of highly-engineered "centers"—meaning in effect: the West versus backward "peripheries" branded as the "Third World." The widespread disbursement of development aid is in some ways an antediluvian remnant of the asymmetric balance of power that marked the colonial era.[3]

This volume has traced different modalities of interaction between Europe and the rest of the world: ecology and agriculture, medicine, interconnections of a political, social, and cultural sort, travel and migration to and from Europe, labor migration, and the ascendency of a world market. Throughout this history, European expansion, long-distance trade, and the international division of labor not only spread Europeans and their technologies across the face of the globe, but also accelerated the import of goods, people, and problems into Europe itself. Globalizing forces inevitably evoked counter-movements and backlash reactions. Nationalistic and even chauvinistic worldviews, for instance, always challenged that which came from outside Europe, and often insisted on raising barriers in order to preserve a homogeneous national or European culture, economy, or even race. So even while Europeans were globalizing they were constantly faced with challenges to their own identity. Euro-centrism was the most common reaction: a distinct historicism and the notion of progress and permanent economic growth. In addition, there was a strong compulsion, bordering on the paranoid, to situate Europe favorably on cartographic and mental maps—generally right in the center with the rest of the world ranged below or to the side.

This was not the least of the reasons for widespread anti-European sentiments: revolutionary and post-colonial intellectuals notoriously criticized Europe's obsession with itself as the center of the universe. Mahatma Gandhi or Jalal Al-e Ahmad diagnosed "Euromania" or "Westoxication" as an illness induced by Western technology and, like the Trojan Horse, capable of infiltrating traditional cultures. At the same time, they were at a loss to find ways to resist the hidden agenda of Western technology. Religious fundamentalism was just one of the possible—though certainly quite endemic—answers, often being mixed with an "Occidentalism" that characterized "the West" as materialistic, decadent, and morally corrupt. The glorification of traditional cultural orientations and reliance on "homemade technologies" were also mobilized as a way of resisting European influences, but in the end none of these responses provided adequate bulwarks against the seductions of Europe's technologies. Almost all attempts to isolate specific cultures or nations from the global circulation of scientific knowledge failed. The promise of "Western" technology proved a lasting temptation; especially *getting connected* to and *being integrated* into the global networks of knowledge.

In this respect, Europe and the material effusions of its lifestyle were as much of an "irresistible empire" as was the United

States with her ostentatious consumer culture. Technology acted as a pre-eminent agent of change and contributed at the most basic level to the shared history of Europe and a globalizing world. The dialectical process of "Europeanizing" the world and "globalizing" Europe was both reciprocal and tortuous. By the mid-nineteenth century, as the world became a European *Terra Nostra*—mastered, exploited, and linked by European technologies—Europe became fundamentally internationalized and receptive to exotic influences. But Europeans could also be reluctant, xenophobic, and even dismissive if the changes came too quickly, were too overwhelming, or appeared to challenge European self-esteem.

In the decades after 1850, the idea of global governance and the rule of international law became ever more widely institutionalized. These ideas had originated in efforts to moderate European conflicts in the Early Modern Era, and they were subsequently enacted in even the remotest regions in order to maintain peace. From its very beginning, international law not only displayed a strong European bias but it also showed a strong tendency to materialize in the form of technical norms and standards, that is, as a Eurocentric "Mechanics of Internationalism." However, these standards were hardly unitary and the worldwide competition among divergent European systems like the British and the French is certainly a key to any imaginable world history of technology in this period of global networking. After the First World War the United States became increasingly influential as a standard-setter, and some Europeans began to fear immanent Americanization. But in reality the U.S. did not dominate international standardization processes until after the Second World War.

While the League of Nations of the interwar years had in effect been a European effort to moderate world politics, the post-Second World War United Nations confronted Europeans with a variety of established and emergent nations (the Global South) under the aegis of two rival superpowers in a truly global effort to manage the perilous new Cold War constellation. As far as Europe itself was concerned, its jump-started postwar reconstruction soon enabled it to resume its former leading position in research and development and in technological mastery—though now clearly tied to the apron strings of the two superpowers. Eastern and Western Europe became de facto satellites and constituted spheres of influence and of collaboration in military and civilian engineering. In practice the highly-dynamic and successful Western sphere quite outpaced the rather slower Eastern one, although it is important here to distinguish consumptive sectors from military and high-tech sectors. In the 1970s, the "shock of the global" impacted on the First, the Second, and the Third Worlds alike.[4] From the 1990s on, the European Union has been seeking to re-establish Europe as a global player, by forging an "innovation union" to meet the challenge of new competitors that have appeared on the global scene.

In the course of 150 years, Europe's protean technologies inspired and underpinned the globalizing ambitions of European nations. We have aimed to show how technology mediated European influence in the rest of the world and how this mediation in turn transformed Europeans. It is clear that the variety of peoples and nations that Europeans encountered was more than reflected in the variety of the outcomes of those encounters. Europeans mapped, they exploited, they exchanged—their interactions ranged from technological and biological genocide to treaties of cooperation and the construction of elaborate colonial infrastructures. It is clear that this staggering variety of technologically-inspired and -enabled interactions over a period of more than 150 eventful years—in the understanding that technology is not only machines, products, systems, and infrastructures, but also the skills, knowledge, and *habitus* that make them work—resists any facile, singular explanation or conclusion. This book is therefore an invitation to other researchers to continue from here.

Quite aside from the enormous variety of political settings, cultures, and colonial programs, we have seen that European interventions in exotic places were unusually vulnerable to being derailed by unintended consequences, which affected all the actors involved. Nonetheless, once established, interrelations created dependencies on both sides. Cultural transfers were rarely unidirectional, and often a kind of Pidgin-knowledge emerged, a hybrid fusion of European and local knowledge and skills.[5] But this very hybridity could once again enhance the feeling of insecurity and foster anxieties.[6] As observers have rightly pointed out, Europe played both the role of "Prometheus unbound" *and* the "Sorcerer's apprentice." Or referring once more to the satirical cartoon of the "civilizing steam engine" that we started with: its blueprints are now more complex than ever, and it is less than ever being operated by Europeans alone.

Endnotes

Introduction

1   J. Lander and R. Lander, *Journal of an Expedition to Explore the Course and Termination of the Niger*, 184.

2   Words by the Portuguese engineer Joaquim José Machado, responsible for the assessment of the Lourenço Marques railway line in Mozambique. See Machado, "Memória ácerca do caminho de ferro de Lourenço Marques à fronteira do Transvaal," 23.

3   Hecht, "Technology, Politics, and National Identity in France."

4   Van Laak, *Imperiale Infrastruktur*.

5   Basalla, "The Spread of Western Science."

6   Headrick, *The Tools of Empire*; Latour, "Drawing Things Together"; Law, "Technology and Heterogeneous Engineering."

7   Gavroglu et al., "Science and Technology in the European Periphery."

8   Carneiroe et al., *Travels of Learning*.

9   Subaltern Studies were founded by Ranajit Guha and a number of collaborators, including David Arnold. They proposed an approach to postcolonial studies based on Antonio Gramsci's concept of "subaltern." Postcolonial Studies addresses the topic of post-colonial identity. The most visible representatives are: Frantz Fanon (colonialism as a source of physical and mental violence), Edward Saïd (the concept of orientalism and the Western misrepresentations of the non-Western), and Dipesh Chakrabarty (the concept of "provincializing" Europe in the sense of placing it on an equal level with other provinces).

10 Howe, *The New Imperial Histories Reader*.

11 Pratt, *Imperial Eyes*, 6–7; Raj, *Relocating Modern Science*, 232–33; Szasz, *Between Indian and White Worlds*; Bénat Tachot and Gurzinski, *Passeurs Culturels*, xi–xii; Middell and Engel, "Bruchzonen der Globalisierung," 15. Misa and Schot, "Inventing Europe," 10; Middell and Naumann, "Global History and the Spatial Turn."

12 Edgerton, "Creole Technologies and Global Histories."

13 Ambirajan, "Science and Technology Education in South India."

14 "Bhadralok" is a Bengali term used to classify the upper and middle classes that gained power during colonial times, composed mainly of English-educated Indians living in Bengal.

15 Letter of the Chief Engineer of the *Companhia Real dos Caminhos de Ferro atravez d'Africa* (Royal Railway Company across Africa), 1888.

16 Machado, "Memória ácerca do caminho de ferro de Lourenço Marques à fronteira do Transvaal," 5.

17 Using Victoria de Grazia's book title, *Irresistible Empire*.

18 Noble, *The Religion of Technology*.

19 Headrick, *The Tools of Empire*.

20 Using as inspiration of Dipesh Chakrabarty's concept in *Provincializing Europe*. See also John MacKenzie's and Andrew Thompson's "Studies in Imperialism" series and their general concept of "imperialism as a cultural phenomenon had as significant an effect on the dominant as on the subordinate societies."

21 Galison, *Einstein's Clocks, Poincare's Maps*, 283.

22 Adas, *Machines as the Measure of Men*.

23 Mudimbe, *The Invention of Africa*, x, 189–90.

24 Le Vaillant, *Voyage de M. Levaillant dans l'intérieur de l'Afrique par le cap de Bonne Esperance dans les années 1780, 1781, 1782, 1783, 1784, 1785*, as quoted from the English version, Vaillant, *Travels into the Interior Parts of Africa by the Way of the Cape of Good Hope in the Years of 1780, 81, 82, 83, 84, and 85*, 167–68.

25 Barth, *Travels and Discoveries in North and Central Africa*, 202.

26 Stanley, *In Darkest Africa*, 38.

27 Quatermain was loosely based on the very real British game hunter and conservationist Frederick Courtney Selous.

28 In an initial phase of conceptualization, Matthias Middell was particularly helpful and inspiring, especially with his broad knowledge of international and global history as well as methodology. We owe him very much.

29 This is David Edgerton's term. See his *The Shock of the Old*, xiv.

1 Europeans Mapping & being Mapped

1 Liu Xihong, "Private Aufzeichnungen ueber England" as quoted in Chen, *Die Entdeckung des Westens*, 45. Later Liu Xihong became the first Chinese ambassador to Germany.

2 Ibid., quoted on 141–42.

3 Weisshaupt, *Europa sieht sich mit fremdem Blick*. According to Osterhammel, *Europa um 1900*, about 150 such eyewitness texts about foreigners fabricated by Europeans were in existence by the eighteenth century.

4 See also: Beasley, *Japan Encounters the Barbarian*; Osterhammel, *Die Entzauberung Asiens*; Mungello, *The Great Encounter of China and the West*; Sen, *Travels to Europe*; Chatterjee and Hawes, *Europe Observed*.

5 As quoted in Harbsmeier, "*Schauspiel Europa*," 341. See also: Raychaudhuri, *Europe Reconsidered*.

6 Weiß, "Die Kolonien in Europa."

7 Osterhammel, *Europa um 1900*, 18–19. See also: Carneiro et al., *Travels of Learning*; Gavroglu et al., "Science and Technology in the European Periphery."

8 Osterhammel, *Europa um 1900*, 20.

9 Sarda, *Hindu Superiority*. See also: Fischer-Tiné, "'Deep Occidentalism'?" For a discussion of influences see: Hobson, *The Eastern Origins of Western Civilisation*.

10 Wintle, *The Image of Europe*, 72. See also: Trakulhun, "Kulturwandel durch Anpassung?"

11 Ibid., 71.

12 Böckelmann, *Die Gelben, die Schwarzen, die Weissen*, 54.

13 Zeng Jize, as quoted in Chen, *Die Entdeckung des Westens,*, 142–43.

14 On Otlet see: Rayward, *The Universe of Information*. On Ostwald see: Krajewski, *Restlosigkeit*, 73–140. One could also refer to Vannevar Bush as a pioneer of the Internet.

15 See: Haeckel, *Die Weltraetsel*.

16 Akerman and Karrow, *Maps*.

17 Leitão, "Um Mundo Novo e uma Nova Ciência," 19.

18 Named after Alberto Cantino, an agent for the Duke of Ferrara who smuggled the map from Portugal to Italy in 1502.

19 Adas, *Machines as the Measure of Men*; Conklin, *A Mission to Civilize*; Fischer-Tiné and Mann, *Colonialism as Civilizing Mission*; Barth and Osterhammel, *Zivilisierungsmissionen*; Osterhammel, *Europe, the "West" and the Civilizing Mission*; Costantini, *Mission civilisatrice*.

20 Alatas, *The Myth of the Lazy Native*.

21 Schultz, "Europa: (k)ein Kontinent?"

22 Eckert, "Globalisierung," 11. See also the chapter "East and West: China (Africa, India, Islam) and the transmission of knowledge from East to West," in Selin, *Encyclopedia of the History of Science, Technology, and Medicine in Non-Western Cultures*; Aust and Schoenpflug, *Vom Gegner lernen*.

23 Wintle, *The Image of Europe*, 349. See also: Osterhammel, *Die Entzauberung Asiens*.

24 Bayly, *The Birth of the Modern World*, 476.

25 Chakrabaty, *Provincializing Europe*.

26 Edelmayer, "The 'Legenda Negra' and the Circulation of Anti-Catholic and Anti-Spanish Prejudices."

27 Schmale, "Die Konstruktion des Homo Europaeus"; Baum, *The Rise and Fall of the Caucasian Race*.

28 Lipphardt, "Von der 'europaeischen Rasse' zu den 'Europiden'"; Lipphardt and Patel, "Auf der Suche nach dem Europaeer."

29 Said's findings have generated numerous works defending the West and stressing European achievements. Ibn Warraq, for example, points out that no other culture has any tradition similar to our biblical criticism (historical-critical exegesis). The West alone has remained intellectually open and inquisitive; it has constantly practiced self-criticism as evidenced by demands to abolish slavery, racism, and colonialism and replace them with universal rights and values. (see: Warraq, *Defending the West*). Headley also argues that human rights and democracy are gifts that Europe has given the world (Headley, *The Europeanization of the World*).

30 Downs and Stea, *Maps in Minds*; Schenk, "Mental Maps."

31 Excerpt from the editorial in the *Delhi Gazette*, New Delhi, India, January 5, 1856, as quoted in Christensen et al., *Europe Meets the World*, 277–78.

32 Coudenvove-Kalergi, *Revolution durch Technik*, 18–23.

33 Ibid., 101.

34 As quoted in Kumar, "Reconstructing India," 241.

35 Weber, *Peasants into Frenchmen*.

36 See, for instance, Ruehling, "'Bilder vom Norden'."

37 For example, Jones, *The European Miracle*, or Landes, *The Wealth and Poverty of Nations*.

38 Maurer, "Europaeische Geschichte."

39   For the Scandinavians' (esp. Carl Linnaeus') share in this endeavor see: Soerlin, "Ordering the World for Europe."

40   Wintle, "Europe as Seen from the Outside," 23; Schultz, *Europa als geographisches Konstrukt*.

41   Ingebritsen, "The Scandinavian Way and Its Legacy in Europe."

42   At almost the same time (1539) the archbishop of Uppsala, Olaus Magnus, depicted a separate animated "Carta Marina" of Scandinavia. See: Magnus, *Die Wunder des Nordens*.

43   Schroeder, " 'Leitbild Norden' statt 'Leitbild Europa'?," 199–200.

44   Ibid. See also: Turner and Nordquist, *The Other European Community*, and Hecker-Stampehl, *Vereinigten Staaten des Nordens*.

45   In recent exhibitions ("Europe Meets the World," Copenhagen 2012) and under Danish EU leadership (second half of 2012), it was presented the other way round: the Vikings became pioneers of European expansion, and the medieval Hansa came to be the first common European market. Likewise, but from the Southern perspective, an exhibition on the role played by Iberian science and technology in the making of Modern Europe was held in Lisbon (Calouste Gulbenkian Foundation, Lisbon, 2013).

46   Stråth, "Multiple Europes," 410.

47   Stråth, " 'Norden' as European Region: Democration and Belonging."

48   Tomasson, "Iceland as 'The First New Nation'"; Tomasson, *Iceland: The First New Society*.

49   Schenk and Winkler, *Der Sueden*.

50   Raj, "Mapping," 688.

51   Goettsch-Elten, "Populaere Bilder vom Norden im 19. und 20. Jahrhundert."

52   See: Ehn et al., *Foersvenskningen av Sverige*.

53   Nagel, *Die Welteislehre*.

54   Ross, *Mit Kind und Kegel in die Arktis*, See also: Milde, "Lernen von den Eskimos."

55   Advertising sheet from the 1930s (owned by Dirk van Laak).

56   Ross, *Mit Kind und Kegel in die Arktis*, 198.

57   For one of the rare turns in perspective see: Harbsmeier, *Stimmen aus dem aeussersten Norden*.

58   Ross, *Mit Kind und Kegel in die Arktis*, 5.

59   Ibid., 6.

60   Ibid., 116.

61   Hønneland and Stokke, *International Cooperation and Arctic Governance*.

62   Fogg, *A History of Antarctic Science*; Larson, *An Empire of Ice*.

63   Peterson, *Managing the Frozen South*; Launius et al., *Globalizing Polar Science*; Roberts, *The European Antarctic*. For a history of the South Pole see: http://www.south-pole.com.

2   *Europe's Significant Others*

1   *Welt am Montag*, November 21, 1898, as quoted in Gründer, "*...da und dort ein junges Deutschland gruenden*," 210.

2 Rohrbach, "Die Bedeutung der Bagdadbahn," 805–6.

3 McMurray, *Distant Ties*. For a more recent record of all the political implications see: McMeekin, *The Berlin-Baghdad Express*.

4 Kohn, *Die Europaeisierung des Orients*, 261.

5 Borchart, *Der europaeische Eisenbahnkoenig*; Meyer, *Mitteleuropa in German Thought and Action, 1815–1945*, 76–77.

6 During the 1930s, the Germans opened a Turkish branch of Junker's airplane factory instead.

7 On the notion of "technocratic internationalism" see also: volume 4 of the Making Europe Series, Kaiser and Schot, *Writing the Rules for Europe*.

8 Said, *Orientalism*. See also: Konrad, "From the 'Turkish Menace' to Exoticism and Orientalism."

9 Hanioğlu, *A Brief History of the Late Ottoman Empire*, 138, who adds, 207, that the late Ottoman Empire just barely succeeded in gaining "second-class membership in the European club." Martykánová, *Reconstructing Ottoman Engineeers*.

10 Kaiser, "The Bagdad Railway and the Armenian Genocide."

11 In his ambition to "turkify" the Ottoman society and to elaborate it into a pan-Turkish vision, Mustapha Kemal was influenced by sociologist Ziya Goekalp. See: Parla, *The Social and Political Thought of Ziya Goekalp*.

12 Kreiser, *Atatuerk*, 245.

13 For the role of émigré experts as agents of Turkish change, see: Kohlrausch and Trischler, *Building Europe on Expertise*, 187–95.

14 For the interwar years see: Mangold-Will, *Begrenzte Freundschaft*.

15 Jaeckh, *Der goldene Pflug*, 219.

16 See: discussions in Schaebler and Stenberg, *Globalization and the Muslim World*.

17 Bellaigue, *Patriot of Persia*.

18 From a similar diagnosis of Western technology, Mahatma Gandhi had claimed the spinning wheel to be a symbol of Indian independence.

19 Achmad, *Occidentosis*, 66–70.

20 Ibid., 71–75. For an alternative perspective compare Marjane Satrapi's autobiographic graphic novel *Persepolis* (2003).

21 Conrad, " 'Europa' aus der Sicht nichtwestlicher Eliten, 1900–1930."

22 Buruma and Margalit, *Okzidentalismus*; Mishra, *From the Ruins of Empire*.

23 See: Aydin, *The Politics of Anti-Westernism in Asia*.

24 Conrad, " 'Europa' aus der Sicht nichtwestlicher Eliten, 1900–1930," 164. In China, a comparable mixture was named *ti-yong*. Osterhammel, *Europa um 1900*, 16.

25 As quoted in Schultz, "Europa: (k)ein Kontinent?," 229.

26 Gouldner, "Stalinism."

27 Ibid., 13.

28 Seegel, *Mapping Europe's Borderlands*.

29 Thum, " 'Europa' im Ostblock"; Faraldo et al., *Europa im Ostblock*.

30 Lemberg, "Zur Entstehung des Osteuropabegriffs im 19. Jahrhundert."

31 Wolff, *Inventing Eastern Europe*; Neumann, *Uses of the Other*.

32 See: Bremer, "Der 'Westen' als Feindbild im theologisch-philosophischen Diskurs der Orthodoxie." See also: Bracewell and Drace-Francis, *A Bibliography of East European Travel Writing on Europe*; Gouzevitch, "Les corps d'ingénieurs comme forme d'organisation professionnelle en Russie."

33   Troitskiy, "Statue of Liberty Made of Russian Copper?"

34   Laurelle, "Die europaeischen Urspruenge des Eurasianismus."

35   Schenk, *Mental Maps*, 502.

36   Beljakow, *Elektrizitaet rings um uns*, 55.

37   Ibid., 80.

38   Salay, *The Soviet Union River Diversion Project*.

39   India and other nations "began to view events in Russia, rather than in the US, as standing at the forefront of the movement of global progress" (Manela, *The Wilsonian Moment*, 174).

40   Josephson, " 'Projects of the Century' in Soviet History." Gestwa, *Die Stalinschen Grossbauten des Kommunismus*. For the perspective of users, see volume 1 of the Making Europe series: Oldenziel and Hård, *Consumers, Tinkerers, Rebels*, 222ff.

41   Graham, *The Ghost of the Executed Engineer*; Schattenberg, *Stalins Ingenieure*.

42   Hughes, *American Genesis*; Rassweiler, *The Generation of Power*.

43   Reynolds, "Science, Technology, and the Cold War."

44   Lincoln, *Die Eroberung Sibiriens*; Breyfogle et al., *Peopling the Russian Periphery*.

45   Rosenberg, *Weltmaerkte und Weltkriege 1870–1945*, 535.

46   Ibid., 320; Sperling, *Der Aufbruch der Provinz*.

47   Lincoln, *The Conquest of a Continent*.

48   Marks, *Road to Power*.

49   Grützmacher, *Die Baikal-Amur-Magistrale*.

50   Richthofen, "Ueber den natuerlichen Weg fuer eine Eisenbahnverbindung zwischen China und Europa."

51   Nansen, *Sibirien, ein Zukunftsland*; Heller, *Sibirien*.

52   As quoted in Rosenbladt, *Der Osten ist gruen?*, 143.

53   Josephson, *New Atlantis Revisited*.

54   For example, Högselius, *Red Gas*. Also Etkind, *Internal Colonization*; Bonhomme, *Russian Exploration*.

55   Ziegler, *Der achte Kontinent*; Naumov and Collins, *The History of Siberia*.

56   Wood, *Russia's Frozen Frontier*.

57   Crosby, *Ecological Imperialism*, 75. Crosby has described the "Columbian exchange" thus: the Europeans exported deadly bacteria that decimated the natives, and imported crops from the New World—maize and later the potato—that contributed to population growth at home.

58   McNeill, *Something New Under The Sun*; Grove, *Ecology, Climate, and Empire*.

59   Crosby, *Ecological Imperialism*, 7. See also: Diamond, *Guns, Germs, and Steel*; McNeill, *Mosquito Empires*.

60   Wendt, *Vom Kolonialismus zur Globalisierung*, 208.

61   Ibid., 187ff. See also: Pilcher, *The Oxford Handbook of Food History*.

62   For a broader record see: volume 1 of the Making Europe series, Oldenziel and Hård, *Consumers, Tinkerer, Rebels*, chapter 2.

63   Wendt, *Vom Kolonialismus zur Globalisierung*, 190.

64   See, for instance, Mintz, *Sweetness and Power*; Milton, *Nathaniel's Nutmeg*; Dash, *Tulipomania*; Kurlansky, *Salt*; Bosma et al., *Sugarlandia Revisited*.

65   Brockway, *Science and Colonial Expansion*; See also: Schiebinger, *Plants and Empire*.

66   Osborne, "Acclimatizing the World," 136.

67 Ibid., 142. See also: Bennett and Hodge, *Science and Empire*.

68 Ibid., 148.

69 Beckert, *Empire of Cotton*; Beckert, "Von Tuskagee nach Togo"; Zimmerman, *Alabama in Africa*.

70 Topik et al., *From Silver to Cocaine*.

71 Osterhammel, "Forschungsreise und Kolonialprogramm," 173. See: story in Judel, "Die Geschichte von Liebigs Fleischextrakt." See also: Edgerton, *The Shock of the Old*, 171.

72 In Article 25 of the Argentine Constitution of 1853, the focus is explicitly placed on creative minds on the subject of immigration: "The Federal Government will encourage European immigration, and it will not restrict, limit or burden with any taxes the entrance into Argentine territory of foreigners who come with the goal of working the land, improving the industries and teaching the sciences and the arts."

73 Brown and Paquelle, *Connections After Colonialism*.

74 Fuentes, "Lateinamerika – die Utopie Europas."

75 Jünger, *Kampf um Kautschuk*, 71.

76 Morel, *Red Rubber*; Stanfield, *Red Rubber*.

77 See: Hochschild, *Bury the Chains*.

78 Pfeisinger et al., *Kolonialwaren*, 126.

79 Grandin, *Fordlandia*.

80 See: Zischka, *Wissenschaft bricht Monopole*; Erker, *Vom nationalen zum globalen Wettbewerb*.

81 Smil, *Enriching the Earth*.

82 Weightman, *The Industrial Revolutionaries*, 363.

83 Holtorf, *Der erste Draht zur Neuen Welt*; Finn and Yang, *Communications under the Seas*; Wenzlhuemer, *Connecting the Nineteenth-Century World*.

84 Holtorf, "Die Modernisierung des nordatlantischen Raumes."

85 Zweig, *Sternstunden der Menschheit*, 129.

86 As quoted in Standage, *The Victorian Internet*, 81.

87 Smil, *Creating the Twentieth Century*.

88 Benjamin, *The Atlantic World*; Winseck and Pike, *Communication and Empire*.

89 Fisch, *Europa zwischen Wachstum und Gleichheit 1850–1914*, 336.

90 Bavaj, " 'The West': A Conceptual Exploration."

91 Andersen, *Intellectual Property Rights: Innovation, Governance and the Institutional Environment*.

92 Compare with volume 2 of the Making Europe series, Kohlrausch and Trischler, *Building Europe on Expertise*.

93 Zweig, "Die Monotonisierung der Welt."

94 Between the 1830s and the 1930s almost 40 million people migrated to the Indian subcontinent, almost 50 million from Russia and Northeast Asia to Siberia and Manchuria, almost 20 million Chinese to South East Asia (see: Eckert, "Globalisierung," 15).

95 Noble, *The Religion of Technology*.

96 Rosenberg, *Spreading the American Dream*, 6–8.

97 Ibid., See also: Ninkovich, *Global Dawn*.

98 For a brief summary see: Klimke, "America."

99 Stead, *The Americanization of the World*.

100   On another occasion, Stead proposed an infrastructurally-integrated Europe as an alternative draft to the United States; see: Stead, "The United States of Europe."

101   Doering-Manteuffel, "Amerikanisierung und Westernisierung."

102   See: Weightman, *The Industrial Revolutionaries.*

103   See, for instance, Schroeter, *Americanization of the European Economy.*

104   Kolnai, *The War Against the West.*

105   Buruma and Margalit, *Occidentalism*, 4.

3   Wars & Peace at Home & Abroad

1   Keegan, *A History of Warfare*; Hacker, "The Machines of War."

2   Berz, *08/15: Ein Standard des 20. Jahrhunderts.*

3   Rabinbach, *The Human Motor.*

4   Jordan, "Society Improved the Way You Can Improve a Dynamo."

5   See: Ellis, *The Social History of the Machine Gun*; Pick, *War Machine.*

6   Hacker, "The Machines of War," 2; Hacker, "Engineering a New Order."

7   Koller, *Von Wilden aller Rassen niedergemetzelt.*

8   Cited from McDonald, *The Collected Works of Florence Nightingale, Vol. 14: The Crimean War*, 561–63. The report being announced here was "Notes on the Care and Treatment of Sick and Wounded during the late War in the East and on the Sanitary Requirements of the Army Generally. Presented by Request to the Secretary of State for War."

9   For example, Royle, *Crimea: The Great Crimean War 1854–1856*; Figes, *Crimea*; Ponting, *The Crimean War.*

10   Morus, "The Nervous System of Britain," 475. See also: Standage, *The Victorian Internet.* Lahiri Choudhury argues that "the newness of 'New Imperialism' lay not in its ideology but its technology" (*Telegraphic Imperialism*, 122).

11   Haller, "Die Erfindung der Heimatfront." See also: Pearton, *The Knowledgeable State.*

12   Daniel, "Der Krimkrieg 1853–1856 und die Entstehungskontexte medialer Kriegsberichterstattung," 46.

13   Baly, *Florence Nightingale and the Nursing Legacy*; Bostridge, *Florence Nightingale: The Woman and her Legend.*

14   Emig, "Die Domestizierung des Krieges"; Emig, "Bringing the War Home."

15   Lien, "Diet and Nutrition." On the same occasion "meat biscuits" were supplied to French soldiers.

16   Desrosières, *La politique des grands nombres*; Tooze, *Statistics and the German State, 1900–1945*; Headrick, *When Information Came of Age.*

17   See: Simms and Trim, *Humanitarian Intervention.*

18   See: Mazumdar, *Insurgent Sepoys.*

19   Gourlay, *Florence Nightingale and the Health of the Raj.*

20   Castro-Santos, "Nursing."

21   Vasold, *Florence Nighingale.*

22   Hutchinson, "Rethinking the Origins of the Red Cross," 578.

23   As quoted in ibid., 572.

24   Dülffer, *Frieden stiften*.

25   Weisbrode, "War." See also: Kalshoven and Zegveld, *Constraints on the Waging of War*.

26   Perrin, *Modern Warfare*.

27   Weisbrode, "War."

28   Edgerton, *The Shock of the Old*.

29   Smil, *Creating the Twentieth Century*.

30   Weisbrode, "War." See also: Löhr, *Die Globalisierung geistiger Eigentumsrechte*.

31   Nye, *Technology Matters*, 174.

32   Ibid., 170.

33   Bashford, "Science and Technology." See also: Headrick, *Power Over Peoples*, 257–301.

34   Chew, *Arming the Periphery*.

35   See: discussion in Reinkowski and Thum, *Helpless Imperialists*.

36   See: examples in Kuß, *Deutsches Militär auf kolonialen Kriegsschauplaetzen*.

37   Adas, *Prophets of Rebellion*.

38   See: Shelton, *Encyclopedia of Genocide and Crimes against Humanity*.

39   See: Kittler, "Zur Theoriegeschichte von Information Warfare."

40   Klein and Schumacher, *Kolonialkriege*; Thomas, *Violence and Colonial Order*.

41   Lindquist, *"Exterminate All the Brutes."*

42   Stuchtey, *Die europäische Expansion und ihre Feinde*.

43   Morel, *Red Rubber*.

44   The pattern of colonial scandals are discribed by Bösch, *Öffentliche Geheimnisse*, 225–327. See also: Taithe, *The Killer Trail*.

45   Hochschild, *King Leopold's Ghost*.

46   Mattelart, *Networking the World 1794–2000*, 25.

47   Maag et al., *Der Krimkrieg als erster europaeischer Medienkrieg*; Keller, *The Ultimate Spectacle*.

48   Knightley, *The First Casualty: The War Correspondent as Hero and Myth-Maker from the Crimea to Iraq*.

49   Mattelard, *Networking the World 1794–2000*, 50.

50   Ibid., 52.

51   Ibid., 2, 4.

52   Herren, *Internationale Organisationen seit 1865*, 12–13.

53   Geyer and Paulmann, *The Mechanics of Internationalism*.

54   Hevia, *The Imperial Security State*.

55   Booth, *In Darkest England, and the Way Out*, 14.

56   Fischer-Tiné, "Global Civil Society and the Forces of Empire," 30.

57   Valverde, "The Dialectic of the Familiar and the Unfamiliar," 495–500.

58   Fischer-Tiné, "Global Civil Society and the Forces of Empire," 40.

59   Booth, *In Darkest England, and the Way Out*, 143.

60   Sengoopta, *Imprint of the Raj*.

61   Chew, *Arming the Periphery*.

62   Simonsen, "Accelerating Modernity," 102.

63   Wohl, *A Passion for Wings* and *The Spectacle of Flight*; Schnuerer, "Aeroplanes and Airships as National and Collective Symbols."

64 Staniland, *Government Birds*; Kranakis, "European Civil Aviation in an Era of Hegemonic Nationalism."

65 Conway, *High-Speed Dreams*, 11. See also: Kranakis, "The 'good miracle'."

66 Vleck, *Empire of the Air*.

67 Sampson, *Empires of the Sky*.

68 Butlin, *Geographies of Empire*, 494.

69 Ibid., 495–96.

70 Brancker, "Imperial Communications," 57, as quoted in Schlimm, *Ordnungen des Verkehrs*, 271.

71 The *Deutsche Lufthansa* flew across the Atlantic to South America via Gambia, and the airline always feared catching malaria and yellow fever (information provided by Marion Hulverscheidt, Kassel). See also: Huber, "The Unification of the Globe by Disease?"

72 Bhimull, "Empire in the Air"; Bhimull, "Contribution to Colonialism, Decolonization, and Aid Theme."

73 Thomas, *Violence and Colonial Order*.

74 Black, *War in the Early Modern World*, 252; Crary, *Techniques of the Observer*; Kehrt, *Moderne Krieger*.

75 Omissi, *Air Power and Colonial Control*, 132.

76 Ibid., 211.

77 Clayton, "Deceptive Might," 290.

78 Art. 22 of the Versailles Treaty created mandates of ex-Ottoman and ex-German dependencies, "which are inhabited by peoples not yet able to stand by themselves under the strenuous conditions of the modern world," as quoted in Fieldhouse, *Western Imperialism in the Middle East 1914–1958*, 69.

79 Butlin, *Geographies of Empire*, 496; Kiernan, *Colonial Empires and Armies 1815–1960*, 194–97; Tanaka, "British 'Humane Bombing' in Iraq During the Interwar Era."

80 Fieldhouse, *Western Imperialism in the Middle East 1914–1958*, 114.

81 Killingray, "A Swift Agent of Government"; Süss, *Tod aus der Luft*, 36.

82 Selassie, "Appeal to the League of Nations," June 1936, http://www.mtholyoke.edu/acad/intrel/selassie.htm, accessed September 8, 2011. For historical contexts see: Brogini Künzi, *Italien und der Abessinienkrieg 1935/36*; Mattioli, *Experimentierfeld der Gewalt*.

83 Jonas, *The Battle of Adwa*.

84 Mattioli, *Experimentierfeld der Gewalt*.

85 Spengler, *Der Mensch und die Technik*, 59–60 (our translation).

86 Esposito, *Mythische Moderne*.

87 Hotta, "Pan-Isms," 807, identifies two main features of these macro-nationalisms: "First is the grand scale of their aspired unions, and second their desire to combine separate and already existing arrangements by appealing to a yet greater cohesive factor."

88 Fritzsche, *A Nation of Fliers*, S. 37; Robinson, *The Zeppelin in Combat*.

89 Introduction to Jünger, *Luftfahrt ist Not!*, 11.

90 Oldenziel, "Islands."

91 Pahl, *Das politische Antlitz der Erde*, 31, who contrasts German aviation to British "highways of Empire," ibid., 16–17. Pahl also published *Die Luftwege der Erde*.

92 Poturzyn, "Koloniale Luftpolitik," 119. For Britain see: McCormack, "Aviation and Empire."

93 Johnston, "Libya 1911." See also: Lindquist, *A History of Bombing*.

94 Segre, "Italo Balbo and the Colonisation of Libya"; Taylor, *Fascist Eagle*.

95 Zischka, *Italien in der Welt*, 188. He refers to the 1908 H.G. Wells novel *The War in the Air, and Particularly How Mr. Bert Smallways Fared While it Lasted*.

96 Nye, *Technology Matters*, 161–74.

97 At the beginning of spaceship-building, constructors like Francesco Lana de Terzi in 1670 had predicted that airships would enable attacks on fortified cities from above: "Who does not recognize that no city would be safe from attacks, with spaceships emerging at every given hour and the crews being deboarded and let down." Cited in Gethmann, "Fahrten im Luftmeer," 130.

98 Hugill, *Global Communications since 1844*; Blumentritt et al., *Tracking the History of Radar*.

99 Edgerton, *The Shock of the Old*.

100 Recherche International / Rheinisches Journalistenbuero, "Unsere Opfer zaehlen nicht."

4 *Scrambling for Eurafrica: Resources & Axes of Infrastructure*

1 Pakenham, *The Scramble for Africa, 1876–1912*.

2 The term "all-red" refers to infrastructures, which were 100 percent under British control, as for instance "all-red" cables, or "all-red" railroad lines. It is interesting to note that neither the telegraph cables nor the railroads projected as "all-red" were, in fact, "all-red." The telegraph cables of the British Empire landed on foreign territory (Portugal) and the British railroad from Cairo to Cape Town was never built. For "all-red" cables see: Silva and Diogo, "From Host to Hostage."

3 Fetherling, *The Gold Crusades*; Clay, "Gold Rushes"; Reeves et al., "Integrating the Historiography of the Nineteenth-Century Gold Rushes."

4 Russel-Wood, "Technology and Society."

5 See: Nightingale, *Segregation*, who stresses the importance of local agents like land registry offices and land development companies in enforcing the "color line."

6 Knight, *The Anglo-Zulu War*; Colenso and Durnford, *History of the Zulu War and its Origin*.

7 Butlin, *Geographies of Empire*, 566.

8 Korieh, "Diamonds," 301. See also: Newbury, *The Diamond Ring*.

9 Iliffe, *Geschichte Afrikas*, 242.

10 Bessant, "South Africa," 700.

11 Davidson and English, *Cecil Rhodes and his Time*; Ziegler, *Legacy*.

12 Marx, *Das Kapital I*, 779, as quoted in Pfeisinger, "Kolonialwaren in der Weltwirtschaft," 12.

13 Rosenberg, "Einleitung," 18.

14 Korieh, "Diamonds," 303.

15 Zukas, "Segregation, racial, Africa." See also: Lapping, *Apartheid: A History*; Beinart and Dubow, *Segregation and Apartheid in Twentieth-Century South Africa*.

16 Korieh, "Diamonds," 303. See also: Headrick, *Power Over Peoples*, 257–301.

17 Portugal lost a great deal in the Berlin Conference: the effective occupation rule was imposed against the historical rights defended by Portugal, the free navigation of African rivers was also imposed (Portugal lost its rights on the Congo, Zambezi, and Rovuma rivers), as well as the Congo river estuary territories.

18 Machado, "Caminho de Ferro de Mossamedes ao Bihé," 219.

19 Ibid., 220.

20 Ibid., 233.

21 Bellefonds, *Mémoires Sur Les Principaux Travaux D'utilité Publiqué Éxécutés En Egypte Depuis La Plus Haute Antiquité Jusqu'à Nos Jours*.

22 Karabell, *Parting the Desert*; Piquet, *Histoire du canal de Suez*; Montel, *Le Chantier Du Canal de Suez (1859–1869*; Lessep, *The History of the Suez Canal*.

23 An overview of the broad impact of the Suez Canal is Huber, *Channelling Mobilities*.

24 Webster, *Fachoda*.

25 Wesseling, *Divide and Rule*; Wesseling, *Expansion and Reaction*.

26 Webster, *Fachoda*.

27 See: Sèbe, *Heroic Imperialists in Africa*.

28 Churchill, *The River War*, 163.

29 Ibid., 312.

30 Ibid., 318.

31 *Die Zeit*, July 14, 1949.

32 Heffernan, "Bringing the Desert to Bloom."

33 Shere, "Frank Shuman"'s Solar Arabian Dream."

34 Ibid.

35 Gall, "Atlantropa"; Van Laak, *Weisse Elefanten*.

36 Bloch, *Das Prinzip Hoffnung*, 775.

5   *From the Raj to the Yellow Peril*

1 Interview with Shyam Kumar, General Manager of the Eastern Railway, in *Frontline*, Volume 21, Issue 17, accessed June 10, 2011.

2 Shridharani, *Story of the Indian Telegraphs*.

3 Nehru, Preface to Shridharani, *Story of the Indian Telegraphs*, as quoted in Ghosh, "The Utilitarianism of Dalhousie and the Material Improvement of India," 97.

4 Mill, *The History of British India*, Vol. II, ch. X, 143.

5 Ghosh, *The Utilitarianism of Dalhousie*, 97.

6 Chapman, *Letter to the Shareholders of the G.I.P.R.*, as quoted in Headrick, *The Tools of Empire*, 182.

7 Kerr, "John Chapman and the Promotion of the Great Indian Peninsula Railway, 1842–1850."

8 Prakash, *Another Reason*, 165.

9 Racknitz, *Die Plünderung des Yuanming Yuan*; Bastid et al., *China from the Opium Wars to the 1911 Revolution*.

10 Hsü, *The Rise of Modern China*.

11 Lovell, *The Opium War*.

12 Beeching, *The Chinese Opium Wars*, 38 and 42. A chest of opium carried about a picul, an Asian unit of weight equivalent to around 60 kg.

13 Lovell, *The Opium War*.

14 Pong, *Shen Pao-chen and China's Modernization in the Nineteenth Century*.

15 Leibo, *Transferring Technology to China*.

16 Gordon, *A Modern History of Japan*.

17 Morris-Suzuki, *The Technological Transformation of Japan*.

18 Clulow, *The Company and the Shogun*.

19 Odagiri et al., *Technology and Industrial Development in Japan*.

20 Gordon, *A Modern History of Japan*, 272–78; Macpherson, *The Economic Development of Japan 1868–1941*.

21 Avakian, *The Meiji Restoration and the Rise of Modern Japan*.

22 Weightman, *The Industrial Revolutionaries*, 285–98.

6   A New World Order & the Collapse of Colonialism

1 Barraclough, "The Revolt against the West."

2 Hirschhausen and Patel, "Europeanization in History," 10.

3 Herren, *Internationale Organisationen seit 1865*, 4. See also: Rittberger and Zangl, *International Organization*; Reinalda, *Routledge History of International Organizations*.

4 Geyer and Paulmann, *The Mechanics of Internationalism*; Porter, *Trust in Numbers*. See also: Ashworth, "Measurement."

5 Cioc, *The Rhine*; Blackbourn, *The Conquest of Nature*; Disco, "Accepting Father Rhine?"

6 Mattelart, *Networking the World, 1794–2000*, 7.

7 Schieder, *Handbuch der europaeischen Geschichte*, Vol. 6, 131.

8 See: Greenaway, *Science International*, on the history of the International Association of Academies (1899–1919), the International Research Council (1919–31), and the International Council of Scientific Unions (since 1931).

9 Puffert, *Tracks Across Continents, Paths Through History*, 240.

10 See also: Laborie, *L'Europe mise en réseaux*.

11 See: contributions of Weiss, Ewing, and Mueller-Pohl in Loehr and Wenzlhuemer, *The Nation State and Beyond*.

12 Vec, "Der Welt Standard," 341.

13 Blaise, *Time Lord*.

14 Herren, *Internationale Organisationen seit 1865*.

15 Galison, *Einstein's Clocks, Poincaré's Maps*.

16 Geyer and Paulmann, *The Mechanics of Internationalism*.

17 Herren, *Hintertueren zur Macht*.

18 Boettger, "Internationalismus und Kolonialismus—Ein Werkstattbericht zur Geschichte des Bruesseler." See also: Stuchtey, *Science Across the European Empires*; Feuerhahn and Rabault-Feuerhahn, *La Fabrique Internationale de la Science*; Bennett and Hodge, *Science and Empire*.

19 The term "world economy" (*Weltwirtschaft*) was coined in 1776 by Johann Heinrich Merck (1741–91). See: http://hsozkult.geschichte.hu-berlin.de/termine/id=11953, accessed August 14, 2012.

20 Bush, *Standards*.

21 Murphy and Yates, *ISO, The International Organization for Standardization*, 107.

22 Ibid., 106. Winseck and Pike, *Communication and Empire*.

23 Herren, *Internationale Organisationen seit 1865*, 73–79.

24 Wilson's speech may be accessed at http://en.wikisource.org/wiki/Fourteen_Points_Speech.

25 Winkler, *Nexus*.

26 Manela, *The Wilsonian Moment*, 51.

27 Ibid., xi.

28 Ibid., xi.

29 Ibid., 219–21.

30 Rothermund, *The Routledge Companion to Decolonization*, 243–57.

31 Hale, *Radio Power*.

32 Edgerton, "Creole Technologies and Global Histories."

33 Le Corbusier, *Feststellungen zu Architektur und Staedtebau*, 152, as quoted in Huse, *Le Corbusier*, 49–50.

34 Schot and Lagendijk, "Technocratic Internationalism in the Interwar Years"; Kaiser and Schot, *Writing the Rules for Europe*.

35 Mazower, *Governing the World*.

36 Laqua, *Internationalism Reconfigured*; Gorman, *The Emergence of International Society in the 1920s*.

37 See: Wenzlhuemer, "The History of Standardisation in Europe."

38 Iryie, *Global Community*; MacKenzie, *A World Beyond Borders*.

39 Clavin and Patel, "The Role of International Organizations in Europeanization," 111–12.

40 Shinohara, "League of Nations System."

41 Schipper et al., "New Connections for an Old Continent."

42 Maul, *Human Rights, Development and Decolonization*.

43 "International Container Bureau."

44 See: Schipper, *Driving Europe*, 146.

45 "International Container Bureau."

46 Donovan and Bonney, *The Box That Changed the World*; Levinson, *The Box*.

47 McPherson, "Containerization."

48 Miller (in *Europe and the Maritime World*) argues that after 1960 global "containerization" still remained a European field of dominance.

49 Herren, "Der Voelkerbund."

50 In 1927, Le Corbusier's plan for Geneva, which was to become the prototype of all future United Nations buildings, had been eliminated from the competition on the grounds of not having been drawn up in Indian ink as the rules of the competition specified (*Encyclopedia Britannica 2005*).

51 Cited in Evenson, *Chandigarh*, 6.

52 See: Ingimundarson, "The Geopolitics of Arctic Natural Resources." See also: http://www.geopoliticsnorth.org.

53 Hannemann, " 'North to the Future' – die Arktis und die Medien."

54 Ross, *Mit Kind und Kegel in die Arktis*, 199.

55 Ibid., 211.

56 Ibid., 214.

57 *Encyclopedia Britannica 2005*, Article "Arctic."

58 Bessant, "South Africa," 700–701.

59 Marx, "Europeans Only."

60 Prashad, *The Darker Nations*.

61 Akurang-Parry, "Anticolonial Movements, Africa," 79.

62 Derrick, *Africa's "Agitators."*

63 See: Young, *Postcolonialism: An Historical Introduction*, 232.

64 "Palestine: Information with Provenance."

65 Marx, "Europeans Only," 156.

66 Biko, "Some African Cultural Concepts," 46–47, as quoted in Marx, "Europeans Only," 167.

67 Lucius, "Am Oranjefluss geht es um Diamanten, Gas und Trinkwasser."

68 Chande, "World War II, Africa," 1143.

69 Global Witness, "Broken Vows."

70 Mollin, *Die USA und der Kolonialismus*; Witte, *The Assassination of Lumumba*. For a more recent record see: Hecht, *Being Nuclear*.

71 Campbell et al., *Regulating Mining in Africa*; Werthmann and Grätz, *Mining Frontiers in Africa*.

72 Campbell, *Blood Diamonds*.

73 "The objective of EUREKA is to raise, through closer cooperation among enterprises and research institutes in the field of advanced technologies, the productivity and competitiveness of Europe's industries and national economies on the world market, and hence strengthen the basis for lasting prosperity and employment." "Hannover Declaration," November 6, 1985.

74 Liniger-Goumaz, *L'Eurafrique: Utopie ou réalité?*; Ageron, "L'idée d'Eurafrique et le dèbat colonial franco-allemand de l'entre-deux-guerres"; Deschamps, "Quelle Afrique pour une Europe unie? L'idée d'Eurafrique à l'aube des années trentes"; Moser, *Europaeische Integration, Dekolonisation, Eurafrika*; Bitsch and Bossuat, *L'Europe unie et l'Afrique*; Van Laak, "Detours around Africa."

75 Edgerton, *The Shock of the Old*, esp., xiv, his concept of "Creole" technology.

76 de Vries, "The Industrious Revolution and the Industrial Revolution."

77 For example, Meredith, *The State of Africa*.

78 Hecht, *The Radiance of France*.

79 This concept, as well as the role of developing countries in the capitalist world system, is strongly discussed among economists. See: Leys, *The Rise and Fall of Development Theory*; Bell, "Development Economies."

80 As referred to by Rothermund, *The Routledge Companion to Decolonization*, 273.

81 Jawaharlal Nehru to U Nu, September 6, 1955.

82 Perkovich, *India's Nuclear Bomb*, 67.

83 Ibid.

84 India and the USSR had signed the Science and Technology Agreement in 1972, which was further strengthened with the Integrated Long Term Programme (ILTP) of Scientific Cooperation.

85 We will use the concept "techno-nationalism" as a national response to the impossibility of importing technology. We are aware, however, that this concept, together with "techno-globalism" is still controversial. See: Edgerton, "The Contradictions of Techno Nationalism and Techno Globalism" and Nakayama, "Techno-nationalism versus Techno-globalism."

86 "The SARAS inaugural flight." *CSIR News*, September 30, 2004.

87 Edgerton, "The Contradictions," 27. For the Royal Riders Club see: http://www.royalridersclub.com/sponsor.php.

88 See: *Hop On Gurls!*, http://www.hopongurls.com.

89 GOELRO is the transliteration of the Russian abbreviation for State Commission for Electrification of Russia.

90 Lenin, "Our Foreign and Domestic Position and Party Tasks" (speech delivered to the Moscow Gubernia Conference Of The R.C.P.(B.), November 21, 1920).

91 Lenin, "Report On The Work Of The All-Russia Central Executive Committee And The Council Of People's Commissars" (speech delivered at the first session of the All-Russia Central Executive Committee, Seventh Convocation, February 2, 1920). The enthronement of electricity is not particular to Lenin, but it was common everywhere in Europe.

92 Womack, *China's Rise in Historical Perspective*.

93 Wei and Brock, *Mr. Science and Chairman Mao's Cultural Revolution*, xv.

94 Needham, "Mao Zedong 1893–1976."

95 Jersild, *The Sino-Soviet Alliance*.

96 DeGlopper, "Science and Technology."

97 Clark, *The Chinese Cultural Revolution*.

98 Mitter, *A Bitter Revolution*.

99 Lewis and Xue, *China Builds the Bomb*; Handberg and Li, *Chinese Space Policy*.

100 Tse-Tung, "We Must Learn to do Economic Work," 241.

101 Vogel, *Deng Xiaoping and the Transformation of China*.

102 Xiaoping, speech at the Opening Ceremony of the National Conference on Science, March 18, 1978.

103 Lardy, "Recasting of the Economic System: Structural Reforms of Agriculture and Industry."

104 Owens and Shaw, *Development Reconsidered*, 3.

105 Edgerton, *The Shock of the Old*, xiv.

106 Bonfiglioli, "Universal Science, Appropriate Technology and Underdevelopment – a Reprise of the Latin America Case."

107 Murteira, *Economia do Conhecimento*.

108 Ha, "International Venture Funding Rose 5 percent in 2008."

7 The Reconstruction Period

1 Grazia, *Irresistible Empire*.

2 Pells, *Not Like Us*.

3 Revel, *L'obsession anti-américaine*; Berman, *Anti-Americanism in Europe*.

4 Rosenberg and Foner, *Spreading the American Dream*, 234; Harsanyi, "Anti-Americanism."

5 Lederer and Burdick, *The Ugly American*.

6 Hochschild, *King Leopold's Ghost*.

7 Witte, *The Assassination of Lumumba*.

8 Mollin, *Die USA und der Kolonialismus*; Reybrouck, *Congo*; Vanthemsche, *Belgium and the Congo*.

9 Westad, *The Global Cold War*; Prashad, *The Darker Nations*; McMahon, *The Cold War in the Third World*.

10 For recent records of "the West" see: Bonnett, *The Idea of the West*; Winkler, *Geschichte des Westens*.

11 Rosenberg and Foner, *Spreading the American Dream*, 4.

12 Schuemer, "Spaghettisiert Euch," 29.

13 Bertaux, "Mutation der Menschheit."

14 See: Polianski and Schwartz, *Die Spur des Sputnik*.

15 Here we reflect the view of Schmidt-Gernig, *Europa als Kontinent der Zukunft*.

16 See: Mazower, *The Dark Continent*; Judt, *Postwar*; Eichengreen, *The European Economy Since 1945*.

17 Van Laak, "Planung, Planbarkeit und Planungseuphorie."

18 Kaiser et al., *The History of the European Union*.

19 Zabusky, *Launching Europe*; Redfield, *Space in the Tropics*.

20 Bade, *Migration in European History*.

21 See: Taylor, "From Trips to Modernity to Holidays in Nostalghia."

22 Mazower, *Der dunkle Kontinent*, 303; Rothermund, *The Routledge Companion to Decolonization*.

23 Kojève, "Kolonialismus in europaeischer Sicht," 128 and 132. The French demographer Alfred Sauvy coined the term "Third World" in 1952 and noted that sooner or later it would marginalize Europe demographically, immediately marking it as a potential threat to Europe. Sauvy demanded programs for family planning and population control as well as the facilitation of European emigration. Europe would be able to aid development while Africa could provide space.

24 Hachtmann, "Fordismus"; Locke and Schoene, *The Entrepreneurial Shift*.

25 Pfister, "Das '1950er' Syndrom."

26 O''Bryan, "Car Culture."

27 Martenstein, "Die Fuersten der Verschwendung," 126.

28 Alber, "Motorization and Colonial Rule."

29 Bourdieu, *La distinction*.

30 Wiersch, *Die Käfer-Chronik.*, 159; Rieger, *The People's Car*.

31 See: http://www.volkswagenag.com/content/vwcorp/info_center/de/news/2008/02/volkswagen_de_mexico_celebrates_ten_years_of_the_new_beetle.html, accessed October 30, 2012.

32 Jones et al., *The Machine that Changed the World*.

33 Poeschke, "Global Sourcing als Instrument zur Nutzung von Kostenoptimierungspotentialen."

34 Juergens, "Approaches towards Integrating Suppliers in Simultaneous Engineering Activities."

35 For a record of the Beetle's international success see: Patton, *Bug*.

36 See: Volkswagen Group Strategy 2018, http://www.volkswagenag.com/content/vwcorp/content/de/the_group/strategy.html.

37 Poser, "Leisure Time and Technology"; Blanchard, *Labour and Leisure in Historical Perspective*; Salazar, *Envisioning Eden*.

38 Scranton, "Preface," vii.

39 Maines, *Hedonizing Technologies*; Koshar, *Histories of Leisure*.

40 "This phrase by Gérard Blitz received considerable endorsement in the postwar context. Freeing people from their restrictions, allowing them to get together, recharge their batteries and return to original pleasures: happiness by Club Med was born." See: Club Méditerranée, "History of the Club Méditerranée."

41 Vidal, "Tourism and Transport in Europe," 483. See also: Dierickx and Lyth, "From Privilege to Popularity"; Lyth, " 'Gimme a ticket on an airoplane'...."

42 Mullen and Munson, *"The Smell of the Continent."*

43 Knoch, "Life on Stage."

44 Gyr, "Geschichte des Tourismus."

45 Deacon et al., *Transnational Lives*.

46 See: Club Méditerranée, "Discover Club Med's History since 1950."

47 Gyr, "Geschichte des Tourismus."

48 Heiss, "Tourism," 1021, who has established that at present, even in the European and American industrial states 10 percent of the gross national product results from tourism.

49 Bausinger, "Grenzenlos... Ein Blick auf den modernen Tourismus."

50 Wendt, *Vom Kolonialismus zur Globalisierung*, 382–85.

51 Vowinckel, *Flugzeugentfuehrungen*.

52 See: Club Méditerranée, "Magical Martinique Holidays."

53 Chimelli, "Europa in der Karibik," 9.

54 Bernau, "Das Wirtschaftswunder im Heiligen Land."

55 Rooney, "How Tel Aviv Became a Tec Hub."

56 Senor and Singer, *Start-Up Nation*. See also: the suggestive 3.18 min video, "Israel – One Hundred Years of Science and Technology," http://www.youtube.com/watch?v=GzilbrH3CcA, accessed September 14, 2013.

57 Heinze-Greenberg, *Europa in Palästina*.

58 H. Davis and D. Davis, *Israel in the World*.

59 See research by Zdeněk Klima, *Radio Prag*, May 19, 2012.

60 See ongoing research on the internationalization of the Tennessee Valley Authority, including the River Jordan, in Lagendijk, "Transnationalising the TVA."

Epilogue: Europeans Globalizing

1 Gilman, *Mandarins of the Future*.
2 Osterhammel and Petersson, *Globalization: A Short History*.
3 See: discussions following, for instance, Kabou, *Et si l'Afrique refusait le développement?*
4 Ferguson et al., *The Shock of the Global*.
5 Fischer-Tiné, *Pidgin-Knowledge* (with respect to medicine).
6 Lindner et al., *Hybrid Cultures—Nervous States*.

Bibliography

Adas, Michael. *Machines As the Measure of Men: Science, Technology, and Ideologies of Western Dominance*. Ithaca: Cornell University Press, 1989.

———. *Prophets of Rebellion: Millenarian Protest Movements against the European Colonial Order*. Chapel Hill: University of North Carolina Press, 1979.

Ageron, Claude-Robert. "L'idée d'Eurafrique et le dèbat colonial franco-allemand de l'entre-deux-guerres." *Revue d'Histoire Moderne et Contemporaine* 22 (1975): 446–75.

Āl-Ahmad, Ğalāl. *Occidentosis: A Plague from the West*. Berkeley, CA: Mizan Press, 1984.

Akerman, James R., and Robert W. Karrow. *Maps: Finding Our Place in the World*. Chicago: University of Chicago Press, 2007.

Akurang-Parry, Kwabena. "Anticolonial Movements, Africa." In *Encyclopedia of Western Colonialism Since 1450*, Vol. 1, edited by Thomas Benjamin, 74–81. Detroit: Macmillan Reference USA, 2007.

Alatas, Hussein. *The Myth of the Lazy Native: A Study of the Image of the Malays, Filipinos and Javanese from the 16th to the 20th Century and Its Function in the Ideology of Colonial Capitalism*. London: F. Cass, 1977.

Alber, Erdmute. "Motorization and Colonial Rule: Two Scandals in Dahomey, 1916." *Journal of African Cultural Studies* 15, no. 1 (2002): 79–92.

Ambirajan, S. "Science and Technology Education in South India." In *Technology and the Raj: Western Technology and Technical Transfers to India: 1700–1947*, edited by R. MacLeod and D. Kumar, 112–33. New Delhi, Thousand Oaks, CA and London: Sage, 1995.

Andersen, Birgitte. *Intellectual Property Rights: Innovation, Governance and the Institutional Environment*. London: Routledge, 2006.

Andreas, Joel. *Rise of the Red Engineers: The Cultural Revolution and the Origins of China's New Class*. Stanford: Stanford University Press, 2009.

Arnold, David. "Europe, Technology, and Colonialism in the 20th Century." *History and Technology* 21, no. 1 (2005): 85–106.

Ashworth, William J. "Measurement." In *The Palgrave Dictionary of Transnational History*, edited by Akira Iriye and Pierre-Yves Saunier, 701–6. Basingstoke: Palgrave Macmillan, 2009.

Aust, Martin, and Daniel Schoenpflug, eds. *Vom Gegner lernen: Feindschaften und Kulturtransfers im Europa des 19. und 20. Jahrunderts*. Frankfurt am Main: Campus Verlag, 2007.

Avakian, Monique. *The Meiji Restoration and the Rise of Modern Japan*. Englewood Cliffs, NJ: Silver Burdett Press, 1991.

Aydin, Cemil. *The Politics of Anti-Westernism in Asia Visions of World Order in Pan-Islamic and Pan-Asian Thought*. New York: Columbia University Press, 2007.

Bade, Klaus J. *Migration in European History*. Malden, MA: Blackwell, 2003.

Baly, Monica E. *Florence Nightingale and the Nursing Legacy*. London: Croom Helm, 1986.

Barraclough, Geoffrey. "The Revolt against the West." In *Decolonization Perspectives from Now and Then*, edited by Prasenjit Duara, 118–30. London: Routledge, 2004.

Barth, Boris, and Jürgen Osterhammel, eds. *Zivilisierungsmissionen: imperiale Weltverbesserung seit dem 18. Jahrhundert*. Konstanz: UVK Verlagsgesellschaft, 2005.

Barth, Heinrich. *Travels and Discoveries in North and Central Africa: Being a Journal of an Expedition Undertaken under the Auspices of H.B.M.'s Government in the Years 1849–1855*. London: F. Cass, 1965.

Basalla, George. "The Spread of Western Science." *Science* 156 (1967): 611–22.

Bashford, Alison. "Science and Technology." In *Encyclopedia of Western Colonialism Since 1450*, edited by Thomas Benjamin, Vol. 3, 991–96. Detroit: Macmillan Reference USA, 2007.

Bastid, Marianne, Marie-Claire Bergere, and Jean Chesneaux. *China from the Opium Wars to the 1911 Revolution*. New York: Pantheon, 1976.

Bates, Darrell. *The Fashoda Incident of 1898: Encounter on the Nile*. Oxford: Oxford University Press, 1984.

Baum, David. *The Rise and Fall of the Caucasian Race: A Political History of Racial Identity*. New York: New York University Press, 2006.

Bausinger, Hermann. "Grenzenlos... Ein Blick auf den modernen Tourismus." In *Reisekultur: von der Pilgerfahrt zum modernen Tourismus*, edited by Hermann Bausinger, Klaus Beyrer, and Gottfried Korff, 343–53. Munich: C.H. Beck, 1991.

Bavaj, Riccardo. " 'The West': A Conceptual Exploration." *European History Online* (EGO), Mainz 2011-11-21, urn:nbn:de:0159–2011112107. http://www.ieg-ego.eu/bavajr-2011-en (accessed January 2, 2013).

Bayly, Christopher A. *The Birth of the Modern World, 1780–1914: Global Connections and Comparisons*. Oxford: Blackwell, 2004.

Beasley, William G. *Japan Encounters the Barbarian: Japanese Travellers in America and Europe*. New Haven: Yale University Press, 1995.

Beckert, Sven. "Von Tuskagee nach Togo. Das Problem der Freiheit im Reich der Baumwolle." In *Globalisierungen*, edited by Paul Nolte and Juergen Osterhammel, 505–45. Göttingen: Vandenhoeck & Ruprecht, 2005.

———. *Empire of Cotton: A New History of Global Capitalism*. London: Allen Lane, 2014.

Beeching, Jack. *The Chinese Opium Wars*. New York: Harcourt Brace Jovanovich, 1977.

Beinart, William, and Saul Dubow, eds. *Segregation and Apartheid in Twentieth-Century South Africa*, London and New York: Routledge, 1995.

Beljakow, A.P. *Elektrizitat rings um uns. Uebersetzung aus dem Russischen von Friedrich V. Raupach*. Berlin: Kinderbuchverlag, 1954.

Bell, C. "Development Economies." In *The New Palgrave Dictionary of Economics*, Vol. 1, edited by John Eatwell, Murray Milgate, and Peter Newman, 818 ff. Basingstoke: Palgrave Macmillan, 1987.

Bellefonds, Louis Maurice Adolphe Linant de. *Mémoires Sur Les Principaux Travaux D'utilité Publiqué Éxécutés En Egypte Depuis La Plus Haute Antiquité Jusqu'à Nos Jours: Accompagné D'un Atlas Renfermant Neuf Planches Grand In-Folio Imprimées En Couleur*. n.p.: Ulan Press, 2012 (reprint of the 1923 original).

Bénat Tachot, Louise, and Serge Gruzinski. *Passeurs culturels: mécanismes de métissage*. Paris: Ed. de la Maison des sciences de l'homme, 2002.

Benjamin, Thomas. *The Atlantic World: Europeans, Africans, Indians and Their Shared History, 1400–1900*. Cambridge: Cambridge University Press, 2009.

Bennett, Brett M., and Joseph Morgan Hodge, eds. *Science and Empire: Knowledge and Networks of Science across the British Empire, 1800–1970*. Basingstoke: Palgrave Macmillan, 2011.

Berman, Russell A. *Anti-Americanism in Europe: A Cultural Problem*. Stanford: Hoover Institution Press, 2004.

Bernau, Patrick. "Das Wirtschaftswunder im Heiligen Land." *Frankfurter Allgemeine Sonntagszeitung*, no. 24 (July 20, 2010): 24.

Bertaux, Pierre. "Mutation der Menschheit: 1963/64." *Themenportal Europaeische Geschichte* (2007). http://www.europa.clio-online.de/2007/Article=11 (accessed September 17, 2012).

Berz, Peter. *08/15: ein Standard des 20. Jahrhunderts*. Munich: Fink, 2001.

Bessant, Leslie. "South Africa." In *History of World Trade Since 1450*, edited by John J. McCusker, 698–702. Farmington Hills, MI: Macmillan Reference USA, 2006.

Bethancourt, Francisco, and Kirti Chaudhuri, eds. *História da Expansão Portuguesa*. vols. 4 and 5. Lisbon: Temas e Debates, 1998.

Bhimull, Chandra D. "Contribution to Colonialism, Decolonization, and Aid." *Tensions of Europe* Project (MS, ca. 2002).

———. "Empire in the Air: Speed, Perception, and Airline Travel in the Atlantic World." PhD diss, University of Michigan, Ann Arbor, 2007.

Bitsch, Marie-Thérèse, and Gérard Bossuat, eds. *L'Europe unie et l'Afrique: de l'idée d'Eurafrique à la Convention de Lomé I : actes du colloque international de Paris, 1er et 2 avril 2004*. Brussels: Bruylant, 2005.

Black, Jeremy. *War in the Early Modern World*. London: UCL Press, 1999.

Blackbourn, David. *The Conquest of Nature: Water, Landscape, and the Making of Modern Germany*. New York: Norton, 2006.

Blaise, Clark. *Time Lord: Sir Sandford Fleming and the Creation of Standard Time*. New York: Pantheon Books, 2000.

Blanchard, Ian. *Labour and Leisure in Historical Perspective, Thirteenth to Twentieth Centuries*. Wiesbaden: Steiner, 1994.

Blanchard, Pascal et al, eds. *Zoos humains et exhibitions coloniales. 150 ans d'inventions de l'Autre*. Paris: Éditions La Découverte, 2011.

———, Gilles Boëtsch, and Nanette Jacomijn Snoep, eds. *Exhibitions : L'invention du sauvage*. Paris: Actes Sud Editions, 2011.

Blaut, John M. *The Colonizer's Model of the World: Geographical Diffusionism and Eurocentric History*. New York and London: Guilford Press 1993.

Bloch, Ernst. *Das Prinzip Hoffnung*, Vol 2. Frankfurt am Main: Suhrkamp, 1973.

Blumtritt, Oskar, Hartmut Petzold, and William Aspray, eds. *Tracking the History of Radar*. Piscataway, NJ: IEEE-Rutgers Center for the History of Electrical Engineering, 1994.

Böckelmann, Frank. *Die Gelben, die Schwarzen, die Weissen*. Frankfurt am Main: Eichborn, 1998.

Bösch, Frank. *Öffentliche Geheimnisse: Skandale, Politik und Medien in Deutschland und Grossbritannien 1880–1914*. Munich: Oldenbourg Verlag, 2009.

Boettger, Jan Henning. "Internationalismus und Kolonialismus – Ein Werkstattbericht zur Geschichte des Bruesseler, Institut Colonial International (1894–1948)." *Jahrbuch für Europäische Überseegeschichte* 6 (2006): 165–72.

Bonfiglioli, Alberto. "Universal Science, Appropriate Technology and Underdevelopment – a Reprise of the Latin America Case." In *Appropriate Technology in Third World Development*, edited by Pradip K. Ghosh, 254. Westport, CN and London: Greenwood Press, 1984.

Bonhomme, Brian. *Russian Exploration, from Siberia to Space: A History*. Jefferson, North Carolina: McFarland & Company, Inc., Publishers, 2012.

Bonnett, Alastair. *The Idea of the West: Culture, Politics, and History*. Basingstoke: Palgrave Macmillan, 2004.

Booth, William. *In Darkest England, and the Way Out*. London: Salvation Army, 1890.

Borchart, Joachim. *Der europaeische Eisenbahnkoenig: Bethel Henry Strousberg*. Munich: C.H. Beck, 1991.

Bosma, Ulbe, Juan A. Giusti-Cordero, and G.R. Knight. *Sugarlandia Revisited: Sugar and Colonialism in Asia and the Americas, 1800–1940*. New York: Berghahn Books, 2007.

Bostridge, Mark. *Florence Nightingale: The Woman and Her Legend*. London: Viking, 2008.

Bourdieu, Pierre. *La distinction: critique sociale du jugement*. Paris: Éditions de minuit, 1979.

Bracewell, Wendy, and Alex Drace-Francis. *A Bibliography of East European Travel Writing on Europe*. Budapest: Central European University Press, 2008.

Brancker, Sefton. "Imperial Communications." *Journal of the Institute of Transport* 6 (1924): 51–58.

Brember, Thomas. "Der 'Westen' als Feindbild im theologisch-philosophischen Diskurs der Orthodoxie." *European History Online (EGO)*, Mainz 2012–03–19, urn:nbn:de:0159–2012030774. http://www.ieg-ego.eu/bremert-2012-de (accessed September 15, 2012).

Breyfogle, Nicholas B., Abby M. Schrader, and Willard Sunderland, eds. *Peopling the Russian Periphery: Borderland Colonization in Eurasian History*. London: Routledge, 2008.

Brockway, Lucile. *Science and Colonial Expansion: The Role of the British Royal Botanic Gardens*. New York: Academic Press, 1979.

Brogini Künzi, Giulia. *Italien und der Abessinienkrieg, 1935/36: Kolonialkrieg oder totaler Krieg?* Paderborn: Schöningh, 2006.

Brown, Matthew, and Gabriel B. Paquette, eds. *Connections After Colonialism: Europe and Latin America in the 1820s*. Tuscaloosa: University of Alabama Press, 2013.

Buruma, Ian, and Avishai Margalit. "Occidentalism." *New York Review of Books*, January 17, 2002, 4–7.

———. *Okzidentalismus: der Westen in den Augen seiner Feinde*. Munich: Hanser, 2005.

Busch, Lawrence. *Standards: Recipes for Reality*. Cambridge, MA: MIT Press, 2011.

Butlin, Robin A. *Geographies of Empire: European Empires and Colonies, C. 1880–1960*. Cambridge: Cambridge University Press, 2009.

Cameron, Rondo. "Imperialism and Technology." In *Technology in Western Civilization, Vol 1: The Emergence of Modern Industrial Society: Earliest Times to 1900*, edited by Melvin Kranzberg and Carroll W. Pursell, 692–706, New York: Oxford University Press, 1967.

Campbell, Bonnie et al., eds. *Regulating Mining in Africa: For Whose Benefit?* Uppsala: Nordiska Afrikainstitutet, 2004.

Campbell, Greg. *Blood Diamonds: Tracing the Deadly Path of the World's Most Precious Stones.* Boulder, CO: Westview Press, 2004.

Carneiro, Ana, Maria Paula Diogo, and Ana Simões, eds. *Travels of Learning: Towards a Geography of Science in Europe.* Dordrecht: Kluwer Academic, 2003.

Castro-Santos, Luiz A. "Nursing." In *The Palgrave Dictionary of Transnational History,* edited by Akira Iriye and Pierre-Yves Saunier, 774. Basingstoke: Palgrave Macmillan, 2009.

Chakrabarty, Dipesh. *Provincializing Europe: Postcolonial Thought and Historical Difference.* Princeton: Princeton University Press, 2000.

Chande, Abdin. "World War II, Africa." In *Encyclopedia of Western Colonialism Since 1450,* Vol. 3, edited by Thomas Benjamin, 1140–43. Detroit: Macmillan Reference USA, 2007.

Chatterjee, Kumkum, and Clement Hawes, eds. *Europe Observed: Multiple Gazes in Early Modern Encounters.* Lewisburg: Bucknell University Press, 2008.

Chen, Feng. *Die Entdeckung Des Westens: Chinas Erste Botschafter in Europa 1866– 1894.* Frankfurt am Main: Fischer-Taschenbuch-Verlag, 2001.

Chew, Emrys. *Arming the Periphery: The Arms Trade in the Indian Ocean During the Age of Global Empire.* Basingstoke: Palgrave Macmillan, 2012.

Chimelli, Rudolph. "Europa in der Karibik. Was Frankreich von Deutschland wirtschaftlich unterscheidet: Ein Besuch auf Martinique." *Sueddeutsche Zeitung,* no. 134 (June 15, 2010): 9.

Christensen, Lars K. et al. *Europe Meets the World.* Copenhagen: National Museum of Denmark, 2012.

Churchill, Winston. *The River War: An Account of the Reconquest of the Sudan.* London and New York: Longmans, Green and Co., 1902.

Cioc, Mark. *The Rhine: An Eco-Biography, 1815–2000.* Seattle: University of Washington Press, 2002.

Clark, Paul. *The Chinese Cultural Revolution: A History.* Cambridge: Cambridge University Press, 2008.

Clavin, Patricia, and Kiran Klaus Patel. "The Role of International Organizations in Europeanization: The Case of the League of Nations and the European Economic Community." In *Europeanization in the Twentieth Century: Historical Approaches,* edited by Martin Conway and Kiran Klaus Patel, 110–31. Basingstoke: Palgrave Macmillan, 2010.

Clay, Karen. "Gold Rushes." In *History of World Trade Since 1450,* edited by John J. McCusker, 328–30. Farmington Hills, MI: Macmillan Reference USA, 2006.

Clayton, Anthony. "'Deceptive Might': Imperial Defence and Security, 1900– 1968," In *The Oxford History of the British Empire,* edited by Judith M. Brown and William Roger Louis, Vol. IV, 287–93. Oxford: Oxford University Press, 1999.

Club Méditerranée. "Discover Club Med's History since 1950." http://www. clubmed.co.uk/img/all/flash/history/EN/historique_EN.swf (accessed November 16, 2012).

———. "Magical Martinique Holidays." http://www.clubmed.co.uk/cm/ all-inclusive-martinique-holidays_p-341-l-EN-pa-MARTINIQUE-ac-ad.html (accessed November 16, 2012).

———. "History of the Club Méditerranée." http://www.clubmed-corporate. com/?cat=189 (accessed November 16, 2012).

Clulow, Adam. *The Company and the Shogun: The Dutch Encounter with Tokugawa Japan*. New York: Columbia University Press 2014.

Colenso, Frances, and Edward Durnford. *History of the Zulu War and its Origin*. Cambridge and New York: Cambridge University Press, 2011.

Conklin, Alice. *A Mission to Civilize: The Republican Idea of Empire in France and West Africa 1895–1930*. Stanford: Stanford University Press, 1998.

Conrad, Sebastian. "'Europa' aus der Sicht nichtwestlicher Eliten, 1900–1930." *Journal of Modern European History* 4 (2006): 158–70.

Conway, Erik M. *High-Speed Dreams: NASA and the Technopolitics of Supersonic Transportation, 1945–1999*. Baltimore: Johns Hopkins University Press, 2005.

Costantini, Dino. *Mission civilisatrice: Le rôle de l'histoire coloniale dans la construction de l'identité politique française*. Paris: La Découverte, 2008.

Coudenhove-Kalergi, Richard Nicolaus. *Revolution durch Technik*. Wien: Paneuropa Verlag, 1932.

Crary, Jonathan. *Techniques of the Observer: On Vision and Modernity in the Nineteenth Century*. Cambridge, MA: MIT Press, 1990.

Crosby, Alfred W. *Ecological Imperialism: The Biological Expansion of Europe, 900–1900*. Cambridge: Cambridge University Press, 1986.

CSIR News. "The SARAS inaugural flight." September 30, 2004. http://www. icast.org.in/news/2004/sep04/sep30cn.html (accessed December 3, 2015).

Cullen, L.M. *A History of Japan, 1582–1941: Internal and External Worlds*. Cambridge: Cambridge University Press, 2003.

Curtin, Philip D. *The World and the West: The European Challenge and the Overseas Response in the Age of Empire*. Cambridge: Cambridge University Press, 2000.

Daniel, Ute. "Der Krimkrieg 1853–1856 und die Entstehungskontexte medialer Kriegsberichterstattung." In *Augenzeugen: Kriegsberichterstattung vom 18. zum 21. Jahrhundert*, edited by Ute Daniel, 40–67. Göttingen: Vandenhoeck & Ruprecht, 2006.

Dash, Mike. *Tulipomania: The Story of the World's Most Coveted Flower and the Extraordinary Passions It Aroused*. New York: Crown Publishers, 1999.

Davidson, Apollon, and Christopher English. *Cecil Rhodes and his Time*. Pretoria: Protea Boekhuis, 2003.

Davis, Helen, and Douglas Davis. *Israel in the World: Changing Lives through Innovation*. London: Weidenfeld & Nicolson, 2005.

De Bellaigue, Christopher. *Patriot of Persia: Muhammad Mossadegh and a Tragic Anglo-American Coup*. New York: Harper, 2012.

De Grazia, Victoria. *Irresistible Empire: America's Advance through Twentieth-Century Europe*. Cambridge, MA: Belknap Press of Harvard University Press, 2005.

Deacon, Desley, Penny Russell, and Angela Woollacott, eds. *Transnational Lives: Biographies of Global Modernity, 1700–Present*. Basingstoke: Palgrave Macmillan, 2010.

DeGlopper, Donald D. "Science and Technology." In *China: A Country Study*, edited by Robert L. Worden, Andrea Matles Savada, and Ronald E. Dolan, 371–405. Washington, DC: Federal Research Division, Library of Congress, 1988.

Derrick, Jonathan. *Africa's "Agitators": Militant Anti-Colonialism in Africa and the West, 1918–1939*. New York: Columbia University Press, 2008.

Deschamps, Etienne. "Quelle Afrique pour une Europe unie? L'idée d'Eurafrique à l'aube des années trentes." In *Penser l'Europe à l'aube des années trentes*, edited by Michel Dumoulin and Xavier Dehan, 95–150. Louvain-la-neuve and Bruessel: Collège Érasme, 1995.

Desrosières, Alain. *La politique des grands nombres: histoire de la raison statistique*. Paris: La Découverte, 1993.

Diamond, Jared. *Guns, Germs, and Steel: The Fates of Human Societies*. New York: Norton, 1997.

Dierickx, Marc, and Peter Lyth. "From Privilege to Popularity: The Growth of Leisure Air Travel since 1945." *Journal of Transport History* 15, no. 2 (1994): 97–116.

Dietmar Rothermund, ed. *The Routledge Companion to Decolonization*, Routledge: London 2006.

Diogo, Maria Paula, and Isabel Maria Amaral, eds. *A Outra face do Império. Ciência, Tecnologia e Medicina (sécs.XIX–XX)*. Lisbon: Colibri, 2012.

———, and Ana Carneiro, "A Revista de Obras Públicas e Minas e a engenharia colonial." In *Formas do Império*, edited by Heloísa Meireles Gesteira, Luís Miguel Carolino, and Pedro Marinho, 517–34. São Paulo: Paz e Terra, 2014.

Disco, Cornelis. "Accepting Father Rhine? Technological Fixes, Vigilance, and Transnational Lobbies as 'European' Strategies of Dutch Municipal Water Supplies 1900–1975." *Environment and History*, Vol. 13, no. 4 (2007): 381–411. http://www.environmentandsociety.org/node/3307 (accessed December 30, 2012).

Doering-Manteuffel, Anselm. "Amerikanisierung und Westernisierung," Version: 1.0. *Docupedia-Zeitgeschichte*, 18.1.2011. https://docupedia.de/zg/Amerikanisierung_und_Westernisierung?oldid=84584 (accessed December 2, 2012).

Donovan, Arthur, and Joseph Bonney. *The Box That Changed the World: Fifty Years of Container Shipping – an Illustrated History*. East Windsor, NJ: Commonwealth Business Media, 2006.

Downs, Roger M., and David Stea. *Maps in Minds: Reflections on Cognitive Mapping*. New York: Harper & Row, 1977.

Dülffer, Jost. *Frieden stiften: Deeskalations- und Friedenspolitik im 20. Jahrhundert*. Cologne: Böhlau Verlag, 2008.

Eckert, Andreas. "Globalisierung." In *Europaeische Erinnerungsorte. Vol 3: Europa und die Welt*, edited by Pim de Boer, Heinz Duchhardt, Georg Kreis, and Wolfgang Schmale, 11–18. Munich: Oldenbourg Verlag, 2012.

Edelmayer, Friedrich. "The 'Legenda Negra' and the Circulation of Anti-Catholic and Anti-Spanish Prejudices." *European History Online (EGO)*. http://www.ieg-ego.eu/edelmayerf-2010-en (accessed September 15, 2012).

Edgerton, David. "Creole Technologies and Global Histories: Rethinking How Things Travel in Space and Time." *History of Science and Technology* 1 (2007): 75–112.

———. "The Contradictions of Techno Nationalism and Techno Globalism: A Historical Perspective." *New Global Studies* 1, no. 1 (2007): 1–32.

———. *The Shock of the Old: Technology and Global History Since 1900*. Oxford: Oxford University Press, 2007.

Ehn, Billy, Jonas Frykman, and Orvar Löfgren. *Försvenskningen av Sverige: det natio-nellas förvandlingar*. Stockholm: Natur och Kultur, 1993.

Eichengreen, Barry J. *The European Economy Since 1945: Coordinated Capitalism and Beyond*. Princeton: Princeton University Press, 2007.

Ellis, John. *The Social History of the Machine Gun*. New York: Pantheon Books, 1975.

Emig, Rainer. "Bringing the War Home: The Crimean War, the Telegraph, and Florence Nightingale." In *Wahrheitsmaschinen: der Einfluss technischer Innovationen auf die Darstellung und das Bild des Krieges in den Medien und Künsten*, edited by Claudia Glunz and Thomas F. Schneider, 287–99. Göttingen: V&R unipress, 2010.

———. "Die Domestizierung des Krieges: Florence Nightingales 'Public Relations'-Strategien waehrend des Krimkrieges." In *Medien - Krieg – Geschlecht. Affirmationen und Irritationen sozialer Ordnungen*, edited by Martina Thiele, 279–94. Wiesbaden: VS Verlag für Sozialwissenschaften, 2010.

Encyclopedia Britannica, Inc. *The New Encyclopedia Britannica*. Chicago: Encyclopedia Britannica, Inc., 2005.

Erker, Paul. *Vom nationalen zum globalen Wettbewerb: die deutsche und die amerikanische Reifenindustrie im 19. und 20. Jahrhundert*. Paderborn: Schöningh, 2005.

Espagne, Michel, Michael Geyer, and Matthias Middell, eds. *European History in an Interconnected World: An Introduction to Transnational History*. Basingstoke: Palgrave Macmillan, 2010.

Esposito, Fernando. *Mythische Moderne: Aviatik, Faschismus und die Sehnsucht nach Ordnung in Deutschland und Italien*. Munich: Oldenbourg Verlag, 2010.

Ėtkind, Aleksandr. *Internal Colonization: Russia's Imperial Experience*. Cambridge: Polity Press, 2011.

Evenson, Norma. *Chandigarh*. Berkeley: University of California Press, 1966.

Faraldo, José M., Paulina Gulińska-Jurgiel, and Christian Domnitz, eds. *Europa Im Ostblock: Vorstellungen Und Diskurse, 1945–1991*. Cologne: Böhlau Verlag, 2008.

Ferguson, Niall, Charles S. Maier, and Erez Manela, eds. *The Shock of the Global: The 1970s in Perspective*. Cambridge, MA: Belknap Press of Harvard University Press, 2011.

Fetherling, Douglas. *The Gold Crusades: A Social History of Gold Rushes, 1849–1929*. Toronto: University of Toronto Press, 1997.

Feuerhahn, Wolf, and Pascale Rabault-Feuerhahn. *La fabrique internationale de la science: les congrès scientifiques de 1865 à 1945*. Paris: CNRS Editions, 2010.

Fieldhouse, David K. *Western Imperialism in the Middle East 1914–1958*. Oxford: Oxford University Press, 2006.

Figes, Orlando. *Crimea: The Last Crusade*. London: Allen Lane, 2010.

Finn, Bernard S., and Daqing Yang, eds. *Communications Under the Seas: The Evolving Cable Network and Its Implications*. Cambridge, MA: MIT Press, 2009.

Fisch, Jörg. *Europa zwischen Wachstum und Gleichheit: 1850–1914*. Stuttgart: Ulmer, 2002.

Fische-Tiné, Harald. "Global Civil Society and the Forces of Empire: The Salvation Army, British Imperialism and the 'Pre-History' of NGOs (c. 1880–1920)." In *Competing Visions of World Order Global Moments and Movements, 1880s–1930s*, edited by Sebastian Conrad and Dominic Sachsenmaier, 29–68. New York: Palgrave Macmillan, 2007.

———. "'Deep Occidentalism'? Europa und 'der Westen' in der Wahrnehmung hinduistischer Intellektueller und Reformer ca. 1890–1930." *Journal of Modern European History* 4, no. 2 (2006): 171–203.

———. *Pidgin-Knowledge: Wissen und Kolonialismus*. Zurich and Berlin: Diaphanes Verlag, 2013.

———, and Michael Mann, eds. *Colonialism As Civilizing Mission: Cultural Ideology in British India*. London: Anthem Press, 2004.

Fogg, Gordon. *A History of Antarctic Science*. Cambridge: Cambridge University Press, 1992.

Fritzsche, Peter. *A Nation of Fliers: German Aviation and the Popular Imagination*. Cambridge, MA: Harvard University Press, 1992.

Fuentes, Carlos. "Lateinamerika – die Utopie Europas." *Die Zeit*, No 23, June 4, 1982.

Galison, Peter. *Einstein's Clocks and Poincaré's Maps: Empires of Time*. New York: Norton, 2003.

Gall, Alexander. "Atlantropa: Technological Visions of a United Europe." In *Networking Europe: Transnational Infrastructures and the Shaping of Europe, 1850–2000*, edited by Arne Kaijser and Eric van der Vleuten, 99–128. Sagamore Beach, MA: Science History Publications, 2006.

Gandhi, Mahatma. *Third Class in Indian Railways*. Kindle Edition, 2012.

Gavroglu, Kostas et al. "Science and Technology in the European Periphery: Some Historiographical Reflections." *History of Science* 46, no. 152 (2008): 153–75.

Gestwa, Klaus. *Die Stalinschen Großbauten des Kommunismus: sowjetische Technik- und Umweltgeschichte, 1948–1967*. Munich: Oldenbourg Verlag, 2010.

Gethmann, Daniel. "Fahrten im Luftmeer: Interkontinentale Landnahme per Luftschiff seit Bartholomeu Lourenço de Gusmão (1709)." In *Projektemacher: zur Produktion von Wissen in der Vorform des Scheiterns*, edited by Markus Krajewski, 128–61. Berlin: Kulturverlag Kadmos, 2004.

Geyer, Martin H., and Johannes Paulmann, eds. *The Mechanics of Internationalism: Culture, Society, and Politics from the 1840s to the First World War*. Oxford: Oxford University Press, 2001.

Ghosh, Suresh Chandra. "The Utilitarianism of Dalhousie and the Material Improvement of India." *Modern Asian Studies* 12 (1978): 97–110.

Gilman, Nils. *Mandarins of the Future: Modernization Theory in Cold War America*. Baltimore: Johns Hopkins University Press, 2003.

Global Witness. "Broken Vows: Exposing the 'Loupe' Holes in the Diamond Industry's Efforts to Prevent the Trade in Conflict Diamonds." London, 2004. https://www.globalwitness.org/sites/default/files/library/globalwitness_rpt_web_04.pdf (accessed December 3, 2015).

Goettsch-Elten, Silke. "Populaere Bilder vom Norden im 19. und 20. Jahrhundert." In *Ultima Thule. Bilder des Nordens von der Antike bis zur Gegenwart*, edited by Hannelore Engel-Braunschmidt et al., 123–43. Frankfurt Am Main: Peter Lang, 2001.

Goodman, Grant. *Japan and the Dutch 1600–1853*. Richmond: Routledge Curzon, 2000.

Gordon, Andrew. *A Modern History of Japan: From Tokugawa Times to the Present*. New York: Oxford University Press, 2003.

Gorman, Daniel. *The Emergence of International Society in the 1920s*. Cambridge: Cambridge University Press, 2012.

Gouldner, Alvin W. "Stalinism: A Study of Internal Colonialism." *Telos* 34 (Winter 1977/78): 5–48.

Gourlay, Jharna. *Florence Nightingale and the Health of the Raj*. Aldershot: Ashgate, 2003.

Gouzevitch, Dmitri, and Irina Gouzévitch, "Les corps d'ingénieurs comme forme d'organisation professionnelle en Russie." *Cahiers du monde russe* 41/4 (2000). http://monderusse.revues.org/60 (accessed September 8, 2012).

Graham, Loren R. *The Ghost of the Executed Engineer: Technology and the Fall of the Soviet Union.* Cambridge, MA: Harvard University Press, 1993.

Grandin, Greg. *Fordlandia: The Rise and Fall of Henry Ford's Forgotten Jungle City.* New York: Metropolitan Books, 2009.

Greenaway, Frank. *Science International: A History of the International Council of Scientific Unions.* Cambridge: Cambridge University Press, 1996.

Grove, Richard. *Ecology, Climate, and Empire: Colonialism and Global Environmental History, 1400–1940.* Cambridge: White Horse Press, 1997.

Gründer, Horst, ed. *"...da und dort ein junges Deutschland gründen": Rassismus, Kolonien und kolonialer Gedanke vom 16. bis zum 20. Jahrhundert.* Munich: Deutscher Taschenbuch Verlag, 1999.

Grützmacher, Johannes. *Die Baikal-Amur-Magistrale: vom stalinistischen Lager zum Mobilisierungsprojekt unter Breznev.* Munich: Oldenbourg Verlag, 2012.

Guha, Ramachandra. *India After Gandhi: The History of the World's Largest Democracy.* New York and London: Harper Perennial, 2008.

Gyr, Ueli. "Geschichte des Tourismus: Strukturen auf dem Weg zur Moderne." *Europaeische Geschichte Online (EGO),* Mainz 2010–12–03, urn:nbn:de:0159–20100921237. http://www.ieg-ego.eu/gyru-2010-de (accessed November 4, 2012).

Ha, Anthony. "International venture funding rose 5 percent in 2008." *Venture Beat,* February 18, 2009. http://venturebeat.com/2009/02/18/international-venture-funding-rose-15-percent-in-2008 (accessed November 16, 2013).

Hachtmann, Rüdiger. "Fordismus, Version: 1.0." *Docupedia-Zeitgeschichte,* 27.10.2011. https://docupedia.de/zg/Fordismus?oldid=80661 (accesed October 7, 2012).

Hacker, Barton C. "Engineering a New Order: Military Institutions, Technical Education, and the Rise of the Industrial State." *Technology and Culture* 34, no. 1 (1993): 1–27.

———. "The Machines of War: Western Military Technology 1850–2000." *History and Technology* 21, no. 3 (2005): 255–300.

Haeckel, Ernst. *Die Weltraetsel.* Stuttgart: A. Kroener, 1899.

Hale, Julian Anthony Stuart. *Radio Power: Propaganda and International Broadcasting.* Philadelphia: Temple University Press, 1975.

Haller, Martin. "Die Erfindung der Heimatfront. Ein Stuttgarter Symposion ueber den Krimkrieg als ersten europaeischen Medienkrieg." *Frankfurter Allgemeine Zeitung,* No. 148, June 29, 2005, N3.

Handberg, Roger, and Zhen Li. *Chinese Space Policy: A Study in Domestic and International Politics.* Abingdon: Routledge, 2007.

Hanioğlu, M. Şükrü. *A Brief History of the Late Ottoman Empire.* Princeton: Princeton University Press, 2008.

Hannemann, Matthias. " 'North to the Future' – die Arktis und die Medien." *Aus Politik und Zeitgeschichte,* 5–6, January 31, 2011, 35–38.

"Hannover Declaration." November 6, 1985. *EUREKA Network.* http://www.eurekanetwork.org/c/document_library/get_file?uuid=1b92be16-ec94–4a7e-a8d1–6dd40e4fb318&groupId=10137 (accessed August 11, 2011).

Harbsmeier, Michael. "Schauspiel Europa. Die aussereuropaeische Entdeckung Europas im 19. Jahrhundert am Beispiel afrikanischer Texte." *Historische Anthropologie* 2 (1994): 331–50.

————. *Stimmen aus dem aeussersten Norden. Wie die Groenlaender Europa fuer sich entdeckten*. Stuttgart: Jan Thorbecke Verlag, 2001.

Harsanyi, Doina Pasca. "Anti-Americanism." In *Encyclopedia of Western Colonialism Since 1450*, Vol 1, edited by Thomas Benjamin, 53–57. Detroit: Macmillan Reference USA, 2007.

Headley, John M. *The Europeanization of the World: On the Origins of Human Rights and Democracy*. Princeton: Princeton University Press, 2008.

Headrick, Daniel R. *Power Over Peoples: Technology, Environments, and Western Imperialism, 1400 to the Present*. Princeton: Princeton University Press, 2010.

————. *The Tentacles of Progress: Technology Transfer in the Age of Imperialism, 1850–1940*, New York and Oxford: Oxford University Press, 1988.

————. *The Tools of Empire: Technology and European Imperialism in the Nineteenth Century*. Oxford and New York: Oxford University Press, 1981.

————. *When Information Came of Age: Technologies of Knowledge in the Age of Reason and Revolution, 1700–1850*. Oxford: Oxford University Press, 2000.

Hecht, Gabrielle. "Technology, Politics, and National Identity in France." In *Technologies of Power*, edited by Michael Thad Allen and Gabrielle Hecht, 253–93. Cambridge, MA: MIT Press, 2001.

————. *Being Nuclear Africans and the Global Uranium Trade*. Cambridge, MA: MIT Press, 2012.

————. *The Radiance of France: Nuclear Power and National Identity After World War II*. Cambridge, MA: MIT Press, 2009.

Hecker-Stampehl, Jan. *Vereinigte Staaten des Nordens. Integrationsideen in Nordeuropa im Zweiten Weltkrieg*. Munich: De Gruyter, 2011.

Heffernan, Michael J. "Bringing the Desert to Bloom: French Ambitions in the Sahara Desert During the Late Nineteenth Century – the Strange Case of 'la mer intérieure'." In *Water, Engineering and Landscape: Water Control and Landscape Transformation in the Modern Period*, edited by Denis Cosgrove and Geoff Petts, 94–114. London: Belhaven Press, 1990.

Heinze-Greenberg, Ita. *Europa in Palästina: die Architekten des zionistischen Projekts 1902–1923*. Zurich: Gta-Verlag, 2011.

Heiss, Hans. "Tourism." In *The Palgrave Dictionary of Transnational History*, edited by Akira Iriye and Pierre-Yves Saunier, 1021–4. Basingstoke: Palgrave Macmillan, 2009.

Heller, Otto. *Sibirien, ein anderes Amerika*. Berlin: Neuer deutscher Verlag, 1930.

Herren, Madeleine. "Der Voelkerbund. Erinnerung an ein globales Europa." In *Europaeische Erinnerungsorte, Vol. 3: Europa und die Welt*, edited by Pim de Boer, Heinz Duchhardt, George Reis, and Wolfgang Schmale, 271–80. Munich: Oldenbourg Verlag, 2012.

————. *Hintertüren zur Macht: Internationalismus und modernisierungsorientierte Aussenpolitik in Belgien, der Schweiz und den USA 1865–1914*. Munich: Oldenbourg Verlag, 2000.

————. *Internationale Organisationen seit 1865: eine Globalgeschichte der internationalen Ordnung*. Darmstadt: Wissenschaftliche Buchgesellschaft, 2009.

Hevia, James Louis. *The Imperial Security State: British Colonial Knowledge and Empire-Building in Asia*. Cambridge: Cambridge University Press, 2012.

Hirschhausen, Ulrike von, and Klaus Patel. "Europeanization in History: An Introduction." In *Europeanization in the Twentieth Century: Historical Approaches*, edited by Martin Conway and Kiran Klaus Patel, 1–18. Basingstoke: Palgrave Macmillan, 2010.

Hobson, John M. *The Eastern Origins of Western Civilization*. Cambridge: Cambridge University Press, 2004.

Hochschild, Adam. *Bury the Chains: Prophets and Rebels in the Fight to Free an Empire's Slaves*. Boston: Houghton Mifflin, 2005.

———. *King Leopold's Ghost: A Story of Greed, Terror, and Heroism in Colonial Africa*. Boston: Houghton Mifflin, 1998.

Hodeir, Catherine, and Michel Pierre. *L'exposition coloniale de 1931*. Waterloo (Belgium): André Versaille, 2011.

Högselius, Per. *Red Gas: Russia and the Origins of European Energy Dependence*. Basingstoke: Palgrave Macmillan, 2013.

Holtorf, Christian. "Die Modernisierung des nordatlantischen Raumes. Cyrus Field, Taliaferro Shafner und das submarine Telegraphennetz von 1858." In *Ortsgespräche: Raum und Kommunikation im 19. und 20. Jahrhundert*i, edited by Alexander C. T. Geppert, Uffa Jensen, and Joern Weinhold, 157–78. Bielefeld: Transcript, 2005.

———. *Der erste Draht zur Neuen Welt: die Verlegung des transatlantischen Telegrafenkabels*. Göttingen: Wallstein Verlag, 2013.

Hønneland, Geir, and Olav Schram Stokke. *International Cooperation and Arctic Governance: Regime Effectiveness and Northern Region Building*. London: Routledge, 2007.

Hotta, Eri. "Pan-Isms." In *The Palgrave Dictionary of Transnational History*, edited by Akira Iriye and Pierre-Yves Saunier, 806–11. Basingstoke: Palgrave Macmillan, 2009.

Howe, Stephen. *The New Imperial Histories Reader*. London: Routledge, 2010.

Hsü, Immanuel C.Y. *The Rise of Modern China*. Oxford: Oxford University Press, 1999 (6th edition).

Huber, Valeska. "The Unification of the Globe by Disease? The International Sanitary Conferences on Cholera, 1851–1894." *Historical Journal* 49, no. 2 (2006): 453–76.

———. *Channelling Mobilities: Migration and Globalisation in the Suez Canal Region and Beyond, 1869–1914*. Cambridge: Cambridge University Press, 2013.

Hughes, Thomas Parke. *American Genesis: A Century of Invention and Technological Enthusiasm, 1870–1970*. New York: Viking, 1989.

Hugill, Peter J. *Global Communications Since 1844: Geopolitics and Technology*. Baltimore: Johns Hopkins University Press, 1999.

Huse, Norbert. *Le Corbusier in Selbstzeugnissen und Bilddokumenten*. Reinbek: Rowohlt, 1976.

Hutchinson, John F. "Rethinking the Origins of the Red Cross." *Bulletin of the History of Medicine* 63, no. 4 (1989): 557–78.

Iliffe, John. *Geschichte Afrikas*. Munich: C.H. Beck, 2000.

Ingebritsen, Christine. "The Scandinavian Way and Its Legacy in Europe." *Scandinavian Studies* 74, no. 3 (2002): 255–64.

Ingimundarson, Valur. "The Geopolitics of Arctic Natural Resources." *European Parliament — Directorate-General For External Policies of the Union*, Brussels, 2010. http://tepsa.be/Valur%20Ingimundarson.pdf (accessed December 3, 2015).

"International Container Bureau." *Lonsea.org* — League of Nations Search Engine. http://www.lonsea.de/pub/org/1171 (accessed December 23, 2012).

"Interview with Shyam Kumar, General Manager of the Eastern Railway." *Frontline*, Volume 21, Issue 17 (2004). http://www.frontline.in/static/html/fl2117/stories/20040827004511600.htm (accessed June 10, 2011).

Iriye, Akira. *Global Community: The Role of International Organizations in the Making of the Contemporary World*. Berkeley: University of California Press, 2002.

Jaeckh, Ernst. *Der goldene Pflug: Lebensernte eines Weltbuergers*. Stuttgart: Deutsche Verlags-Anstalt, 1954.

Jersild, Austin. *The Sino-Soviet Alliance: An International History*. Chapel Hill, NC: University of North Carolina Press, 2014.

Johnston, Alan. "Libya 1911: How an Italian pilot began the air war era." *BBC News Europe*, May 10, 2011.. http://www.bbc.co.uk/news/world-europe. (accessed September 20, 2012).

Jonas, Raymond Anthony. *The Battle of Adwa: African Victory in the Age of Empire*. Cambridge, MA: Belknap Press of Harvard University Press, 2011.

Jones, Eric. *The European Miracle: Environments, Economies, and Geopolitics in the History of Europe and Asia*. Cambridge: Cambridge University Press, 2003.

Jones, James A., *Industrial Labor in the Colonial World: Workers of the Chemin de Fer Dakar-Niger, 1881–1963*. Portsmouth, NH: Heinemann, 2002.

Jordan, John M. "'Society Improved the Way You Can Improve a Dynamo': Charles P. Steinmetz and the Politics of Efficiency." *Technology and Culture* 30, no. 1 (1989): 57–82.

Josephson, Paul R. "'Projects of the Century' in Soviet History: Large-Scale Technologies from Lenin to Gorbachev." *Technology and Culture* 36, no. 3 (1995): 519–59.

———. *New Atlantis Revisited: Akademgorodok, the Siberian City of Science*. Princeton: Princeton University Press, 1997.

Judel, Guenther Klaus. "Die Geschichte von Liebigs Fleischextrakt. Zur populaersten Erfindung des beruehmten Chemikers." *Spiegel der Forschung* 20, no. 1 (2003): 6–17.

Judt, Tony. *Postwar: A History of Europe Since 1945*. New York: Penguin Press, 2005.

Juergens, Ulrich. "Approaches towards Integrating Suppliers in Simultaneous Engineering Activities: The Case of Two German Automakers." *International Journal of Automotive Technology and Management* 1, no. 1 (2001): 61–77.

Jünger, Ernst. *Luftfahrt ist not!* Berlin: W. Andermann, 1930.

Jünger, Wolfgang. *Kampf um Kautschuk*. Leipzig: W. Goldmann, 1942.

Kabou, Axelle. *Et si l'Afrique refusait le développement?* Paris: L'Harmattan, 1991.

Kaiser, Hilmar. "The Bagdad Railway and the Armenian Genocide, 1915–16: A Case Study in German Resistance and Complicity." In *Remembrance and Denial: The Case of the Armenian Genocide*, edited by Richard G. Hovannisian, 67–112. Detroit: Wayne State University Press, 1998.

Kaiser, Wolfram, and Johan W. Schot. *Writing the Rules for Europe: Experts, Cartels, and International Organizations*. Basingstoke: Palgrave Macmillan, 2014.

———, Brigitte Leucht, and Morten Rasmussen, eds. *The History of the European Union: Origins of a Trans- and Supranational Polity 1950–72*. New York: Routledge, 2009.

Kalshoven, Frits, and Liesbeth Zegveld. *Constraints on the Waging of War: An Introduction to International Humanitarian Law*. Cambridge: Cambridge University Press, 2011.

Karabell, Zachary. *Parting the Desert: The Creation of the Suez Canal*. New York: Knopf, 2003.

Keegan, John. *A History of Warfare*. New York: Knopf, 1993.

Kehrt, Christian. *Moderne Krieger: die Technikerfahrungen deutscher Militärpiloten 1910–1945*. Paderborn: Schöningh, 2010.

Keller, Ulrich. *The Ultimate Spectacle: A Visual History of the Crimean War*. Amsterdam: Gordon and Breach, 2001.

Kerr, Ian J. "John Chapman and the Promotion of the Great Indian Peninsula Railway, 1842–1850," http://www.docutren.com/HistoriaFerroviaria/Semmering2004/pdf/14.pdf (accessed June 10, 2011).

Kiernan, Victor G. *Colonial Empires and Armies, 1815–1960*. Montreal: McGill-Queen's University Press, 1998.

Kierzkowski, Hendryk, ed. *Europe and Globalization*. Basingstoke: Palgrave Macmillan, 2002.

Killingray, David. "'A Swift Agent of Government': Air Power and British Colonial Africa, 1916–1939." *Journal of African History* 25 (1986): 429–44.

Kittler, Friedrich. "Zur Theoriegeschichte von Information Warfare." *Ars Electronica Archiv* (1998). http://90.146.8.18/de/archives/festival_archive/festival_catalogs/festival_artikel.asp?iProjectID=8439 (accessed December 22, 2012).

Klein, Thoralf, and Frank Schumacher, eds. *Kolonialkriege: militärische Gewalt im Zeichen des Imperialismus*. Hamburg: Hamburger Edition, 2006.

Klimke, Martin. "America." In *The Palgrave Dictionary of Transnational History*, edited by Akira Iriye and Pierre-Yves Saunier, 33–36. Basingstoke: Palgrave Macmillan, 2009.

Knight, Ian. *The Anglo-Zulu War*. Oxford: Osprey, 2004.

Knightley, Phillip. *The First Casualty: The War Correspondent as Hero and Myth-Maker from the Crimea to Iraq*. Baltimore: Johns Hopkins University Press, 2004.

Knoch, Habbo. "Life on Stage. Grand Hotels as Urban Interzones around 1900." In *Creative Urban Milieus: Historical Perspectives on Culture, Economy, and the City*, edited by Martina Hessler and Clemens Zimmermann, 137–58. Frankfurt am Main: Campus Verlag, 2008.

Kolnai, Aurel. *The War Against the West*. New York: Viking Press, 1938.

Kohlrausch, Martin and Helmuth Trischler. *Building Europe on Expertise: Innovators, Organizers, Networkers*. Basingstoke: Palgrave Macmillan, 2014.

Kohn, Hans. *Die Europaeisierung des Orients*. Berlin: Schocken, 1934.

Kojève, Alexandre. "Kolonialismus in europaeischer Sicht." In *Schmittiana: Beiträge zu Leben und Werk Carl Schmitts, Band VII*, edited by Angela Stender and Piet Tomissen, 125–40. Berlin: Duncker & Humblot, 2001.

Koller, Christian. *Von Wilden aller Rassen niedergemetzelt: die Diskussion um die Verwendung von Kolonialtruppen in Europa zwischen Rassismus, Kolonial- und Militärpolitik (1914–1930)*. Stuttgart: Steiner, 2001.

Korieh, Chima J. "Diamonds." In *Encyclopedia of Western Colonialism Since 1450*, Vol 1, edited by Thomas Benjamin, 301–4. Detroit: Macmillan Reference USA, 2007.

Koshar, Rudy. *Histories of Leisure*. Oxford: Berg, 2002.

Krajewski, Markus. *Restlosigkeit: Weltprojekte um 1900*. Frankfurt am Main: Fischer Taschenbuch Verlag, 2006.

Kranakis, Eda. "European Civil Aviation in an Era of Hegemonic Nationalism: Infrastructure, Air Mobility, and European Identity Formation, 1919–1933." In *Materializing Europe: Transnational Infrastructures and the Project of Europe*, edited by Alec Badenoch and Andreas Fickers, 290–326. New York: Palgrave Macmillan, 2010.

———. "The 'good miracle': Building a European Airspace Commons, 1919–1939." In *Cosmopolitan Commons: Sharing Resources and Risks across Borders*, edited by Nil Disco and Eda Kranakis, 57–96. Cambridge, MA: MIT Press, 2013.

Kreiser, Klaus. *Atatuerk: Eine Biographie*. Munich: C.H. Beck, 2008.

Kroll, Stephan. "Völkerrecht und Weltwirtschaft im 19. Jahrhundert." *H-Soz-Kult*, 29/07/2009. http://hsozkult.geschichte.hu-berlin.de/termine/id=11953 (accessed August 14, 2012).

Kumar, Deepak. "Reconstructing India: Disunity in the Science and Technology for Development Discourse, 1900–1947." *Osiris* 15 (2001): 241–57.

Kurlansky, Mark. *Salt: A World History*. New York: Walker and Co., 2002.

Kuß, Susanne. *Deutsches Militär auf kolonialen Kriegsschauplätzen Eskalation von Gewalt zu Beginn des 20. Jahrhunderts*. Berlin: Links, 2010.

Laak, Dirk van. "Detours around Africa: The Connection between Developing Colonies and Integrating Europe." In *Materializing Europe: Transnational Infrastructures and the Project of Europe*, edited by Alec Badenoch and Andreas Fickers, 27–43. New York: Palgrave Macmillan, 2010.

———. *Weiße Elefanten: Anspruch und Scheitern technischer Großprojekte im 20. Jahrhundert*. Munich: DVA, 1999.

———. "Planung, Planbarkeit und Planungseuphorie (Werkstatt-Version)." *Docupedia-Zeitgeschichte*, February 11, 2010. http://docupedia.de/zg/Planung (accessed December 3, 2015).

———. *Imperiale Infrastruktur. Deutsche Planungen fuer eine Erschliessung Afrikas, 1880–1960*. Paderborn: Schoeningh, 2004.

Laborie, Léonard. *L'Europe mise en réseaux: la France et la coopération internationale dans les postes et les télécommunications (années 1850-années 1950)*. Brussels: Peter Lang, 2010.

Ladikas, Miltos et al., eds. *Science and Technology Governance and Ethics: A Global Perspective from Europe, India and China*. Springer Open, 2015.

Lagendijk, Vincent. "Transnationalising the TVA: International River Development in Troubled Waters." http://media.leidenuniv.nl/legacy/gi-paper-lagendi-jk-ttt.pdf (accessed October 2, 2013).

Lahiri Choudhury, Deep Kanta. *Telegraphic Imperialism Crisis and Panic in the Indian Empire, C.1830–1920*. New York: Palgrave Macmillan, 2010.

Lander, John and Richard Lander. *Journal of an Expedition to Explore the Course and Termination of the Niger*. Boston: Adamant Media Corporation, 2002 (Reprint of the 1833 edition).

Landes, David S. *The Wealth and Poverty of Nations: Why Some Are so Rich and Some so Poor*. New York: Norton, 1998.

Lapping, Brian. *Apartheid: A History*. New York: G. Braziller, 1987.

Laqua, Daniel. *Internationalism Reconfigured: Transnational Ideas and Movements between the World Wars*. London: I.B. Tauris, 2011.

Lardy, Nicholas R. "Recasting of the Economic System: Structural Reforms of Agriculture and Industry." In *China in the Era of Deng Xiaoping: A Decade of Reform*, edited by Susan H. Marsh and Michael Ying-mao Kau, 103–19. London: M.E. Sharpe, 1993.

Larson, Edward J. *An Empire of Ice: Scott, Shackleton, and the Heroic Age of Antarctic Science*. New Haven: Yale University Press, 2011.

Latour, Bruno. "Drawing Things Together." In *Representation in Scientific Practice*, edited by Michael Lynch and Steven Woolgar, 19–68. Cambridge, MA: MIT Press, 1990.

Latourette, Kenneth Scott. *Japan: From Ancient Times to 1918*. Kindle Edition, 2012.

Launius, Roger D., James Rodger Fleming, and David H. DeVorkin, eds. *Globalizing Polar Science: Reconsidering the International Polar and Geophysical Years*. New York: Palgrave Macmillan, 2010.

Laurelle, Marlène. "Die europaeischen Urspruenge des Eurasianismus. Russische Emigranten und ihre Kritik am Eurasianismus 1920–1930." In *Vom Gegner lernen: Feindschaften und Kulturtransfers im Europa des 19. und 20. Jahrhunderts*, edited by Martin Aust and Daniel Schoenpflug, 157–78. Frankfurt am Main: Campus Verlag, 2007.

Law, John. "Technology and Heterogeneous Engineering: The Case of Portuguese Expansion." In *The Social Construction of Technological Systems*, edited by Wiebe E. Bijker et al., 105–27. Cambridge, MA: MIT Press, 1987.

Lederer, William J., and Eugene Burdick. *The Ugly American*. New York: Norton, 1958.

Leibo, Steven A. *Transferring Technology to China: Prosper Giquel and the Self-Strengthening Movement*. Berkeley: Institute of East Asian Studies, University of California, Berkeley, Center for Chinese Studies, 1985.

Leitão, Henrique, ed. "Um Mundo Novo e uma Nova Ciência." In *360° · Ciência Descoberta, Catálogo da Exposição*, edited by Henrique Leitão, 16–39. Lisbon: Fundação Calouste Gulbenkian, 2013.

Lemberg, Hans. "Zur Entstehung des Osteuropabegriffs im 19. Jahrhundert. Vom 'Norden' zum 'Osten' Europas." *Jahrbücher für Geschichte Osteuropas* 33, no. 1 (1985): 48–91.

Lenin, Vladimir. "Report On The Work Of The All-Russia Central Executive Committee And The Council Of People's Commissars." Speech delivered at the first session of the All-Russia Central Executive Committee, Seventh Convocation, February 2, 1920. See http://ww.marxists.org/archive/lenin/works/date/1920.htm (accessed December 3, 2015).

———. "Our Foreign and Domestic Position and Party Tasks." Speech Delivered To The Moscow Gubernia Conference Of The R.C.P.(B.), November 21, 1920. https://www.marxists.org/archive/lenin/works/1920/nov/21.htm (accessed December 3, 2015).

Lessep, Ferdinand de. *The History of the Suez Canal: A Personal Narrative*. Whitefish, MT: Kessinger Publishing, 2007 (facsimile reprint of the 1876 original).

Letter of the Chief Engineer of the Companhia Real dos Caminhos de Ferro atravez d'Africa [Royal Railway Company across Africa], 1888. Arquivo Histórico Ultramarino – AHU (Overseas Historical Archive), 2678, Sala 3, Est. 16, Prat. 17, n° 13420.

Levinson, Marc. *The Box: How the Shipping Container Made the World Smaller and the World Economy Bigger*. Princeton: Princeton University Press, 2006.

Lewis, David Levering. *Race to Fashoda*. New York: Henry Holt, 2001.

Lewis, John Wilson, and Litai Xue. *China Builds the Bomb*. Stanford: Stanford University Press, 1988.

Leys, Colin. *The Rise & Fall of Development Theory*. Nairobi: East African Educational Publishers, 1996.

Lien, Mariannne Elisabeth. "Diet and Nutrition." In *The Palgrave Dictionary of Transnational History*, edited by Akira Iriye and Pierre-Yves Saunier, 279. Basingstoke: Palgrave Macmillan, 2009.

Lincoln, W. Bruce. *Die Eroberung Sibiriens*. Munich and Zürich: Piper, 1996.

———. *The Conquest of a Continent: Siberia and the Russians*. New York: Random House, 1994.

Lindner, Ulrike, Maren Moehring, Mark Stein and Silke Stroh, eds. *Hybrid Cultures – Nervous States: Britain and Germany in a (Post)Colonial World*. Amsterdam: Ropopi, 2010.

Lindqvist, Sven. *"Exterminate All the Brutes": One Man's Odyssey into the Heart of Darkness and the Origins of European Genocide*. New York: New Press, 2007.

———. *A History of Bombing*. New York: New Press, 2001.

Liniger-Goumaz, Max. *L'Eurafrique, utopie ou réalité?* Yaoundé: Éditions CLE, 1972.

Lipphardt, Veronika. "Von der 'europaeischen Rasse' zu den 'Europiden'. Wissen um die biologische Beschaffenheit des Europaeers in Sach- und Lehrbuechern, 1950–1989." In *Der Europaeer – ein Konstrukt. Wissensbestaende, Diskurse, Praktiken*, edited Lorraine Blunche, Veronika Lipphardt, and Kiran Klaus Patel, 158–86. Göttingen: Wallstein, 2009.

———, and Kiran Klaus Patel. "Auf der Suche nach dem Europaeer. Wissenschaftliche Konstruktionen des Homo Europaeus." *Themenportal Europaeische Geschichte*. http://www.europa.clio-online.de/2007/Article=204 (accessed January 2, 2013).

Locke, Robert R., and Katja E. Schöne. *The Entrepreneurial Shift: Americanization in European High-Technology Management Education*. Cambridge: Cambridge University Press, 2004.

Löhr, Isabella. *Die Globalisierung geistiger Eigentumsrechte: neue Strukturen internationaler Zusammenarbeit, 1886–1952*. Göttingen: Vandenhoeck & Ruprecht, 2010.

———, and Roland Wenzlhuemer, eds. *The Nation State and Beyond: Governing Globalization Processes in the Nineteenth and Early Twentieth Centuries*. Berlin: Springer, 2013.

Lovell, Julia. *The Opium War: Drugs, Dreams and the Making of China*. London: Picador, 2011.

Low, Morris, ed. *Building a Modern Japan: Science, Technology, and Medicine in the Meiji Era and Beyond*. Basingstoke: Palgrave Macmillan, 2005.

Lucius, Robert von. "Am Oranjefluss geht es um Diamanten, Gas und Trinkwasser." *Frankfurter Allgemeine Zeitung*, No. 303, December 30, 2000, 4.

Lyth, Peter. " 'Gimme a ticket on an airoplane...': The Jet Engine and the Revolution in Leisure Air Travel, 1960–1975." In *Construction d'une industrie touristique aux 19e et 20e siècles: perspectives internationales*, edited by Laurent Tissot, 111–22. Neuchâtel: Éd. Alphil, 2003.

Maag, Georg, Wolfram Pyta, and Martin Windisch, eds. *Der Krimkrieg als erster europaeischer Medienkrieg*. Berlin: LIT Verlag Münster, 2010.

Machado, Joaquim José. "Caminho de Ferro de Mossamedes ao Bihé." *Revista de Obras Públicas* 21, no. 247 and 248 (July–August, 1890): 219–96.

———. "Memória ácerca do caminho de ferro de Lourenço Marques à fronteira do Transvaal." *Revista de Obras Públicas e Minas* 12, no. 445 (1882): 1–57.

MacKenzie, David Clark. *A World Beyond Borders: An Introduction to the History of International Organizations.* Toronto: University of Toronto Press, 2010.

Macpherson, W.J. *The Economic Development of Japan, 1868–1941.* Cambridge: Cambridge University Press, 1995.

Magnus, Olaus. *Die Wunder des Nordens. Erschlossen von Elena Balzamo und Reinhard Kaiser.* Frankfurt am Main: Eichborn, 2006.

Maines, Rachel. *Hedonizing Technologies: Paths to Pleasure in Hobbies and Leisure.* Baltimore: Johns Hopkins University Press, 2009.

Manela, Erez. *The Wilsonian Moment: Self-Determination and the International Origins of Anticolonial Nationalism.* Oxford: Oxford University Press, 2007.

Mangold-Will, Sabine. *Begrenzte Freundschaft: Deutschland und die Türkei 1918–1933.* Goettingen: Wallstein, 2013.

Marks, Steven G. *Road to Power: The Trans-Siberian Railroad and the Colonization of Asian Russia, 1850–1917.* Ithaca: Cornell University Press, 1991.

Martenstein, Harald. "Die Fuersten der Verschwendung: In der Maerchenwelt der Maharadschas." *Geo-Epoche*, no. 41 (2010): 120–9.

Martykánová, Darina. *Reconstructing Ottoman Engineers: Archaeology of a Profession (1789–1914).* Plus (online edition), 2010.

Marx, Christoph. " 'Europeans Only': Europa als Leitbild, Vorbild und Zerrbild in Suedafrika, 1948–2008." In *Bilder von Europa: Innen- und Aussenansichten von der Antike bis zur Gegenwart*, edited by Benjamin Drechsel et al., 155–73. Bielefeld: Transcript, 2010.

Mattelart, Armand. *Networking the World, 1794–2000.* Minneapolis: University of Minnesota Press, 2000.

Mattioli, Aram. *Experimentierfeld der Gewalt: der Abessinienkrieg und seine internationale Bedeutung 1935–1941.* Zürich: Orell Füssli, 2005.

Maul, Daniel R. *Human Rights, Development and Decolonization: The International Labour Organization, 1940–70.* New York: Palgrave Macmillan, 2012.

Maurer, Michael. "Europaeische Geschichte." In *Aufriss der historischen Wissenschaften*, edited by Michael Maurer, Vol. 2, 99–197. Stuttgart: Reclam, 2001.

Mazower, Mark. *Dark Continent: Europe's Twentieth Century.* New York: Knopf, 1999.

———. *Der dunkle Kontinent.* Frankfurt am Main: Fischer, 2002.

———. *Governing the World: The History of an Idea, 1815 to the Present.* New York: Penguin Books, 2013.

Mazumdar, Shaswati. *Insurgent Sepoys: Europe Views the Revolt of 1857.* London: Routledge, 2011.

McCormack, Robert Lewis. "Aviation and Empire: The British African Experience, 1919–1939." PhD diss., Dalhousie University, 1974.

McDonald, Lynn. *Florence Nightingale. Collected Works of Florence Nightingale, Vol 14.* Waterloo: Wilfrid Laurier University Press, 2010.

McMahon, Robert J. *The Cold War in the Third World.* Oxford and New York: Oxford University Press, 2013.

McMeekin, Sean. *The Berlin-Baghdad Express: The Ottoman Empire and Germany's Bid for World Power*. Cambridge, MA: Belknap Press of Harvard University Press, 2010.

McMurray, Jonathan S. *Distant Ties: Germany, the Ottoman Empire, and the Construction of the Baghdad Railway*. Westport, CN: Praeger, 2001.

McNeill, John Robert. *Mosquito Empires: Ecology and War in the Greater Caribbean, 1620–1914*. New York: Cambridge University Press, 2010.

———. *Something New Under the Sun: An Environmental History of the Twentieth-Century World*. New York: Norton, 2000.

McPherson, Kenneth. "Containerization." In *History of World Trade Since 1450*, edited by John J. McCusker, 154–57. Farmington Hills, MI: Macmillan Reference USA, 2006.

Meredith, Martin. *The State of Africa: A History of Fifty Years of Independence*. London: Free Press, 2005.

Meyer, Henry Cord. *Mitteleuropa in German Thought and Action, 1815–1945*. The Hague: Nijhoff, 1955.

Middell, Matthias, ed. "Welt- und Globalgeschichte in Europa." *Historical Social Research* 31, no. 2 (2006).

———, and Katja Naumann. "Global History and the Spatial Turn: From the Impact of Area Studies to the Study of Critical Junctures of Globalization." *Journal of Global History* 5, no. 1 (2010): 149–70.

———, and Ulf Engel, eds. "Bruchzonen der Globalisierung." *Comparativ* 15, no. 5/6 (2005): 5–38.

Milde, David. "Lernen von den Eskimos. Der Weltfahrer Colin Ross zwischen Moderne und Nationalsozialismus." In *Der Technikdiskurs in der Hitler-Stalin-Aera*, edited by Wolfgang Emmerich and Carl Wege, 146–58. Stuttgart: J.B. Metzler, 1995.

Mill, James. *The History of British India in 6 vols.* (third edition) London: Baldwin, Cradock, and Joy, 1826. Volume 2.

Miller, Michael B. *Europe and the Maritime World: A Twentieth-Century History*. Cambridge: Cambridge University Press, 2012.

Milton, Giles. *Nathaniel's Nutmeg, or, The True and Incredible Adventures of the Spice Trader Who Changed the Course of History*. New York: Farrar, Straus and Giroux, 1999.

Mintz, Sidney Wilfred. *Sweetness and Power: The Place of Sugar in Modern History*. New York: Viking, 1985.

Misa, Thomas, and Johan Schot. "Inventing Europe: Technology and the Hidden Integration of Europe." *History and Technology* 21, no. 1 (2005): 1–19.

Mishra, Pankaj. *From the Ruins of Empire: The Intellectuals Who Remade Asia*. New York: Farrar, Straus & Giroux, 2012.

Mitter, Rana. *A Bitter Revolution: China's Struggle with the Modern World*. Oxford: Oxford University Press, 2004.

Mollin, Gerhard. *Die USA und der Kolonialismus: Amerika als Partner und Nachfolger der belgischen Macht in Afrika 1939–1965*. Berlin: De Gruyter, 1996.

Montel, Nathalie. *Le Chantier Du Canal de Suez (1859–1869): Une Histoire des Pratiques Techniques*. Paris: Presses de l'École Nationale des Ponts et Chaussées, 1999.

Morel, Edmund D. *Red Rubber: The Story of the Rubber Slave Trade Flourishing on the Congo in the Year of Grace 1906*. New York: Negro Universities Press, 1969.

Morris-Suzuki, Tessa. *The Technological Transformation of Japan: From the Seventeenth to the Twenty-First Century*. Cambridge: Cambridge University Press, 1994.

Morus, Iwan Rhys. " 'The Nervous System of Britain'. Space, Time and the Electric Telegraph in the Victorian Age." *British Journal for the History of Science* 33, no. 4 (2000): 455–75.

Moser, Thomas. *Europäische Integration, Dekolonisation, Eurafrika: eine historische Analyse über die Entstehungsbedingungen der eurafrikanischen Gemeinschaft von der Weltwirtschaftskrise bis zum Jaunde-Vertrag, 1929–1963*. Baden-Baden: Nomos, 2000.

Mudimbe, Victor Yves. *The Invention of Africa: Gnosis, Philosophy, and the Order of Knowledge*. Bloomington: Indiana University Press, 1988.

Mullen, Richard, and James Munson. *'The Smell of the Continent': The British Discover Europe 1814–1914*. London: Macmillan, 2009.

Mungello, David. *The Great Encounter of China and the West, 1500–1800*. Lanham, MD: Rowman & Littlefield, 1999.

Murphy, Craig, and Jo Anne Yates. *ISO, the International Organization for Standardization: Global Governance through Voluntary Consensus*. London: Routledge, 2008.

Murteira, Mário. *Economia do Conhecimento*. Lisbon: Quimera, 2004.

Nagel, Brigitte. *Die Welteislehre: ihre Geschichte und ihre Rolle im "Dritten Reich."* Stuttgart: Verlag für Geschichte der Naturwissenschaften und der Technik, 1991.

Nakayama, Shigeru. "Techno-nationalism versus Techno-globalism." *East Asian Science, Technology and Society* 6, no. 1 (2012): 9–15.

Nansen, Fridtjof. *Sibirien, ein Zukunftsland*. Leipzig: F.A. Brockhaus, 1922.

Naughton, Barry J. *The Chinese Economy: Transitions and Growth*. Cambridge, MA: MIT Press, 2006.

Naumov, I.V., and David Norman Collins. *The History of Siberia*. London: Routledge, 2006.

Needham, Joseph. "Mao Zedong 1893–1976." *China Now* 65, no. 2 (1976): 2.

Nehru, Jawaharlal. Jawaharlal Nehru to U Nu. New Delhi, September 6, 1955. In *Selected Works of Jawaharlal* Nehru, Vol. 30, edited by Jawahrlal Nehru. http://www.claudearpi.net/maintenance/uploaded_pics/SW30.pdf (accessed December 3, 2015).

Neumann, Iver B. *Uses of the Other: "The East" in European Identity Formation*. Minneapolis: University of Minnesota Press, 1999.

Newbury, Colin W. *The Diamond Ring: Business, Politics, and Precious Stones in South Africa, 1867–1947*. Oxford: Clarendon Press, 1989.

Nightingale, Carl H.. *Segregation: A Global History of Divided Cities*. Chicago and London, University of Chicago Press, 2012.

Ninkovich, Frank A. *Global Dawn: The Cultural Foundation of American Internationalism, 1865–1890*. Cambridge, MA: Harvard University Press, 2009.

Noble, David F. *The Religion of Technology: The Divinity of Man and the Spirit of Invention*. New York: Knopf, 1997.

Nye, David E. *Technology Matters: Questions to Live With*. Cambridge, MA: MIT Press, 2006.

O'Bryan, Scott. "Car Culture." In *The Palgrave Dictionary of Transnational History*, edited by Akira Iriye and Pierre-Yves Saunier, 119. Basingstoke: Palgrave Macmillan, 2009.

Odagiri, Hiroyuki, and Akira Gotō. *Technology and Industrial Development in Japan: Building Capabilities by Learning, Innovation, and Public Policy*. Oxford: Clarendon Press, 1996.

Oldenziel, Ruth. "Islands: The United States as a Networked Empire." In *Entangled Geographies: Empire and Technopolitics in the Global Cold War*, edited by Gabrielle Hecht, 13–41. Cambridge, MA: MIT Press, 2011.

———, and Mikael Hård. *Consumers, Tinkerers, Rebels: The People Who Shaped Europe*. Basingstoke: Palgrave Macmillan, 2013.

Omissi, David E. *Air Power and Colonial Control: The Royal Air Force, 1919–1939*. Manchester: Manchester University Press, 1990.

Osborne, Michael A. "Acclimatizing the World." In *Nature and Empire: Science and the Colonial Enterprise*, edited by Roy MacLeod, 135–51. Chicago: University of Chicago Press, 2000.

Osterhammel, Juergen. "Forschungsreise und Kolonialprogramm. Ferdinand von Richthofen und die Erschliessung Chinas im 19. Jahrhundert." *Archiv fuer Kulturgeschichte* 69 (1987): 150–95.

———. *Die Entzauberung Asiens: Europa und die asiatischen Reiche im 18. Jahrhundert*. München: C.H. Beck, 1998.

———. *Europa um 1900: auf der Suche nach einer Sicht "von außen."* Bochum: Stiftung Bibliothek d. Ruhrgebiets, 2008.

———. *Europe, the "West" and the Civilizing Mission*. London: German Historical Institute, 2006.

———. *The Transformation of the World: A Global History of the Nineteenth Century*. Princeton: Princeton University Press, 2015.

———, and Niels P. Petersson. *Globalization: A Short History*. Princeton: Princeton University Press, 2009.

Owens, Edgar, and Robert Shaw. *Development Reconsidered: Bridging the Gap between Government and People*. Lexington, MA: Lexington Books, 1972.

Pahl, Walther. *Das politische Antlitz der Erde*. Leipzig: Goldmann, 1938.

———. *Die Luftwege der Erde*. Hamburg: Hanseatische Verlagsanstalt, 1936.

Pakenham, Thomas. *The Scramble for Africa, 1876–1912*. New York: Random House, 1991.

"Palestine: Information with Provenance." *PIWP database*. http://cosmos.ucc.ie/cs1064/jabowen/IPSC/php/authors.php?auid=42975 (accessed December 18, 2012).

Parla, Taha. *The Social and Political Thought of Ziya Goekalp, 1876–1924*. Leiden: Brill, 1985.

Patton, Phil. *Bug: The Strange Mutations of the World's Most Famous Automobile*. New York: Simon & Schuster, 2002.

Pearton, Maurice. *Diplomacy, War, and Technology Since 1830*. Lawrence: University Press of Kansas, 1984.

Pells, Richard H. *Not Like Us: How Europeans Have Loved, Hated, and Transformed American Culture Since World War II*. New York: Basic Books, 1997.

Perkovich, George. *India's Nuclear Bomb: The Impact on Global Proliferation*. Berkeley: University of California Press, 1999.

Perrin, Benjamin. *Modern Warfare: Armed Groups, Private Militaries, Humanitarian Organizations, and the Law*. Vancouver: UBC Press, 2012.

Peterson, Mildred J. *Managing the Frozen South: The Creation and Evolution of the Antarctic Treaty System*. Berkeley: University of California Press, 1988.

Pfeisinger, Gerhard, Stefan Schennach, and Martin Frimmel, eds. *Kolonialwaren: die Schaffung der ungleichen Welt*. Göttingen: Lamuv, 1989.

———. "Kolonialwaren in der Weltwirtschaft." In *Kolonialwaren: die Schaffung der ungleichen Welt*, edited by Gerhard Pfeisinger and Stefan Schennach, 12–18. Göttingen: Lamuv, 1989.

Pfister, Christian. "Das '1950er Syndrom': Die umweltgeschichtliche Epochenschwelle zwischen Industriegesellschaft und Konsumgesellschaft." In *Das 1950er Syndrom: der Weg in die Konsumgesellschaft*, edited by Christian Pfister, 51–95. Bern: Haupt, 1995.

Pick, Daniel. *War Machine: The Rationalisation of Slaughter in the Modern Age*. New Haven: Yale University Press, 1993.

Pilcher, Jeffrey M. *The Oxford Handbook of Food History*. Oxford: Oxford University Press, 2012.

Piquet, Caroline. *Histoire du canal de Suez*. Paris: Librairie Académique Perrin, 2009.

Poeschke, Martina. "Global Sourcing als Instrument zur Nutzung von Kostenoptimierungspotentialen – dargestellt am Beispiel der Volkswagen AG." Diploma Thesis Fachhochschule [Master's thesis], Braunschweig/Wolfenbuettel, 2000.

Polianski, Igor J., and Matthias Schwartz, eds. *Die Spur des Sputnik: Kulturhistorische Expeditionen ins kosmische Zeitalter*. Frankfurt am Main: Campus Verlag, 2009.

Pong, David. *Shen Pao-Chen and China's Modernization in the Nineteenth Century*. Cambridge: Cambridge University Press, 1994.

Ponting, Clive. *The Crimean War*. London: Chatto & Windus, 2004.

Porter, Theodore M. *Trust in Numbers: The Pursuit of Objectivity in Science and Public Life*. Princeton: Princeton University Press, 1995.

Poser, Stefan. "Leisure Time and Technology." *European History Online (EGO)*, Mainz 2011–09–26, urn:nbn:de:0159–2011051216. http://www.ieg-ego.eu/posers-2010-en (accessed November 15, 2012).

Poturzyn, Friedrich A. Fischer von. "Koloniale Luftpolitik." In *Die Weltgeltung Der Deutschen Luftfahrt Unter Mitarbeit*, edited by Heinz Orlovius and Ernst Schultze, 115–26. Stuttgart: F. Enke, 1938.

Prakash, Gyan. *Another Reason: Science and the Imagination of Modern India*. Princeton and Oxford: Princeton University Press, 1999.

Prashad, Vijay. *The Darker Nations: A People's History of the Third World*. New York: New Press, 2007.

Pratt, Mary Louise. *Imperial Eyes: Travel Writing and Transculturation*. London: Routledge, 1992.

Puffert, Douglas J. *Tracks Across Continents, Paths Through History: The Economic Dynamics of Standardization in Railway Gauge*. Chicago: University of Chicago Press, 2009.

Rabinbach, Anson. *The Human Motor: Energy, Fatigue, and the Origins of Modernity.* New York: Basic Books, 1990.

Racknitz, Ines Eben von. *Die Plünderung des Yuanming yuan: Imperiale Beutenahme im britisch-französischen Chinafeldzug von 1860.* Stuttgart: Franz Steiner Verlag, 2012.

Raj, Kapil. "Mapping." In *Palgrave Dictionary of Transnational History*, edited by Akira Iriye and Pierre-Yves Saunier, 688. Basingstoke: Palgrave Macmillan, 2009.

———. *Relocating Modern Science: Circulation and the Construction of Knowledge in South Asia and Europe, 1650–1900.* New York: Palgrave Macmillan, 2007.

Rassweiler, Anne D. *The Generation of Power: The History of Dneprostroi.* New York: Oxford University Press, 1988.

Raychaudhuri, Tapan. *Europe Reconsidered: Perceptions of the West in Nineteenth Century Bengal.* Delhi: Oxford University Press, 1988.

Rayward, W. Boyd. *The Universe of Information: The Work of Paul Otlet for Documentation and International Organisation.* Moscow: Published for International Federation for Documentation (FID) by All-Union Institute for Scientific and Technical Information (VINITI), 1975.

Recherche International e.V., and Rheinisches JournalistInnenbüro (Köln). *Unsere Opfer zählen nicht: die Dritte Welt im Zweiten Weltkrieg.* Berlin and Hamburg: Assoziation A, 2005.

Redfield, Peter. *Space in the Tropics: From Convicts to Rockets in French Guiana.* Berkeley: University of California Press, 2000.

Reeves, Keir et al. "Integrating the Historiography of the Nineteenth-Century Gold Rushes." *Australian Economic History Review* 50, no. 2 (2010): 111–28.

Reinalda, Bob. *Routledge History of International Organizations: From 1815 to the Present Day.* London: Routledge, 2009.

Reinkowski, Maurus, and Gregor Thum, eds. *Helpless Imperialists: Imperial Failure, Fear and Radicalization.* Göttingen: Vandenhoeck & Ruprecht, 2013.

Revel, Jean-François. *L'obsession anti-américaine: son fonctionnement, ses causes, ses inconséquences.* Paris: Plon, 2002.

Reybrouck, David van. *Congo: Een geschiedenis.* Amsterdam: De Bezige Bij, 2010.

Reynolds, David. "Science, Technology, and the Cold War." In *The Cambridge History of the Cold War, Volume III: Endings*, edited by Melvyn P. Leffler and Odd Arne Westad, 378–99. Cambridge: Cambridge University Press, 2010.

Richthofen, Ferdinand von. "Ueber den natuerlichen Weg fuer eine Eisenbahnverbindung zwischen China und Europa." *Verhandlungen der Gesellschaft fuer Erdkunde zu Berlin* 4 (1874): 1–14.

Rieger, Bernhard. *The People's Car: A Global History of the Volkswagen Beetle.* Cambridge, MA: Harvard University Press, 2013.

Riffenburgh, Beau. *Mapping the World: The Story of Cartography.* London: Andre Deutsch, 2015.

Rittberger, Volker, and Bernhard Zangl. *International Organization: Polity, Politics And Policies.* Basingstoke: Palgrave Macmillan, 2006.

Roberts, Peder. *The European Antarctic: Science and Strategy in Scandinavia and the British Empire.* New York: Palgrave Macmillan, 2011.

Robinson, Douglas Hill. *The Zeppelin in Combat: A History of the German Naval Airship Division, 1912–1918.* Henley-on-Thames: Foulis, 1971.

Rohrbach, Paul. "Die Bedeutung der Bagdadbahn." *Verhandlungen des Deutschen Kolonialkongresses*, 1902 (1903): 800–807.

Rooney, Ben. "How Tel Aviv Became a Tec Hub." *Wall Street Journal*, January 27, 2012. http://blogs.wsj.com/tech-europe/2012/01/27/how-tel-aviv-became-a-tech-hub/tab/print (accessed October 3, 2012).

Rosenberg, Emily S. "Einleitung." In *Geschichte der Welt*, Bd. 5, edited by Akira Iriye, Jürgen Osterhammel, and Emily S. Rosenberg, 9–32, Munich: C.H. Beck, 2012.

———, and Eric Foner. *Spreading the American Dream: American Economic and Cultural Expansion, 1890–1945*. New York: Hill and Wang, 1982.

Rosenbladt, Sabine. *Der Osten ist grün?: Ökoreportagen aus der DDR, Sowjetunion, Tschechoslowakei, Polen, Ungarn*. Hamburg: Rasch und Röhring, 1986.

Ross, Colin. *Mit Kind und Kegel in die Arktis; mit 50 Abbildungen und einer Karte*. Leipzig: F.A. Brockhaus, 1934.

Rossabi, Morris. *A History of China*. Chichester: Wiley-Blackwell, 2014.

Rothermund, Dietmar, ed. *The Routledge Companion to Decolonization*. London: Routledge, 2006.

Royle, Trevor. *Crimea: The Great Crimean War, 1854–1856*. New York: St. Martin's Press, 2000.

Ruehling, Lutz. "'Bilder vom Norden.' Imagines, Stereotype und ihre Funktionen." In *Imagologie des Nordens: Kulturelle Konstruktionen von Noerdlichkeit in interdisziplinaerer Perspektive*, edited by Astrid Arndt et al., 279–300. Frankfurt am Main: Peter Lang, 2004.

Russel-Wood, Anthony J.R. "Technology and Society: The Impact of Gold Mining on the Institution of Slavery in Portuguese America." In *Technology and European Overseas Enterprise: Diffusion, Adaption, and Adoption*, edited by Michael Adas, 181–206. Aldershot: Variorum, 1996.

Said, Edward W. *Orientalism*. New York: Vintage Books, 1979.

Salay, Jürgen. *The Soviet Union River Diversion Project: From Plan to Cancellation 1976–1986*. Uppsala Papers in Economic History, Research Report 17. Uppsala: Ekonomisk-historiska institutionen, 1988.

Salazar, Noel B. *Envisioning Eden: Mobilizing Imaginaries in Tourism and Beyond*. New York: Berghahn Books, 2010.

Sampson, Anthony. *Empires of the Sky: The Politics, Contests and Cartels of World Airlines*. New York: Random House, 1984.

Sarda, Har Bilas. *Hindu Superiority: An Attempt to Determine the Position of the Hindu Race in the Scale of Nations*. Ajmer: Rajputana Printing Works, 1906.

Schaebler, Birgit, and Leif Stenberg, eds. *Globalization and the Muslim World: Culture, Religion, and Modernity*. Syracuse, NY: Syracuse University Press, 2004.

Schattenberg, Susanne. *Stalins Ingenieure: Lebenswelten zwischen Technik und Terror in den 1930er Jahren*. Munich: Oldenbourg Verlag, 2002.

Schenk, Frithjof Benjamin. "Mental Maps. Die Konstruktion von geographischen Raeumen in Europa seit der Aufklaerung." *Geschichte und Gesellschaft* 28 (2002): 493–514.

Schenk, Frithjof Benjamin, and Martina Winkler, eds. *Der Sueden: Neue Perspektiven auf eine europaeische Geschichtsregion*. Frankfurt am Main: Campus Verlag, 2007.

Schiebinger, Londa L. *Plants and Empire: Colonial Bioprospecting in the Atlantic World*. Cambridge, MA: Harvard University Press, 2004.

Schieder, Theodor. *Handbuch der europaeischen Geschichte*, Vol 6. Stuttgart: Union Verlag, 1968.

Schipper, Frank. *Driving Europe: Building Europe on Roads in the Twentieth Century*. Amsterdam: Aksant, 2008.

———, Vincent Lagendijk, and Irene Anastasiadou. "New Connections for an Old Continent: Rail, Road and Electricity in the League of Nations Organisation for Communications and Transit." In *Materializing Europe: Transnational Infrastructures and the Project of Europe*, edited by Alec Badenoch and Andreas Fickers, 113–43. New York: Palgrave Macmillan, 2010.

Schlimm, Anette. *Ordnungen des Verkehrs Arbeit an der Moderne - deutsche und britische Verkehrsexpertise im 20. Jahrhundert*. Bielefeld: Transcript, 2011.

Schmidt-Gernig, Alexander. "Europa als Kontinent der Zukunft: Pierre Bertaux und die Zeitdiagnostik der 1960er Jahre." *Themenportal Europaeische Geschichte* (2007). http://www.europa.clio-online.de/2007/Article=144 (accessed September 18, 2012).

Schnuerer, Florian. "Aeroplanes and Airships as National and Collective Symbols in Western Europe before the First World War (1908–1914)." *Memoria y Civilización* 12 (2009): 155–89.

Schot, Johan, and Vincent Lagendijk. "Technocratic Internationalism in the Interwar Years: Building Europe on Motorways and Electricity Networks." *Journal of Modern European History* 6, no. 2 (2008): 196–216.

Schroeder, Stephan Michael. "'Leitbild Norden' statt 'Leitbild Europa'? Die Gruende der nordeuropaeischen Europaskepsis." In *Leitbild Europa? Europabilder und ihre Wirkungen in der Neuzeit*, edited by Juergen Elvert and Juergen Nielsen-Sikora, 193–207. Stuttgart: Steiner, 2009.

Schröter, Harm G. *Americanization of the European Economy: A Compact Survey of American Economic Influence in Europe since the 1880s*. Dordrecht: Springer, 2015.

Schuemer, Dirk. "Spaghettisiert euch: Alle Welt beklagt den amerikanischen Einfluss, doch die globale Leitkultur kommt aus Italien." *Frankfurter Allgemeine Zeitung*, No. 226, September 28, 2002, 29.

Schultz, Hans-Dietrich. "Europa: (k)ein Kontinent? Das Europa deutscher Geographen." In *Welt-Raeume: Geschichte, Geographie und Globalisierung um 1900*, edited by Iris Schroeder and Sabine Hoehler, 203–31. Frankfurt am Main: Campus Verlag, 2005.

———. *Europa als geographisches Konstrukt*. Jena: Friedrich-Schiller-Univ., Institut für Geographie, 1999.

Scranton, Philip. Preface to *The Business of Tourism: Place, Faith, and History*, edited by Janet F. Davidson and Philip Scranton, vii. Philadelphia: University of Pennsylvania Press, 2007.

Sèbe, Berny. *Heroic Imperialists in Africa: The Promotion of British and French Colonial Heroes, 1870–1939*. Manchester: Manchester University Press, 2013.

Seegel, Steven. *Mapping Europe's Borderlands: Russian Cartography in the Age of Empire*. Chicago: University of Chicago Press, 2012.

Segre, Claudio G. "Italo Balbo and the Colonisation of Libya." *Journal of Contemporary History* 7, no. 3 (1972): 141–55.

Selassie, Haile. "Appeal to the League of Nations," June 1936.. http://www.mtholyoke.edu/acad/intrel/selassie.htm (accessed September 8, 2011).

Selin, Helaine, ed. *Encyclopedia of the History of Science, Technology, and Medicine in Non-Western Cultures*. Boston: Kluwer Academic, 1997.

Sen, Simonti. *Travels to Europe: Self and Other in Bengali Travel Narratives, 1870–1910*. New Delhi: Orient Longman, 2005.

Sengoopta, Chandak. *Imprint of the Raj: How Fingerprinting Was Born in Colonial India*. London: Macmillan, 2003.

Senor, Dan, and Saul Singer. *Start-Up Nation: The Story of Israel's Economic Miracle*. New York: Twelve, 2009.

Shelton, Dinah, ed. *Encyclopedia of Genocide and Crimes against Humanity*. Detroit: Macmillan Reference, 2005.

Shere, Jeremy. "Frank Shuman's Solar Arabian Dream." In *Renewable: A Book-in-Progress about Renewable Energy*. http://renewablebook.com/chapter-excerpts/350–2/ (accessed December 18, 2012).

Shinohara, Hatsue. "League of Nations System." In *The Palgrave Dictionary of Transnational History*, edited by Akira Iriye and Pierre-Yves Saunier, 645–49. Basingstoke: Palgrave Macmillan, 2009.

Silva, Ana Paula, and Maria Paula Diogo. "From Host to Hostage." In *Networking Europe: Transnational Infrastructures and the Shaping of Europe, 1850–2000*, edited by Arne Kaijser and Eric van der Vleuten, 51–69. Sagamore Beach, MA: Science History Publications, 2006.

Simms, Brendan, and D.J.B. Trim. *Humanitarian Intervention: A History*. Cambridge: Cambridge University Press, 2011.

Simões, Ana, Ana Carneiro, and Maria Paula Diogo, eds. *Travels of Learning: A Geography of Science in Europe*. Dordrecht: Kluwer Academic, 2003.

Simonsen, D.G. "Accelerating Modernity: Time-Space Compression in the Wake of the Aeroplane." *Journal of Transport History* 26, no. 2 (2009): 98–117.

Smil, Vaclav. *Creating the Twentieth Century: Technical Innovations of 1867–1914 and Their Lasting Impact*. Oxford: Oxford University Press, 2005.

———. *Enriching the Earth. Fritz Haber, Carl Bosch, and the Transformation of World Food Production*. Cambridge, MA: MIT Press, 2001.

Soerlin, Sverker. "Ordering the World for Europe: Science as Intelligence and Information as Seen from the Northern Periphery." *Osiris* 15 (2000): 51–69.

Spengler, Oswald. *Der Mensch und die Technik. Ein Beitrag zu einer Philosophie des Lebens*. Munich: Beck, 1931.

Sperling, Walter. *Der Aufbruch der Provinz: Die Eisenbahn und die Neuordnung der Räume im Zarenreich*. Frankfurt am Main: Campus Verlag, 2011.

Shridharāni, Krishnalal. *Story of the Indian Telegraphs: A Century of Progress*. New Delhi: Posts and Telegraphs Dept, 1953.

Standage, Tom. *The Victorian Internet: The Remarkable Story of the Telegraph and the Nineteenth Century's On-Line Pioneers*. New York: Walker and Co., 1998.

Stanfield, Michael Edward. *Red Rubber, Bleeding Trees: Violence, Slavery, and Empire in Northwest Amazonia, 1850–1933*. Albuquerque, NM: University of New Mexico Press, 1998.

Staniland, Martin. *Government Birds: Air Transport and the State in Western Europe*. Lanham, MD: Rowman & Littlefield, 2003.

Stanley, Henry M. *In Darkest Africa: Or, The Quest, Rescue and Retreat of Emin, Governor of Equatoria*. New York: Chas. Scribner's Sons, 1890.

Stead, William Thomas. "The United States of Europe." *Review of Reviews*, July 1897, 18–29.

———. *The Americanization of the World, or, The Trend of the Twentieth Century*. New York: H. Markley, 1902.

Stråth, Bo. "Multiple Europes: Integration, Identity and Demarcation to the Other." In *Europa and the Other and Europa as the Other*, edited by Bo Stråth, 385–420. Brussels: Peter Lang, 2010.

———. "'Norden' as European Region: Democration and Belonging." In *Domains and Divisions of European History*, edited by Johann P. Arnason and Natalie J. Doyle, 198–215. Liverpool: Liverpool University Press, 2010.

Stuchtey, Benedikt. *Die europaeische Expansion und ihre Feinde. Kolonialismuskritik vom 18. bis in das 20. Jahrhundert*. Munich: Oldenbourg Verlag, 2010.

———., ed. *Science Across the European Empires, 1800–1950*. Oxford: Oxford University Press, 2005.

Süss, Dietmar. *Tod aus der Luft: Kriegsgesellschaft und Luftkrieg in Deutschland und England*. Munich: Siedler, 2011.

Szasz, Margaret. *Between Indian and White Worlds: The Cultural Broker*. Norman: University of Oklahoma Press, 1994.

Taithe, Bertrand. *The Killer Trail: A Colonial Scandal in the Heart of Africa*. Oxford: Oxford University Press, 2009.

Tanaka, Yuki. "British 'Humane Bombing' in Iraq during the Interwar Era." In *Bombing Civilians: A Twentieth-Century History*, edited by Yuki Tanaka and Marilyn B. Young, 8–29. New York: New Press, 2009.

Taylor, Blaine. *Fascist Eagle: Italy's Air Marshal Italo Balbo*. Missoula, MN: Pictorial Histories Pub. Co., 1996.

Taylor, Karin. "From Trips to Modernity to Holidays in Nostalghia – Tourism History in Eastern and South Eastern Europe." *Tension of Europe/Inventing Europe*, Working Paper Nr. WP_2011_01. http://www.tensionsofeurope.eu/www/nl/files/get/publications/WP_2011_01_Taylor.pdf (accessed March 1, 2011).

Tharoor, Shashi. *The Elephant, The Tiger, and the Cell Phone: Reflections on INDIA – The Emerging 21st-Century Power*. Kindle Edition, 2011.

Thomas, Martin. *Violence and Colonial Order: Police, Workers and Protest in the European Colonial Empires, 1918–40*. Cambridge: Cambridge University Press, 2012.

Thum, Gregor. "'Europa' im Ostblock. Weisse Flecken in der Geschichte der europaeischen Integration." *Zeithistorische Forschungen* [Studies in Contemporary History] 1, No. 3 (2004). http://www.zeithistorische-forschungen.de/3-2004/id=4732 (accessed December 3, 2015).

Tomasson, Richard F. "Iceland as 'The First New Nation'." *Scandinavian Political Studies* 10 (1975): 33–51.

———. *Iceland, the First New Society*. Minneapolis: University of Minnesota Press, 1980.

Tooze, J. Adam. *Statistics and the German State, 1900–1945: The Making of Modern Economic Knowledge*. Cambridge: Cambridge University Press, 2001.

Topik, Steven, Carlos Marichal, and Zephyr L. Frank, eds. *From Silver to Cocaine: Latin American Commodity Chains and the Building of the World Economy, 1500–2000*. Durham, NC: Duke University Press, 2006.

Trakulhun, Sven. "Kulturwandel durch Anpassung? Matteo Ricci und die Jesuitenmission in China." In *Zeitenblicke* 11, no. 1. http://www.zeitenblicke.de/2012/1/Trakulhun/index_html (accessed November 22, 2012).

Troitskiy, Valeriy. "Statue of Liberty Made of Russian Copper?" *Russian-American Business.* http://www.russianamericanbusiness.org/EN/web_CONTENT/articles/2005.01.20/group_05/2_articl/articl.shtml (accessed August 19, 2012).

Tse-Tung, Mao. "We Must Learn to do Economic Work." In *Selected Works of Mao Tse-Tung*, Vol 3, 239–246. Peking: Foreign Languages Press, 1965. http://www.marxists.org/reference/archive/mao/selected-works/volume-3/mswv3_22.htm (accessed December 3, 2015).

Turner, Barry, and Gunilla Nordquist, eds. *The Other European Community: Integration and Cooperation in Nordic Europe.* New York: St. Martin's Press, 1982.

Vaillant, Franc. *Travels into the Interior Parts of Africa by the Way of the Cape of Good Hope in the Years 1780, 8l, 82, 83, 84, and 85.* London: G.G.J. and J. Robinson, 1790.

Valverde, Maria. "The Dialectic of the Familiar and the Unfamiliar: 'The Jungle' in Early Slum Travel Writing." *Sociology* 30, no. 3 (1996): 493–509.

Vanthemsche, Guy. *Belgium and the Congo, 1885–1980.* Cambridge: Cambridge University Press, 2012.

Vasold, Manfred. *Florence Nightingale: eine Frau im Kampf für die Menschlichkeit.* Regensburg: Pustet, 2003.

Vec, Miloš. "Der Welt Standard: Bureau International des Poids et Mesures, Sèvres." In *Mekkas der Moderne: Pilgerstaetten der Wissensgesellschaft*, edited by Hilmar Schmundt, Miloš Vec, and Hildegard Westphal, 339–42. Köln: Boehlau, 2010.

Vidal, Javier. "Tourism and Transport in Europe, 1930–2000." In *Neue Wege in ein neues Europa: Geschichte und Verkehr im 20. Jahrhundert*, edited by Ralf Roth and Karl Schloegel, 476–90. Frankfurt am Main: Campus Verlag, 2009.

Vleck, Jenifer Van. *Empire of the Air: Aviation and the American Ascendancy.* Cambridge, MA: Harvard University Press, 2013.

Vogel, Ezra F. *Deng Xiaoping and the Transformation of China.* Cambridge, MA: Belknap Press of Harvard University Press, 2011.

Vowinckel, Annette. *Flugzeugentfuehrungen: Eine Kulturgeschichte.* Goettingen: Wallstein, 2011.

Vries, Jan de. "The Industrious Revolution and the Industrial Revolution." *Journal of Economic History* 54 (1994): 249–70.

Warraq, Ibn. *Defending the West: A Critique of Edward Said's Orientalism.* Amherst, NY: Prometheus Books, 2007.

Weber, Eugen. *Peasants into Frenchmen: The Modernization of Rural France, 1870–1914.* Stanford: Stanford University Press, 1976.

Webster, Paul. *Fachoda: La Bataille pour le Nil.* Paris: Éditions du Félin, 2001.

Wei, Chunjuan Nancy, and Darryl E. Brock. *Mr. Science and Chairman Mao's Cultural Revolution: Science and Technology in Modern China.* Lexington, MA: Lexington Books, 2013.

Weightman, Gavin. *The Industrial Revolutionaries: The Making of the Modern World 1776–1914.* New York: Grove Press, 2007.

Weisbrode, Kenneth. "War." In *The Palgrave Dictionary of Transnational* History, edited by Akira Iriye and Pierre-Yves Saunier, 1090. Basingstoke: Palgrave Macmillan, 2009.

Weiß, Andreas. "Die Kolonien in Europa." *Berliner Journal für Soziologie* 18, no. 4 (2008): 663–77.

Weisshaupt, Winfried. *Europa sieht sich mit fremdem Blick: Werke nach dem Schema der. "Lettres persanes" in der europäischen, insbesondere der deutschen Literatur des. 18. Jahrhunderts*. Frankfurt am Main: Peter Lang, 1979.

Wendt, Reinhard. *Vom Kolonialismus zur Globalisierung: Europa und die Welt seit 1500*. Paderborn: Schöningh, 2007.

Wenzlhuemer, Roland. "The History of Standardisation in Europe." *European History Online (EGO)*, December 3, 2010. urn:nbn:de:0159–20100921441. http://www.ieg-ego.eu/wenzlhuemerr-2010-en (accessed January 2, 2013).

———. *Connecting the Nineteenth-Century World: The Telegraph and Globalization*. Cambridge: Cambridge University Press, 2013.

Werthmann, Katja, and Tilo Grätz. *Mining Frontiers in Africa: Anthropological and Historical Perspectives*. Cologne: Rüdiger Köppe Verlag, 2012.

Wesseling, Hendrik L. *Divide and Rule: The Partition of Africa, 1880–1914*. Westport, CN: Praeger, 1996.

———. *Expansion and Reaction: Essays on European Expansion and Reaction in Asia and Africa*. Leiden: Leiden University Press, 1978.

Westad, Odd Arne. *The Global Cold War: Third World Interventions and the Making of Our Times*. Cambridge: Cambridge University Press, 2005.

Wiersch, Bernd. *Die Käfer-Chronik: die Geschichte einer Autolegende*. Bielefeld: Delius Klasing, 2005.

Wilson, Woodrow. "Fourteen Points Speech." Speech presented to a joint session of the United States Congress on January 8, 1918. http://en.wikisource.org/wiki/Fourteen_Points_Speech (accessed December 3, 2015).

Winkler, Heinrich August. *Geschichte des Westens*. 4 Vols. Munich: C.H. Beck, 2009–2015.

Winkler, Jonathan Reed. *Nexus Strategic Communications and American Security in World War I*. Cambridge, MA: Harvard University Press, 2013.

Winseck, Dwayne Roy, and Robert M. Pike. *Communication and Empire: Media, Markets, and Globalization, 1860–1930*. Durham, NC: Duke University Press, 2007.

Wintle, Michael. "Europe as Seen From the Outside. A Brief Visual Survey." In *Imagining Europe. Europe and European Civilisation as Seen from its Margins and by the Rest of the World, in Nineteenth and Twentieth Centuries*, edited by Michael Wintle, 23–48. Brussels: Peter Lang, 2008.

———. *The Image of Europe: Visualizing Europe in Cartography and Iconography Throughout the Ages*. Cambridge: Cambridge University Press, 2009.

Witte, Ludo de. *The Assassination of Lumumba*. London: Verso, 2001.

Wohl, Robert. *A Passion for Wings: Aviation and the Western Imagination, 1908–1918*. New Haven: Yale University Press, 1994.

———. *The Spectacle of Flight: Aviation and the Western Imagination, 1920–1950*. New Haven: Yale University Press, 2005.

Wolff, Larry. *Inventing Eastern Europe: The Map of Civilization on the Mind of the Enlightenment*. Stanford: Stanford University Press, 1994.

Wolfgang, Schmale. "Die Konstruktion des Homo Europaeus." *Comparative European History Review* 1 (2001): 165–84.

Womack, Brantly. *China's Rise in Historical Perspective*. Lanham, MD: Rowman & Littlefield, 2010.

Womack, James P., Daniel T. Jones, and Daniel Roos, eds. *The Machine That Changed the World: Based on the Massachusetts Institute of Technology 5-Million Dollar 5-Year Study on the Future of the Automobile.* New York: Rawson Associates, 1990.

Wood, Alan. *Russia's Frozen Frontier: A History of Siberia and the Russian Far East, 1581–1991.* London: Bloomsbury Academic, 2011.

Xiaoping, Deng. Speech at the Opening Ceremony of the National Conference on Science, March 18, 1978. http://www.china.org.cn/english/features/dengxiaoping/103390.htm (accessed December 3, 2015).

Young, Robert. *Postcolonialism.* Oxford: Oxford University Press, 2003.

Zabusky, Stacia E. *Launching Europe: An Ethnography of European Cooperation in Space Science.* Princeton: Princeton University Press, 1995.

Zachariah, Benjamin. *Nehru.* London: Routledge, 2004.

Ziegler, Gudrun. *Der achte Kontinent: Die Eroberung Sibiriens.* Berlin: Ullstein, 2005.

Ziegler, Philip. *Legacy: Cecil Rhodes, the Rhodes Trust and Rhodes Scholarships.* New Haven: Yale University Press, 2008.

Zimmerman, Andrew. *Alabama in Africa: Booker T. Washington, the German Empire, and the Globalization of the New South.* Princeton: Princeton University Press, 2010.

Zischka, Anton. *Italien in der Welt.* Leipzig: W. Goldmann, 1937.

———. *Wissenschaft bricht Monopole: Der Forscherkampf um neue Rohstoffe und neuen Lebensraum.* Leipzig: Wilhelm Goldmann Verlag, 1936.

Zukas, Lorna Lueker. "Segregation, racial, Africa." In *Encyclopedia of Western Colonialism Since 1450*, Vol. 3, edited by Thomas Benjamin, 1001–2. Detroit: Macmillan Reference USA, 2007.

Zweig, Stefan. "Die Monotonisierung der Welt" (1925). http://www.cicero.de/berliner-republik/die-monotonisierung-der-welt/36657 (accessed December 24, 2012).

———. *Sternstunden der Menschheit: Zwölf historische Miniaturen.* Frankfurt am Main: Fischer Taschenbuch, 1964.

Illustration Credits

Permissions to reproduce the illustrations and photographs in this book have generously been granted by the institutions and collections named here, and are gratefully acknowledged by the editors, the authors, the Foundation for the History of Technology, and Palgrave Macmillan.

The Foundation for the History of Technology has carefully tried to locate all rights holders. Parties who despite this feel that they are entitled to certain rights are requested to contact the Foundation for the History of Technology (www.histech.nl).

Cover: Original caption: "Civilisations-Dampf-Maschine auf der Londoner Industrie-Ausstellung." Illustration from *Kladderadatsch*, SH. 3, S.20 (1848–51). Permission by Deutsches Historisches Museum, Berlin.

Making Europe:
Series Acknowledgements

Making Europe is the result/product of an unusual collaboration among a host of individuals and organizations. The *Making Europe* authors and series editors feel extremely fortunate to be working with them. We list here individuals and organizations who contributed to the entire series. Each volume in the series also has its own separate acknowledgements.

Making Europe was initiated by:

- Foundation for the History of Technology (www.histech.nl)

Making Europe is sponsored by:

- Eindhoven University of Technology (www.tue.nl)
- Fonds 21 (www.fonds21.nl) – formerly known as SNS REAAL Fonds
- Next Generations Infrastructures (www.nextgenerationsinfrastructures.eu)
- Foundation for the History of Technology Corporate Program (www.histech.nl) that includes:
  - DSM (www.dsm.com)
  - EBN (www.ebn.nl)
  - FrieslandCampina(www.frieslandcampina.com)
  - Philips (www.philips.nl)
  - SIDN (www.sidn.nl)
  - TNO (www.tno.nl)

Making Europe has been made possible thanks to:

- Tensions of Europe Network (www.tensionsofeurope.eu)
- European Science Foundation EUROCORES Programme Inventing Europe – Technology and the Making of Europe, 1850 to the Present (www.esf.org)
- Research Theme Group Grant of the Netherlands Institute of Advanced Studies (NIAS) in 2010–11 (www.nias.nl)
- European University Institute in Florence, Italy for providing the support in developing the series and for the founding workshop (3–6 July 2008) (www. eui.eu)

Making Europe benefited from the feedback of a community of scholars who have been involved in the series from the start:

Håkon With Andersen, Alec Badenoch, Robert Bud, David Burigana, Cornelis Disco, Paul Edwards, Valentina Fava, Karen Johnson Freeze, Andrea Guintini, Gabrielle Hecht, Rüdiger Klein, Eda Kranakis, John Krige, Leonard Laborie, Vincent Lagendijk, Suzanne Lommers, Slawomir Lotysz, Dagmara Jaješniak-Quast, Karl-Erik Michelsen, Matthias Middell, Thomas J. Misa, Dobrinka Parusheva, Kiran Patel, Pierre-Yves Saunier, Emanuela Scarpellini, Frank Schipper, Michael Strang, Ivan Tchalakov, Frank Trentmann, Aristotle Tympas, Hans Weinberger

Making Europe relied on the unflagging support of:

Picture editors

- Katherine Kay-Mouat
- Giel van Hooff
- Jan Korsten (Management)
- Camiel Lintsen – Kade 05 Eindhoven (graphs and maps)

Text editors

- Lisa Friedman
- James Morrison

PalgraveMacmillan

- Jenny McCall (Publisher)
- Jade Moulds (Editorial Assistant)
- Philip Hillyer (Copy-Editor and Editorial Services Consultant)
- Susan Boobis (Indexer)

Office Foundation for the History of Technology:

- Sonja Beekers (Secretarial Support)
- Jan Korsten (Business Director)
- Loek Stoks (Bookkeeping)
- Henk Treur (Volunteer)
- Erik van der Vleuten (Scientific Director)

Board Foundation for the History of Technology:

- Emmo Meijer (chair)
- Jacques Joosten (treasurer)
- Dirk van Delft
- Herman de Boon
- Eric Fischer
- Frans Greidanus
- Michiel Westermann
- Ernst Homburg (advisor)
- Harry Lintsen (advisor)
- Johan Schot (advisor)
- Martin Schuurmans (advisor)

Index